Design of CMOS RF Integrated Circuits and Systems

Design of CMOS RF Integrated Circuits and Systems

Kiat Seng YEO
Manh Anh DO
Chirn Chye BOON
Nanyang Technological University, Singapore

World Scientific

NEW JERSEY • LONDON • SINGAPORE • BEIJING • SHANGHAI • HONG KONG • TAIPEI • CHENNAI

Published by

World Scientific Publishing Co. Pte. Ltd.
5 Toh Tuck Link, Singapore 596224
USA office: 27 Warren Street, Suite 401-402, Hackensack, NJ 07601
UK office: 57 Shelton Street, Covent Garden, London WC2H 9HE

British Library Cataloguing-in-Publication Data
A catalogue record for this book is available from the British Library.

DESIGN OF CMOS RF INTEGRATED CIRCUITS AND SYSTEMS

ISBN-13 978-981-4271-55-4
ISBN-10 981-4271-55-1

Printed in Singapore.

CONTENTS

PREFACE

High demand for wireless products in the mass consumer markets is driving the semiconductor industry towards total integration and implementation of a complete transceiver (analog and digital signal processing) on a single chip. With the rapid evolution of deep sub-micron radio frequency (RF) CMOS technologies, it is no longer impossible to have exceptionally compact, high-performance and low-cost RF integrated circuits and systems. This book provides the most comprehensive and in-depth coverage of the latest RF integrated circuit design developments in CMOS technology. It is a practical and cutting-edge guide, packed with tested circuit techniques and innovative design methodologies for solving challenging problems associated with RF integrated circuits and systems. This invaluable resource features a collection of the finest design practices that may soon drive the system-on-chip revolution. Using this book's state-of-the-art design techniques, you can apply existing technologies in novel ways and to create new circuit designs for the future.

This book begins with an introduction on the basic concepts of radio frequency integrated circuits (RFIC) that contains multitude topics of essential fundamentals in RFIC such as the principles of communication engineering, transceiver architectures and wireless standards. These fundamental concepts serve to fill the gap between traditional low frequency IC design and RFIC design.

Chapter 2 focuses on RFIC technology and the recent development of process design kits (PDKs) that revolutionized the semiconductor industry in shortening the design cycle. The design and optimization of RF transistors for a broad range of bias conditions and operating frequencies are discussed. Modeling of RF MOSFETs to accurately predict the RF characteristics at high frequencies is also given. In addition, this chapter covers the design and analysis of passive devices such as on-chip inductors, baluns and varactors and how they are employed in realizing low-noise and low-power RFIC's. For the first time, RF interconnects and its loss mechanisms, the relevant parasitics and their effects on RF circuits' performances are discussed in great details.

Chapter 3 devotes solely to describing different types of low noise amplifiers (LNAs) for RF applications. LNA is one of the most important building blocks in the communication system. Given the importance of input matching, various input networks are analyzed. An innovative parallel *LC* network to replace the large gate inductor, which is normally required for normal input architecture LNA, is presented. A 2.4GHz low noise amplifier is described as a design example.

Chapter 4 presents common configurations of active and passive mixers. A high gain active mixer with current booster, an image-reject mixer for low intermediate frequency (IF) architectures, and a double balanced passive mixer with zero power consumption and low noise figure are described. As leakages of high frequency signals from one port to another are unavoidable due to magnetic and capacitive couplings, conduction through substrate, and unbalanced structures of circuits, this chapter also addresses the importance of port isolation and DC offset in direct conversion mixers and proposes solutions to overcome them.

Chapter 5 gives the basic concepts and terminologies used in RF voltage-controlled oscillator (VCO). VCO is a critical component in a frequency synthesizer, which is used to generate certain frequencies. In this chapter, various VCO architectures are presented and supported by design examples. Associate noise sources of a LC tank circuit, linear time variant phase noise analysis and phase noise calculation are discussed. Emphasis is also placed on the practical aspects and detailed descriptions of a methodology for the design of an LC VCO, which are not reported in any other book yet.

Apart from LNAs, mixers and VCOs, phase-locked loops (PLLs) and prescalers, which are extensively covered in chapters 6 and 7 respectively, are requisite attributes in contributing to a large fraction of the total power consumption on a single chip. While the application of the PLL in the demodulation of frequency modulated (FM) signals has never been popular in analog transceivers, its application in modern data recovery circuits for data rates of several gigabits per second in optical communications has pushed the design of the phase-frequency detector into the RF range, making the implementation of CMOS PLL one of the most challenging tasks in RFIC design. The most apparent feature that differentiates a PLL frequency synthesizer from other phase-locked loops lies in the frequency divider. In chapter 7, much attention and efforts have been given to this circuit block to minimize its power consumption and size and to improve its performance.

Finally, this book endeavors to bridge the gap between a typical textbook-style RFIC book and a design-oriented book that contains few insights to the actual theory. It is one of the few rare books that focus entirely on CMOS technology. We anticipate to position this book at a level suitable for senior undergraduate courses, and to provide an excellent guide to RFIC for graduate students, researchers and engineers. It is also a good reference for professors, circuit designers, scientists and instructors working in the area of RF integrated circuit and system design.

ACKNOWLEDGEMENTS

This book would not have been made possible without the help and support of many.

Kiat Seng YEO is thankful to his wife Shannen for her unwavering support, understanding and love and their three daughters Shun Yuan, Shun Yi and Shun Yu for lighting up his life.

Manh Anh DO dedicates his work to Marcella for her love, care, friendship and companionship for the last 32 years.

Chirn Chye BOON dedicates this book to his loving mother Siew Kiat, lovely wife Xiaoling, family and friends who are most dear to him.

We are truly thankful to Ms Chelsea Chin, Desk Editor of World Scientific and her team for the professional production of this book. We wish to thank them for all the help in bringing this book into reality.

We wish to express our appreciation to Nanyang Technological University for creating an excellent environment for education and research.

Finally, the success of this book would not have been possible without the kind assistance from our researchers and students: Choon Beng SIA, Ah Fatt TONG, Chee Chong LIM, Lih Chieh PNG, Lin JIA, Linggajaya KAUFIK, Jingjing LIU, Shouxian MOU and Xiaopeng YU.

Knowledge is limited to what we know whereas imagination embraces everything we do not know.

Kiat Seng YEO

CHAPTER 1

RF CMOS Systems on Chips

The design of RF (radio frequency) system on a single chip demands from the designers a good understanding of multiple disciplines including the principles of communication engineering, transceiver architectures, wireless standards, signal processing, integrated circuit (IC) design techniques, device physics, Electronic Design Automation (EDA) tools, device modeling, and semiconductor technologies, etc. While the main focus of this book is on the design techniques used in various RF-IC building blocks, this first chapter introduces the various RF technologies which affect the design specifications of individual circuits and consequently the system specifications. For a certain set of system specifications, trade-offs among specifications of individual circuits are possible. Hence the circuit designers need to some basic understanding of common RF system architectures and the interrelation of individual block's performances.

1.1 Modern RF Mobile Technologies

The recent convergence of the four technologies, namely, wireless communications, microelectronics, computers, and information technology, has propelled a phenomenon expansion of the RF mobile communications industry worldwide over the last decade. The new generation of mobile phones and data terminals opens a new horizon for communications and information technologies handling voice, data and video and offers a wide range of services. Today the major applications of the modern RF technology are operating in the frequency bands described below under numerous industrial standards:

Cordless phones:	49MHz, 900MHz
Cellular phones:	850MHz, 900MHz, 1.8GHz and 1.9GHz
3G and UTMS:	1.8-2.2GHz
Wireless LAN:	900MHz, 2.4GHz, 5.3GHz and 5.8GHz
Broadband Wireless Access:	2-11GHz
RFID Systems:	900MHz, 2.4GHz and 5.8GHz
GPS:	1.5GHz
Satellite television:	10GHz
UWB:	3.1-10.6GHz

1.2 The RF Transceiver System

RF communications is the most popular form of wireless communications due to its ability of providing services to users during mobility. Examples include pagers, broadcasting, HF and VHF radio telephones, cordless phones, and cellular phones. In order to send voice and / or video information or data, which will be referred to as baseband signal, through a RF wireless link, a modulator is employed in the transmitter. In the receiver, the baseband signal is recovered by a demodulator.

Figure 1.1 illustrates this basic concept. The principles of modern RF transceiver circuits and systems are well reviewed in [1-3]. In the modulator, the baseband signal modulates a high frequency carrier to produce a passband signal for transmission.

The conversion to a high frequency allows the implementation of a smaller and more efficient antenna, and the efficient management of the radio spectrum and allocation of RF channels. The shape and width of the passband signal spectrum is dependent on the type of modulation and its generalized expression is given by equation (1.1):

$$S(t) = a(t)\cos[\omega_c t + \theta(t)] \tag{1.1}$$

$a(t)$ and $\theta(t)$ modulate the amplitude and the phase angle of the carrier $A\cos\,\omega_c t$. The instantaneous frequency is thus equal to $\omega = \omega_c + d\theta / dt$. The receiver demodulates $S(t)$ to extract the original baseband signal.

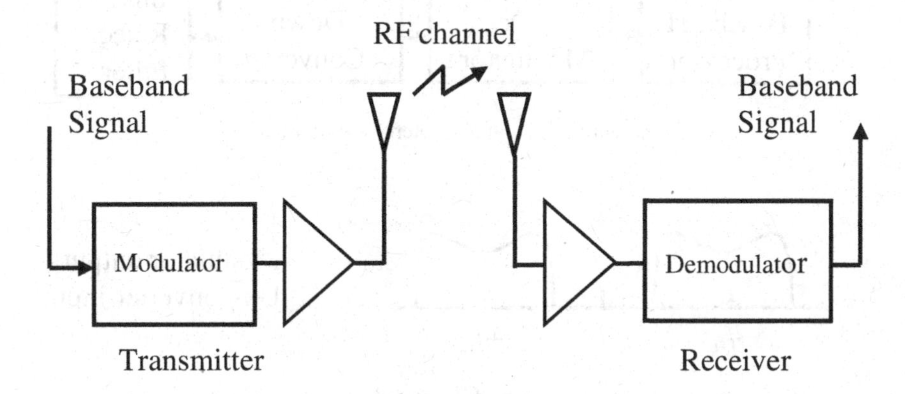

Figure 1.1 A basic communication system.

To provide the channel selection capability for a multi-channel transceiver system, modulators/demodulators are often designed at a standard intermediate frequency (IF), and up/down converters are used to shift the passband signal to or from the RF channels (Figure 1.2). This transceiver architecture is referred to as the heterodyne architecture. Receivers using down-conversions twice are referred to as double conversion receivers.

The up-conversion from an intermediate frequency to a radio frequency is commonly implemented by mixing the IF signal with a sine wave from a local oscillator (LO). This mixing process involves the addition and subtraction of input frequencies, and produces image signals as shown in Figure 1.3.

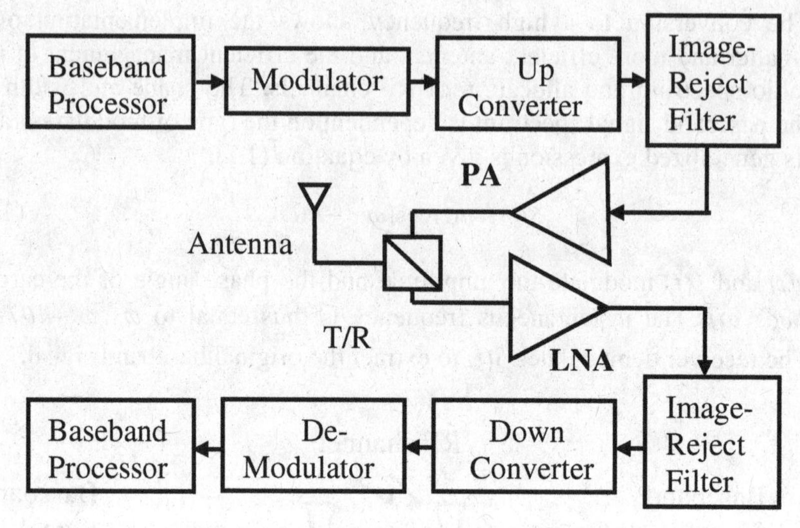

Figure 1.2 A typical heterodyne transceiver.

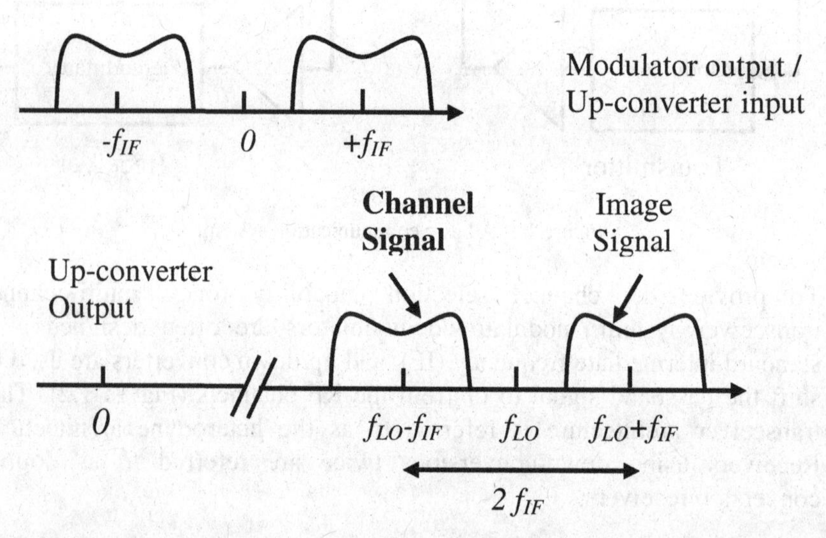

Figure 1.3 IF to RF up-conversion with interchangeable channel and image signals.

In Figure 1.3, the wanted channel is located at f_{LO} - IF so the image signal located at f_{LO} + IF has to be filtered out before transmission. The process is described as using a high injection of the local oscillator. Vice versa, for a low injection of the LO, the channel signal is located at f_{LO} + IF and the image signal at $f_{LO} - IF$.

At the receiver end, if the high injection method is used, $f_{channel} = (f_{LO} - f_{IF})$ is down-converted to f_{IF} as shown in Figure 1.4.

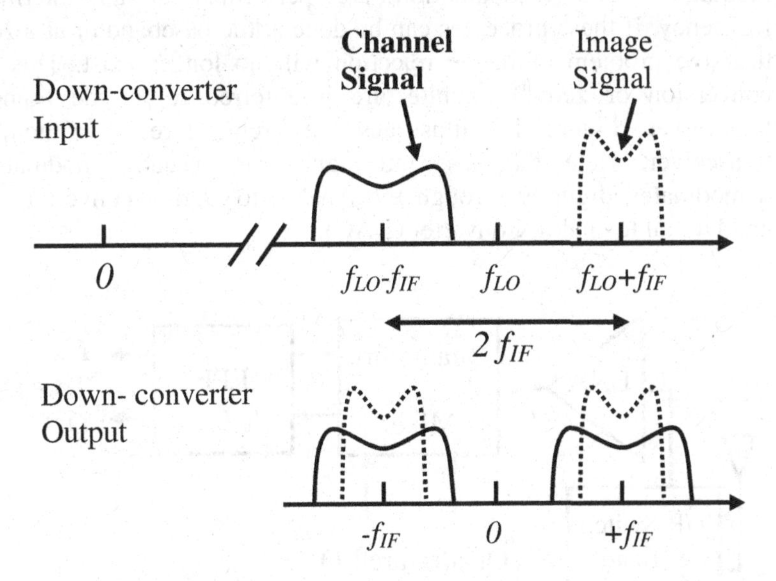

Figure 1.4 Interference by image signal in the down-conversion from RF to IF.

Figure 1.4 illustrates that if $f_{image} = (f_{LO} + f_{IF})$ also exists at the mixer input, an interference will be produced at the output. Hence the image has to be filtered out before the down-conversion. The image-reject filters in Figure 1.2 usually present a bottle neck in the implementation of a fully integrated transceiver on a chip due to the difficulty in implementing high-Q on-chip inductors.

The off-chip solutions require SAW (surface acoustic wave) filters, or ceramic filters, or high-Q LC filters. In order to reduce the requirements of the image reject filter, a high IF is commonly used to separate the wanted signal and the image. A high IF will, however, lead to high power consumption in the IF circuits, thus a compromise between the image rejection and the power consumption is necessary.

The discussion so far is based on the heterodyne architecture where the modulation and demodulation are performed at an intermediate frequency. If these processes can be done at the baseband or at a zero IF then the problem of image rejection will no longer exist. This direct conversion or zero-IF architecture is referred to as the homodyne transceiver. Figure 1.5 illustrates the architecture of a homodyne transceiver. Here the baseband signals are usually modulated or demodulated digitally through the Analog to Digital converter (ADC) and Digital to analog Converter (DAC).

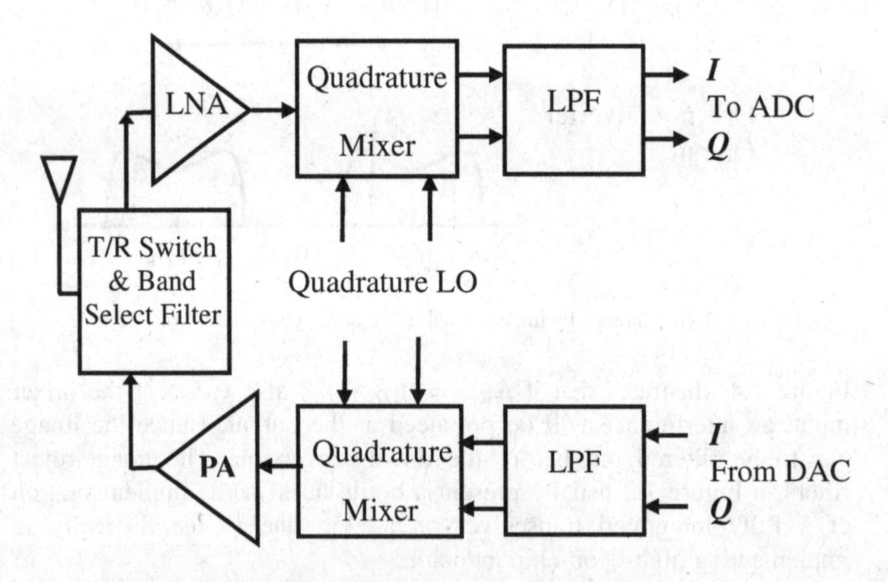

Figure 1.5 Simplified block diagram of the homodyne transceiver front-end.

The conversion of baseband signals to and from RF signals requires, however, a quadrature topology due to the asymmetrical spectra of the modulated signals. In the transmitter, the modulated baseband signal has the I and Q components, which will be respectively mixed with two 90° phase shifted LO signals then combined before applied to the power amplifier. In the receiver, the RF signal is equally splitted and respectively mixed with two 90° phase shifted LO signals to produce the I and Q components of the modulated baseband signal, which is usually demodulated digitally.

The direct translation of the spectrum to and from the zero frequency creates new problems. In the reception, severe DC Offsets of baseband signals can occur due to the leakage of the LO in the RF path or the leakage of RF signal in the LO path. This DC offset can saturate the baseband stages prohibiting the detection of the desired signal. For certain semiconductor technologies, the high baseband noise level could affect the receiver sensitivity.

For the direct conversion transmitter, the main problem is the coupling of the strong modulated signal at the PA output to the LO of the same center frequency. The LO signal can be corrupted by the transmitted signal through a mechanism called injection pulling. The quadrature conversion can be implemented with a 90° phase shift in the RF path or the LO path. Mismatches in gain and phase shift of I and Q signals increases errors in the receiver. In spite of these problems, the advantage of removing the off-chip image reject filter is attractive and the direct conversion techniques have gained popularity in recent years [4-8].

In certain applications where the requirements for image rejection in the receiver are less stringent, the low IF techniques can be used to avoid the problem of off-chip image reject filter in the heterodyne architecture, and the problem of DC offset and baseband noise in the homodyne architecture.

The IF is usually selected from a few hundred kHz to less than 2MHz. The low IF architecture is quite similar to that of Figure 1.5, except for the replacement of the low pass filter (LPF) with the bandpass filter followed by a quadrature phase shifter or polyphase filter. The I and Q outputs are then subtracted to cancel out the image signal. An image rejection of more than 30dB can be achieved with this technique. These image reject techniques are described in [2,9]. As the IF is relatively low, the architecture of Figure 1.5 can also be used directly, and the quadrature phase shift and image subtraction can carried out digitally after the analog to digital conversion to obtain better image rejection.

1.3 Modulation and Demodulation Techniques

Equation (1.1) indicates that the informative baseband signal can be incorporated in $a(t)$ or $\theta(t)$ to modulate the amplitude or the phase angle of the carrier $A_c \cos \omega_c t$ generated in the transmitter. In the receiver, the original baseband signal is extracted from the received signal in a demodulation process. An amplitude-modulated (AM) signal is described by equation (1.2):

$$S_{am}(t) = A_c[1 + m(t)]\cos(\omega_c t) \tag{1.2}$$

where $m(t)$ represents the modulating signal in the baseband, and $|m(t)|$ is the modulation index. The AM process is just a multiplication of the carrier and the baseband signal with a certain amount of dc offset, and can be implemented by a mixer as in the frequency conversion described in the last section. The shape and bandwidth of the signal spectrum is not changed. In the receiver, the demodulation can be a down conversion of the RF signal to the baseband (coherent detection) or a detection of the envelope of the RF signal. A small drop of the Signal to Noise Ratio (*SNR*) occurs in the demodulation.

As the instantaneous frequency of an angle modulated signal is equal to $\omega = \omega_c + d\theta / dt$, the angle modulation process is commonly implemented through either the phase modulation (PM) or frequency modulation (FM) and described by equations (1.3) and (1.4) respectively:

$$S_{pm}(t) = A_c \cos(\omega_c t + k_p m(t)) \tag{1.3}$$

$$S_{fm}(t) = A_c \cos(\omega_c t + k_f \int m(t)dt) \quad (1.4)$$

The modulation index is given by $\beta_p = k_p\, m(t)$ for a PM signal, and $\beta_f = k_f\, m(t)/(2\pi f_m) = \Delta f / f_m$ for an FM signal, where f_m is the frequency of $m(t)$ and $\Delta f = k_f\, m(t)/2\pi$ is the frequency deviation. The instantaneous frequency of an FM signal, $\omega = \omega_c + k_f\, m(t)$ is linearly related to the modulating signal. For several decades, the frequency modulation is more widely used due to its simpler implementation in analog transceivers for both narrow band ($\beta_f \ll 1$) and wideband ($\beta_f \gg 1$) applications.

The FM signal bandwidth is governed by the modulation index according to the Carson's rule $W = 2(\beta_f + 1)B_{bb}$, where B_{bb} is the baseband bandwidth or the maximum value of f_m. The demodulation of a wideband FM signal gives an improvement of the output *SNR* by a factor $3(\beta_f + 1)\beta_f^2$. The basic principles of modulation and demodulation are described in [3].

The migration of communication systems from analog to digital during the 80s gave the parallel definitions of Amplitude Shift Keying (ASK), Frequency Shift Keying (FSK), and Phase Shift Keying (PSK) to the digital modulation process. For the simple binary modulation, ASK is referred to as On-Off keying (OOK), while FSK and PSK are referred to as BFSK and BPSK respectively. The modulating signal $m(t)$ is a string of data:

$$m(t) = \sum n b_n p(t - nT_b) \quad (1.5)$$

where b_n is the bit value over the duration T_b, having two possible values, e.g. 0 and 1, or -1 and +1, and $p(t)$ is a pulse shape function, e.g. rectangular, raised-cosine, etc. The optimum detection of $m(t)$ is thus obtained by a correlation of the received signal with the reference pulses, e.g. $p_1(t)$ for bit '0' and $p_2(t)$ for bit '1'.

The optimum detection is also referred to as coherent detection, and the correlator is also referred to as the matched filter. The mathematical treatments given in [3,10] show that a maximum *SNR* obtained for the optimum detection is dependent only on the energy of each bit regardless of its pulse shape and bandwidth. The detection performance of the type of modulation is measured by its Bit Error Rate (BER), which is dependent solely on the ratio E_b/N_o, where E_b is the average bit energy, and $N_o/2$ is the two sided power density of additive white noise. E_b is proportional to the signal power and the bit duration T_b. A better or lower BER is achievable by increasing the signal power or T_b. OOK and BFSK are found to have the same performance for the same average bit energy. BPSK can have the same performance as OOK and BFSK for half of the later average bit energy. The ratio E_b/N_o is related to the *SNR* by the expression: $E_b/N_o = SNR \times (B_{bb} / R)$, where the data rate $R = 1/T_b$, $E_b = SB_{bb}T_b$, and the *SNR* is defined as the ratio of the signal power density S to the noise density N_o.

In order to increase the data rate without increasing the signal bandwidth, many data bits can be grouped together in one "symbol". Multilevel digital signals are referred to as M-ary signaling, e.g. MASK, MFSK and MPSK. In this case, b_n can assume more than two values in equation (5), and each transmitted level is called a symbol. The term "symbol rate" thus replaces the term bit rate. The Quadrature Amplitude Modulation (QAM) and Quadrature Phase-Shift Keying (QPSK) are the commonly used signaling schemes with M = 4, where two bits are transmitted simultaneously by quadrature carriers. For a selected modulation type, the required *SNR* is proportional to the number of bits per symbol over the required E_b/N_o. For examples, the 4-QAM transmitting 2 bits per symbol will require an *SNR* of 13.6dB or 3dB higher than the required E_b/N_o of 10.6dB for a BER of 10^{-6} for $B_{bb}T_b = 1$.

Similarly, the 16-PSK transmitting 4 bits per symbol will require an *SNR* of 24.3dB compared to the required E_b/N_o of 18.3 B. The early MFSK signaling scheme is not bandwidth efficient due to the wide spacing of multiple frequency components. This problem is overcome in the Orthogonal-Frequency Division Multiplexing (OFDM) technique.

An OFDM signal of duration T consists of a set of multiple complex exponential sub-carriers arranged with a frequency spacing of $1/T$ so that the peak of one sub-carrier spectrum occurs at the zero notch of the other spectra of all other sub-carriers to maintain their orthogonality [11]. Equations (1.3) and (1.4) are applicable to digital modulated signals and can be rewritten as:

$$S_{pm}(t) = A_c \cos(k_p m(t))\cos(\omega_c t) - A_c \sin(k_p m(t))\sin(\omega_c t) \quad (1.6)$$

$$S_{fm}(t) = A_c \cos(k_f \int m(t)dt)\cos(\omega_c t) - A_c \sin(k_f \int m(t)dt)\sin(\omega_c t) \quad (1.7)$$

Hence all PSK and FSK signals can be generated by the computation of the I and Q baseband components, e.g. $\cos [k_p m(t)]$ and $\sin [k_p m(t)]$, then up-convert them to the RF frequency with the quadrature carriers $A_c \cos \omega_c t$ and $A_c \sin \omega_c t$. The demodulation is simply the quadrature down conversion process to recover the I and Q baseband components then to compute $m(t)$. Digital signal processors (DSP) are used in modern transceivers to compute the I and Q components from $m(t)$ or vice versa. High speed Digital to Analog Converters (DAC) and Analog-to-Digital Converters (ADC) are the interface circuits between the DSP and the RF front end.

1.4 Multiple Access Techniques

Two common techniques are used to provide the simultaneous two-way communications. In Time-Division Duplexing (TDD), the antenna is switched between the transmitter and the receiver.

In Frequency Division Duplexing (FDD), the transmitter and the receiver operate on different frequency bands, and two bandpass filters are combined as a duplexer filter to interface the transmitter and the receiver with the antenna. FDD and Frequency-Division Multiple Access (FDMA) were used in early analog cellular phones,

Advanced Mobile Phone Service (AMPS). Transmitting & receiving bands of 824-849 & 869-894MHz respectively are partitioned into 830 channels of 30kHz width, distributed over K cells. A few channels are reserved for supervisory signals to initiate, maintain and terminate a call. Time-Division Multiple Access (TDMA) allows a multiple access of the same channel at different times.

For a time-frame T_F, each user will access to the channel within a time-slot T_{sl}. Digital signals are compressed, coded, and transmitted only in allocated slots. Examples are the North America Digital Cellular (NADC) system, the Digital European Cordless Phone (DECT), and the Global System For Mobile Communications (GSM). Code-Division Multiple Access (CDMA) is a spread spectrum technique, allowing multiple users to access to the same frequency band at the same time. In the Direct-Sequence Spread Spectrum (DSSS) CDMA technique (also called DS-CDMA), the baseband data of each user is multiplied with a unique coded sequence of shorter pulses called chips. Thus, if the data bit duration is n times the chip duration, the baseband bandwidth is spread by n times after being coded. Usually, bipolar data and sequences are used for spreading. The received signal can only be de-spread by a multiplication with the same code.

The performance is dependent on the orthogonal property of the code, i.e. if the received signal and the reference signal are of different codes, the multiplication will yield a zero. The IS-95 standard employs DS-CDMA in mobile phones with the data rate of 9.6kbps, while the IEEE 802.11b uses this technique in LNA with data rates up to 11Mbps (Mega bits per second). Frequency Hoping (FH) CDMA can be viewed as FDMA with pseudo-random channel allocation. The carrier frequency of the transmitter is hopped according to a channel code. This technique is currently popular in wireless LAN (Local Area Network) applications with moderate data rate, e.g. the Blue tooth standard uses FH for data rate up to 1Mbps. High data rates are usually achieved through the M-ary modulation schemes.

OFDM is the latest technology employed in wireless LAN for high data rates up to 54Mbps under the IEEE 802.11a & g standards [11]. Most modern communication systems employ digital modulation and demodulation techniques combined with the multiple access techniques which are complex and more suitable to be implemented at the baseband by the digital signal processor. These digital baseband techniques are well described in [3, 10].

1.5 Receiver Sensitivity and Linearity

The receiver sensitivity is the minimum input signal level that can be detected by the receiver with a certain performance quality measured at the output, e.g. *SNR*, E_b/N_o, and *BER*. For a chosen type of modulation, the E_b/N_o, and the *BER* can be determined from the *SNR*. To compare the *SNRs* at the input and output of a circuit or system, the noise factor is defined as $F = SNR_{in} / SNR_{out}$ and the noise figure $NF = 10 \log F$ is a measure of the degradation of the *SNR* from the input to the output in dB.

The conventional approach for the calculation of the noise factor is based on the power matching of the circuit or system input to the signal/noise source, so the input noise power is simply dependent on the bandwidth concerned B, $N_{in} = kTB$, where $k = 1.38 \times 10^{-23}$ J/K and T is the absolute temperature. The channel bandwidth is usually used as B in this calculation. If G is the available power gain of the circuit or system then $F.\ kTB = N_{out} / G$ is call input referred noise power, where N_{out} is the output noise power. From the definition of *NF*, the receiver sensitivity can be calculated as the minimum input signal power for a certain requirement of the output *SNR*:

$$p_{in,\min} = NF + 10\log(KTB) + SNR_{out} \tag{1.8}$$

where $10\log kT = -174$ dBm/Hz is the single sided thermal noise power spectral density. The calculation of the noise factor of a system formed by multiple cascaded stages can be facilitated by Friis' equation:

$$F = F_1 + \frac{F_2 - 1}{G_1} + \ldots + \frac{F_n - 1}{G_{n-1}} \tag{1.9}$$

where F_i and G_i represent the noise factor and available power gain of stage i respectively. Friis' equation is only valid when the output of one stage is power matched to the input of following stage, usually at a low impedance value, e.g. 50Ω. However, in a fully integrated receiver, the power matching of one stage to the following stage is not implemented hence the concepts of power gain and noise figure have to be modified to suit the voltage mode of operation. The noise factor is defined by IEEE as the ratio of the total output noise to the input noise due to source. Usually, the mean square value of spot noise voltage is used in lieu of noise power, and the mean square voltage of spot noise due to the source resistance R_s is 4 kTR_s. Hence the noise factor is given by [1]:

$$F = \frac{V_{n,out}^2}{4kTR_s A_{vs}^2} \tag{1.10}$$

where the voltage gain A_{vs} is the ratio of the output voltage to the source voltage. To ease the calculation of the noise factor of a cascade system, the loaded voltage gain A_{vl} is defined the ratio of the voltage at the load to the voltage at the input of the circuit or system [12]. Hence, the total voltage gain will be the product of individual circuit blocks' gains, where the output of each circuit block is loaded by the input impedance of the next circuit block. The overall noise factor of a cascade system is then given by:

$$F = 1 + \left(\frac{R_s + R_{in}}{R_{in}}\right)^2 \frac{V_{n,1}^2 + \frac{V_{n,2}^2}{A_{vl1}^2} + \frac{V_{n,3}^2}{A_{vl2}^2}}{4kTR_s} \tag{1.11}$$

where R_{in}, $V_{n,i}^2$ and A_{vli} represent the input impedance of the first stage, the mean square value of input referred noise and the loaded voltage gain of the building block i respectively. The sum $V_{n,in}^2 = V_{n,1}^2 + \frac{V_{n,2}^2}{A_{vl1}^2} + \frac{V_{n,3}^2}{A_{vl1}^2 A_{vl2}^2} + \ldots$ is the total input referred noise of the system. The input referred noise of each individual stage is referred back to the input of the first stage so its value is divided by the gain of all preceding stages.

The linearity of a receiver is described by the maximum input signal level beyond the distortion of the input signal will cause problems for the reception or detection of the signal. The two commonly used specifications are the 1-dB compression point, and the input referred third order intercept point (*IIP3*). *IIP3* is the input power at which the third order intermodulation output power (P_{IM3}) is equal to the output power of the fundamental component. The 1-dB compression point is the input power at which the available power gain is dropped by 1dB. It can be shown that the 1-dB compression point is 9.6dB less than *IIP3* [1]. For a cascade system, we have:

$$\frac{1}{IIP3} = \frac{1}{IIP3_1} + \frac{G_1}{IIP3_2} + \frac{G_1 G_2}{IIP3_3} + .. \qquad (1.12)$$

However, for the same reason as mentioned before regarding the voltage mode operation of the integrated receiver, the total mean square value V_{IIP3}^2 of the *IIP3* are given by [12]:

$$\frac{1}{V_{IIP3}^2} = \frac{1}{V_{IIP3,1}^2} + \frac{A_{vl1}^2}{V_{IIP3,2}^2} + \frac{A_{vl1}^2 A_{vl2}^2}{V_{IIP3,3}^2} + .. \qquad (1.13)$$

where $V_{IIP3,i}^2$ is the mean square voltage of *IIP3* of stage *i*.

As the input impedance of a receiver is usually designed at 50Ω to match R_s, we can calculate $IIP3 = V_{IIP3}^2 / 50$ and determine the maximum allowable input power to a receiver, $p_{in,\max}$ according to equation (1.14) [1]:

$$p_{in,\max} = \frac{2IIP3 + NF + 10\log kTB}{3} \qquad (1.14)$$

At this input level, the input referred P_{IM3} equates to (NF + 10 log kTB), the input referred noise power. The difference between $P_{in,\max}$ and $P_{in,\min}$ is the spurious free dynamic range of the receiver. A systematic approach for using equations (1.8-1.14) is described in [12] with numerical examples.

1.6 On-chip Power Amplifier

Many current single chip transceivers designed for short range communications are implemented with on-chip power amplifiers (PA). The current deep submicron technology operates on a low supply voltage and has a low breakdown voltage so most existing designs of PAs have an output power less than 1W or 30dBm. If the antenna impedance is 50Ω then the required voltage swing is 20Vpp while a supply voltage of 3V or less is usually preferable in mobile applications.

Although passive impedance transformation circuits can be used to substitute the required high voltage swing with a high current swing in the PA, these impedance transformation circuits will cause loss so the output power will be much less than 1W. Furthermore, in most applications, linear PAs are required so low efficiency PA topologies such as class A and class AB have to be used. As the efficiency is usually less than 30% for class A PAs and is about or less than 50% for class AB PAs [13-15], the power to be dissipated through the chip is more than 1W. Consequently, the performance of other circuits on the chip may be affected by the higher operating temperature. This problem can be avoided if the output power of on-chip PAs is limited to around 20-26dBm for applications with operating distances within a few hundreds meters. For longer operating distances, off-chip high power PAs can be used as power boosters.

To improve the efficiency of the PA, off-chip high Q inductors have been commonly used in PA design. The requirement of fully integrated PA with good efficiency is still a challenge for many designers. The maximum input power to the PA is usually related to the nonlinearity of the PA and the 1-dB compression point and *IIP3* as described in the previous section. The adjacent channel power ratio (ACPR) is also a commonly used parameter describing the linearity of the PA. ACPR is defined as the ratio of the output power in the desired channel to the spurious power in the adjacent channel measured over the same bandwidth.

1.7 The Cellular Phone Concept

Early two-way radio-telephone services are mainly used in commercial applications in simplex mode due to the limited number of available channels. Examples of the simplex or push-to-talk mode are VHF radios used by taxis and police.

The operating distance of the system is mainly determined by the received *SNR*. A less sensitive receiver can be compensated by a higher power transmitted signal. The introduction of cellular phones during the 80's is based on the frequency reuse concept to provide a wider service to public. Figure 1.6 shows the common 7-cell structure. Different frequencies are used in each cell, but the same frequency pattern can be repeated in the adjacent 7-cells. Other common reuse patterns are with K = 4, 12, and 19. The relationship between the cell radius and the frequency reuse distance is [2]:

$$q = D/R = \sqrt{3K} \tag{1.15}$$

e.g. $q = D/R = 4.6$ for $K = 7$

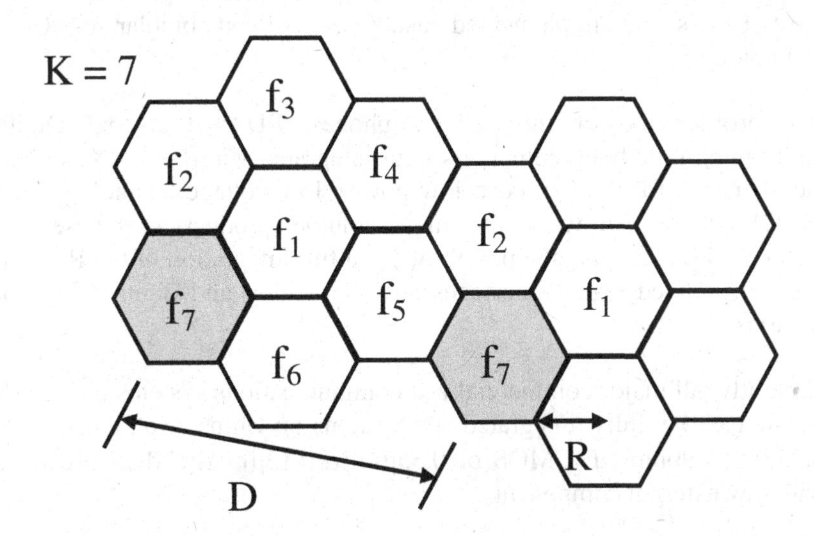

Figure 1.6 Principle of 7-cell frequency reuse.

In a cellular phone system, co-channel interference is caused by the frequency reuse. The carrier/interference ratio is, however, independent of the transmitted power of each cell. For 6 sources of interference, we have:

$$\frac{C}{I} = \frac{R^{-\gamma}}{6D^{-\gamma}} = \frac{q^{\gamma}}{6} \tag{1.16}$$

where γ is the propagation loss slope determined by the environment, e.g. $\gamma = 4$ in the urban environment, thus for the 7-cell structure, $q = 4.6$ and C/I = 18.7dB. Hence, there is no advantage to increase the transmitted power, or to improve the receiver sensitivity to obtain the *SNR* much higher than the *C/I* ratio determined by the K-cell structure.

1.8 The CMOS RF Technology

RF integrated circuits (RFICs) are made up of high speed transistors and passive components like resistors, inductors, capacitors and varactors. The common performance parameters are the operating frequency, frequency range, gain, loss, linearity and power efficiency. Until recently, most RFICs are implemented using the silicon bipolar or GaAs technology [16].

The proliferation of the cellular phones, PDAs (Personal Digital Assistants), note book computers with integrated wireless LAN devices, etc. demands for the low cost, low power, low voltage technology such as CMOS or BiCMOS for high volume production. These two technologies offer also the possibility of a full integration of the RF front ends, the mixed signal interfaces and the analog and digital baseband sections.

Currently, all major commercialized communications systems mentioned above can be fully integrated as Systems-on-Chips (SoC) using the 0.25μm to 90nm RF CMOS or 0.35μm to 0.13μm BiCMOS processes with few external components.

The selection of the technology for the RF front end is usually decided by the operating frequency and f_T and f_{max} of the technology. The scale down of the device size has increased f_T from around 30GHz for the 0.25μm process to around 80GHz for the 0.13μm process [17]. Linear RF circuits are commonly designed to operate at less than 1/10 of f_T. For example, the 0.35μm or 0.25μm CMOS technology is usually chosen for applications at less than 2.5GHz. The 0.25μm or 0.18μm CMOS technology would be selected for application in the ISM 2.4GHz band, while the 0.18μm or 0.13μm CMOS technology for the higher operating frequencies. The BiCMOS SiGe (Silicon Germanium) technology offers a better performance in terms of f_T, noise, and power handling, but at a significantly higher cost.

To the IC industry, the cost of the chip is the most crucial factor in the selection of the technology, the partition of the system, and the level of integration. The manufacturing price difference from one technology generation to the next is quite substantial that the single chip solution is not always the best solution for a specific application. For example, the wafer price for the 0.18μm CMOS process is more than twice the price for the 0.35μm CMOS process. If the baseband and mixed signal portion of the circuit is implementable in the 0.35μm technology and is significantly larger than the RF portion of the circuit, then the decision may be to implement the RF portion separately in the 0.18μm technology. The packaging price and electrical characteristics of the package also influence the decision of chip partition.

For example, the single chip solution results in the requirement of a large package for the large die. The large package has longer bond wires which could affect the performance of the RF circuits. This problem could be resolved by using more expensive chip scale packaging (CSP) types such as the flip chip. Until recently, most RF power amplifiers (PA) and Transmitter / Receiver (T/R) switches have been implemented as separate chips using the GaAs technology. CMOS PAs and T/R switches handling up to 30dBm have now been developed for short range wireless applications up to several hundred meters [15, 18]. Single chip CMOS RF transceivers including PAs and T/R switches are becoming a reality.

References

[1] Behzad Razavi "RF Microelectronics", Prentice Hall, New Jersey, USA, 1998.

[2] William C.Y. Lee "Mobile Cellular Telecommunications – Analog and Digital Systems", McGraw-Hill, New York, 1995.

[3] Leon W. Couch "Digital and Analog Communication Systems", Prentice Hall, London, 1997.

[4] B. Lindquist et al., "A New Approach to Eliminate the DC Offset in TDMA Direct Conversion Receiver", *IEEE Vehicular Technology Conference*, pp. 754-757, 1993.

[5] J. K. Cavers, M. W. Liao, "Adaptive Compensation for Imbalance and Offset Losses in Direct Conversion Transceivers", *IEEE Transaction on Vehicular Technology*, vol. 42, no. 4, pp. 581-588, 1993.

[6] M. Lehne, J. T. Stonick, U. Moon, "An Adaptive Offset Cancellation Mixer for Direct Conversion Receivers in 2.4GHz CMOS", *ISCAS 2000,* May 28-31, 2000.

[7] Zhaofeng Zhang; Tsui, L.; Zhiheng Chen; Lau, J.; "A CMOS Self-mixing-free Front-end for Direct Conversion Applications", *Circuits and Systems, ISCAS 2001. The 2001 IEEE International Symposium on*, Vol. 4, 6-9 May 2001 pp. 386 –389.

[8] J. Pihl, K. T. Christensen, E. Bruun, "Direct Downconversion with Switching CMOS Mixer", *The 2001 IEEE International Symposium on Circuits and Systems, ISCAS 2001*, vol. 1, 6-9 May 2001.

[9] M.A. Do, K. Linggajaya, K.S. Yeo, and J.G. Ma "Broadband Image Reject Down-converters", *2003 International Symposium On VLSI Technology, Systems, and Applications (VLSI-TSA),* 23-25 April 2003, Hsinchu, Taiwan, Invited Paper.

[10] Bernard Sklar "Digital Communications – Fundamentals and Applications", Prentice Hall, New Jersey, 1988.

[11] Bahai, Ahmad R.S. "Multi-carrier Digital Communications: Theory and Applications of OFDM", Kluwer Academic/Plenum, 1999.

[12] W. Sheng, A. Emira, E. Sanchez-Sinencio, "CMOS RF Receiver System Design: A Systematic Approach", *IEEE Transactions on Circuits and Systems—I Regular Papers*, Vol. 53, No. 5, May 2006.

[13] J. Griffiths and V. Sadhir, "A Low Cost 3.6V Single-Supply GaAs Power Amplifier For The 1.9-GHz DECT System," *Conference Proceedings, IEEE NTC '95 The Microwave Systems Conference*, Orlando, May 17-19, 1995, pp. 37-40.

[14] Y.J.E. Chen, M.Hamai, D.Heo, A. Sutono, S.Yoo, and J.Laskar, "RF Power Amplifier Integration in CMOS Technology", *IEEE MTT-S Digest* pp.545-548, 2000.

[15] Cheng-Chi Yen and Huey-Ru Chuang, "A 0.25-_m 20-dBm 2.4-GHz CMOS Power Amplifier With an Integrated Diode Linearizer', *IEEE Microwave and wireless components letters*, vol. 13, no. 2, pp. 45-47, February 2003.

[16] Larson, L.E. "Integrated Circuit Technology Options for RFIC's – Present Status and Future Directions", *IEEE J. of Solid State Circuits*, Vol. 33, pp. 387-399, 1998.

[17] Tajinder Manku "Microwave-CMOS – Device Physics and Design", *IEEE J. of Solid State Circuits*, Vol. 34, No. 3, 1999.

[18] Pradeep B. Khannur,"A CMOS Power Amplifier with power control and T/R switch for 2.45-GHz Bluetooth/ISM band application," *2003 IEEE Radio Frequency Integrated Circuits Symposium*, pp.145-148.

When knowledge ends, imagination begins.

Kiat Seng YEO

CHAPTER 2

RF CMOS Devices and Process Design Kits

2.1 Introduction

CMOS technology [1] has emerged as the top solution for wireless applications due to its cost advantage, performance improvement, and ease of integration for high performance digital circuits and high-speed analog/RF circuits. This is especially true if one wants to achieve the ultimate goal of full integration: the complete transceiver system on a single chip, both the analog front-end and the digital demodulator implemented on the same die [2]. This can only be achieved in either a CMOS or a BiCMOS technology. Each different component on the same chip has to be suitably optimized such that every component's performance complements each other and functions effectively as a whole.

A critical issue for the development of microwave circuits using MOS transistor [3] is the availability of a compact MOS model that is valid for a broad range of bias conditions and operating frequencies. An important issue for RF devices is the availability of compact RF models of both the active devices and the passive devices to accurately predict the RF characteristics at high frequencies, including the millimeter wave range [4]. For modeling of RF MOSFETs, the substrate-signal coupling network model and non-quasi-static effect (NQS) are important at high frequency close to and higher than f_T.

With the increasing demand of low-cost, low-noise, and low-power wireless communication systems, on-chip inductors play an extremely significant role in the design of radio frequency (RF) front-end circuitry

[5]. RF inductors have been widely used in matching networks, LC tanks, passive filters, transformers, etc, and have recently been introduced as tuning elements for RF applications [6]. Variation in inductance of the tunable RF inductors allow for adjusting the impedance matching as well as the frequency selection to optimize the circuit performance.

Passive transformers [7] are major components in RF transceivers that are used in applications such as impedance transformation to achieve efficient power transfer, for isolation of DC currents and voltages while maintaining efficient AC transmission as well as for balun applications to obtain single-ended-to-differential functionality and thereby improve noise suppression. The key to accurate modeling is the ability to identify the transformer loss mechanisms, the relevant parasitic, and their effects.

The decrease in minimum feature size of devices has led to a proportional decrease in interconnect cross-sectional area and pitch [8]. The parasitic resistance, capacitance, and inductance associated with interconnects are beginning to influence the circuit performance and have increasingly become one of the primary showstoppers in the evolution of deep submicron technology. For on-chip communication [9], traditional interconnects transmit baseband signals and are *RC* dominant. They have inherent signal distortion and large *RC* delay and the delay cannot scale as well as transistor speed.

Within the VCO, the inductor and varactor [10] capacitance, which lie at the core, set the oscillation frequency and their losses largely determine the performance of the VCO itself. As frequencies have increased, it has become increasingly difficult to route high-frequency signals to external components. In addition, the benefits of system integration in terms of size and cost reduction have meant that both the varactor and the inductor are now commonly integrated onto the same substrate as the remaining electronics. It is, therefore, desirable to be able to characterize and to optimize the quality factor of the varactor as well as that of the inductor.

On-chip RF capacitor [11] is one of the key components for RF integrated circuit (RFIC) designs such as filters, phase shifters, and oscillators. Geometry design variables include number of fingers, finger length, finger width, spacing, dielectric constant, which would affect the quality factor and capacitance of the device. Poor Q factor is primary due to the resistive losses in the plates and contacts and due to the parasitic capacitance between the passive component and the lossy silicon substrate.

A process design kit (PDK) is a collection of verified data files that are used by a set of custom IC design EDA tools to provide a complete analog/mixed-signal/RF design flow. Optimized model files ensure the accuracy and consistency of device behavior at the backend so that the front-end circuit simulations can be correct. A good PDK has a vast dataset and ruleset for schematic simulation in the front end which works under a strong set of physical backend SPICE models.

2.2 RF Transistors

MOSFETs [12][13] are one of the three basic types of field effect transistors besides MESFETs and JFETs. All three have the same operational principle where the resistance of a conducting channel is modulated by applying a signal voltage called gate-source voltage. MOSFETs based silicon are used extensively in integrated circuits, the excellent insulating properties of silicon dioxide playing an important part here. The low power consumption of the MOSFET makes the technology suitable for portable devices.

The MOSFET consists of four terminals: gate, drain, source, and body. Normally, the source and the body terminals are connected to each other, so that it becomes like a three-terminal device. But as the operating frequency increases, the substrate current will increase and flows through the distributed RC network in the substrate. This will caused the potential of the intrinsic substrate node to be different from the source and the extrinsic body terminal.

Therefore to model a four-terminal MOSFET, additional DC measurements must be made to characterize the biasing effect on the body structure. Since the biasing effect of the body structure can be characterized in the DC measurement, the source and the body terminals can be shorted together and measured as a two-port network during the RF measurement so as to extract its characteristic and the RF parasitics of the device layout.

2.2.1 BSIM3v3 Model

BSIM3v3 [14][15] is a dc scalable MOSFET model. It is considered a physical model because most of its parameters have strong correlation to the process and device structure design. But there are also some parameters that have weak physical meaning and are only introduced for the model fitting purposes. BSIM3v3 has been widely used by the industry and it is initially developed for analog and digital circuit simulation. However, for the RF simulation, it lacks of some important models that are required to predict the RF parasitic and therefore, sub-circuit components are added to the core model to simulate the RF parasitic effects [16] [17] using the macro modeling approach.

In the core model, the gate resistance R_g is not included and hence it is unable to accurately predict the input admittance. At DC and low frequency region, R_g is purely the gate electrode resistance. But as the frequency increases, the gate electrode must be treated as a distributed transmission line [18] so as to model R_g accurately. By treating the gate terminal as a distributed transmission line, the gate impedance can be derived:

$$Z_{in} = \frac{N}{j\omega C_p w} + \frac{\rho_{poly} w / L}{3N} + \frac{j\omega L_s w}{3N}, \quad \text{Single contacted gate} \tag{2.1}$$

$$Z_{in} = \frac{N}{j\omega C_p w} + \frac{\rho_{poly} w / L}{12N} + \frac{j\omega L_s w}{12N}, \quad \text{Double contacted gate} \tag{2.2}$$

The variable N is defined as the number of finger, C_p is the gate capacitance, ρ_{poly} is the gate resistivity, L and w is the length and width of a single finger, and L_s is the series inductance in the gate terminal. A detailed derivation for (2.1) and (2.2) can be obtained in [18]. Besides the distributed gate resistance which is important in wide transistors, another physical effect is the distributed or non-quasi-static (NQS) effect in the channel [16]. The common approach for RF modeling for BSIM3v3 is by adding sub-circuit components to model R_g and R_{sub} resistances [18] and several other RF parasitic effects such as deep n-well biasing. Other models with different sub-circuit components are reported in [19-24].

2.2.2 BSIM4 Model

The R_g effect is included into its core model in BSIM4 [14]. This resistance is being separated into gate electrode resistance and channel-reflected resistance. As the operating frequency increases, the gate electrode is being treated as a distributed resistance so that it can be modeled more accurately. The gate electrode resistance is derived in (2.1) and (2.2). The channel-reflected resistance is not a physical resistance. It is the resistance as "seen" by the gate signal and is a function of biasing [25]. This channel-reflected gate resistance is one of the various approaches to account for the Non-Quasi-Static (NQS) effect [25]. Quasi-Static (QS) [14] is defined as the ability for the charges in the channel to respond immediately to the biasing and most of the commercially available models are QS models. When the RF MOSFET is operated at high frequency, the response speed of the device to the input biasing may not be fast enough and the device will show some signal delay - this is the NQS effect. In BSIM4, different configurations of the R_g model can be selected via the parameter RGATEMOD.

In BSIM3v3, there is no internal substrate resistance. Hence, the model is unable to fit the output admittance of the RF MOSFET. In BSIM4, the substrate resistances are now being modeled by a five resistors network. The substrate network can be selected by the parameter RBODYMOD. Note that this substrate network is not scalable, since they have no geometry dependence parameter in the model [15].

In BSIM4, the inclusion of the gate and substrate resistance model has made BSIM4 another alternative for RF modeling. But as the proposed substrate network in BSIM4 is not scalable, it is unable predict the RF characteristics for different geometrical sizes [15]. Similarly, for BSIM3v3 model, the sub-circuit components that are used to model substrate resistances are also not scalable with geometry. Therefore, there is a need to improve the model so as to achieve RF scalability.

2.2.3 Figure of Merit

In multi-fingered transistor design, the size of the transistor is mainly controlled by the finger number (N_f), unit width (W_f) and channel length (L_g) of the transistor. For most RF circuit design, transistors with smallest gate length are always used due to their fast response and high drain current. Therefore, the RF circuit designer will need to select transistors based on either N_f or W_f. By optimizing per finger unit width with respect to f_T, f_{MAX}, minimum noise figure (NF_{min}) and flicker noise, the best W_f of the transistor can be selected to be used in a specific circuit application, such as the low noise amplifier (LNA), voltage controlled oscillator (VCO), and mixer.

The following figure of merits f_T, f_{MAX}, NF_{min} and flicker noise spectral density have been commonly used in [26], [27], [28], [29] to characterize the performance for the RFCMOS transistor, but these FOMs are normally presented with respect to the change in technology or channel length.

It is important to study these 4 FOMs with respect to W_f on different transistor sizes so as to obtain the optimized width per finger based on either one or all the 4 FOMs.

2.2.3.1 f_T definition and extraction

f_T is defined as the unity current gain frequency at which the short circuit current gain of the transistor becomes unity and is shown in Eqn. (2.3).

$$\omega_T = 2 \cdot \pi \cdot f_T = \frac{g_m}{C_g} \tag{2.3}$$

$$C_g = C_{gs} + C_{gb} + C_{gd} \tag{2.4}$$

$$H_{21} = \frac{I_{out}}{I_{in}} \tag{2.5}$$

The short circuit current gain in (2.5) is used for the extraction of f_T. It can be easily obtained by performing a 2-port conversion into H-parameters from the measured de-embedded S-parameters of the transistor.

2.2.3.2 f_{MAX} definition and extraction

f_{MAX} is defined as the frequency at which the ratio of the load power to the input power becomes unity. From [30], it is derived as

$$f_{MAX} \cong \sqrt{\frac{f_T}{8\pi R_g C_{gd}}} \tag{2.6}$$

$$GU = \frac{0.5 \cdot \left|\frac{S_{21}}{S_{12}} - 1\right|^2}{k \cdot \left|\frac{S_{21}}{S_{12}}\right| - real\left(\frac{S_{21}}{S_{12}}\right)} \tag{2.7}$$

$$k = \frac{\left(1 - |S_{11}|^2 - |S_{22}|^2 + |S_{11} + S_{22} - S_{12} \cdot S_{21}|^2\right)}{2 \cdot |S_{12}||S_{21}|} \tag{2.8}$$

The extraction of f_{MAX} is done using the unilateral power gain (GU) as shown in (2.7). This power gain can be obtained when the input of the transistor is conjugate-matched to the input signal source, the load is also conjugate-matched with the transistor output impedance, and an appropriate network is used to cancel the effect of feedback from the output to the input [31].

Tong [32] developed a new FOM to study the transistor's W_f effect on its HF noise performance:

$$FOM = F_{\min} R_n \tag{2.9}$$

$$R_n = \frac{\gamma g_{d0}}{g_m^2} = \frac{\gamma}{\alpha} \frac{1}{g_m} \tag{2.10}$$

$$\alpha = \frac{g_m}{g_{d0}} \tag{2.11}$$

where F_{min} is in the minimum noise factor:

$$F_{\min} \approx 1 + \frac{f}{f_{\max}} \sqrt{P + R - 2C\sqrt{RP}} \times \sqrt{1 + \left(2\frac{f_{\max}}{f_c}\right)^2 g_m (R_g + R_s + R_i)} \tag{2.12}$$

$$P = \frac{\overline{i_{nd}^2}}{4kTg_m\Delta f} \tag{2.13}$$

$$R = \frac{\overline{i_{ng}^2}}{4kT\left(\frac{\omega C_{gs}^2}{g_m}\right)\Delta f} \tag{2.14}$$

$$C = \operatorname{Im}\left(\frac{\overline{i_{ng} i_{nd}^*}}{\sqrt{\overline{i_{ng}^2} * \overline{i_{nd}^2}}}\right) \tag{2.15}$$

$$f_c = \frac{g_m}{2\pi C_{gs}} \tag{2.16}$$

$$i_{nd}^2 = 4kT\gamma g_{d0}\Delta f \tag{2.17}$$

$$i_{ng}^2 = 4kT\delta\left(\frac{\omega^2 C_{gs}^2}{5gd0}\right)\Delta f \tag{2.18}$$

$$C \equiv \frac{i_{ng} i_{nd}^*}{\sqrt{\overline{i_{ng}^2} * \overline{i_{nd}^2}}} \tag{2.19}$$

γ is the noise factor. R_n is the normalized noise resistance defined in a linear two-port noisy network [33]. In [34], the parameter R_n can be derived and simplified as shown in (2.10). For long channel devices, the parameter α is equal to unity and it will gradually decrease as the channel length reduces [34] in the expressions (2.9) to (2.11).

By multiplying the parameter F_{min} and R_n together, the resultant equation is obtained as follows:

$$FOM = \frac{\gamma}{\alpha}\left(\frac{1}{g_m}\right) + f\,K_1\frac{\gamma}{\alpha}\sqrt{\left(\frac{16\pi^2}{g_m^3}\right)\left(R_g C_{gd} C_g + C_{gs}^2\left(R_g + R_s + R_i\right)\right)} \tag{2.20}$$

$$K_1 = \sqrt{P + R - 2C\sqrt{RP}} \tag{2.21}$$

By extracting the small-signal parameters for all the devices under comparison and substituting them back into (2.20), the calculated FOM can be obtained.

2.2.4 RF Parasitics in MOSFETs

In the DC and low frequency MOSFET models, the parasitic at the gate and substrate structures are normally neglected as their effects are usually small in that frequency range. The source and drain parasitic resistances are also treated as "virtual" components and are only included to model the voltage drop in the *I-V* equation [35]. However, in radio frequency range, the effect of these parasitics can no longer be neglected as they affect the device performance significantly such that they must be considered in the equivalent circuit of the device.

2.2.5 Scalable RF CMOS Transistor Modeling

The relentless downscaling of CMOS technologies has greatly improved the RF performance of MOSFET. It has been reported that for a technology node of 90nm, high f_t of 209GHz, and f_{max} of 248GHz are achieved [36]. Furthermore, the scaling down of the transistor has brought about lower NF_{min} and it is now comparable to the reported SiGe BJT process [36][37]. The improved RFCMOS performance coupled with its lower cost has motivated circuit designers to integrate digital, mixed-signal, and RF transceiver blocks into a single chip [38]-[42].

Most of the RF models developed today are based on the macro modeling approach. In this approach, sub-circuit components are added to the transistor's core model to model the RF parasitics of MOSFET structures [16][43]. The core model used are usually the commercially available models such as BSIM3v3 [15] and BSIM4 [14][44]. The sub-circuit components are extracted from the measured S-parameters of the transistor; however, the extracted values of these RF components can be different when different extraction techniques are used. All existing RF parameter extraction techniques are based on the transistor's small-signal equivalent circuit analysis. Therefore, to characterize an RF MOSFET, all its RF parasitic elements must be included into the small-signal equivalent circuit. Although it has been demonstrated that including the sub-circuit components into the core model can accurately simulate the transistor's RF characteristics, such methods are normally for discrete transistor sizes. In order to generate a geometry scalable RFCMOS model, the extracted sub-circuit component values must be studied for its geometry dependency in order to formulate equations to capture their physical effects at high frequency regions. A physical scalable RFCMOS model can thus be generated. Presently, some publications are reported on scalable RF MOSFET modeling [45]-[47] but these publications [45][47] do not show all the geometry scalable equations of the sub-circuit components. In [46], the formulated equations for these sub-circuit components are empirical and have no physical meaning. Furthermore, only one device size of f_T and f_{MAX} plot are presented.

2.2.5.1 RF MOSFET model

Figure 2.1 shows the proposed RF equivalent sub-circuit model [4]. All the sub-circuit components are physically based and can be used for the transistor that has the source and body terminal tied together and grounded.

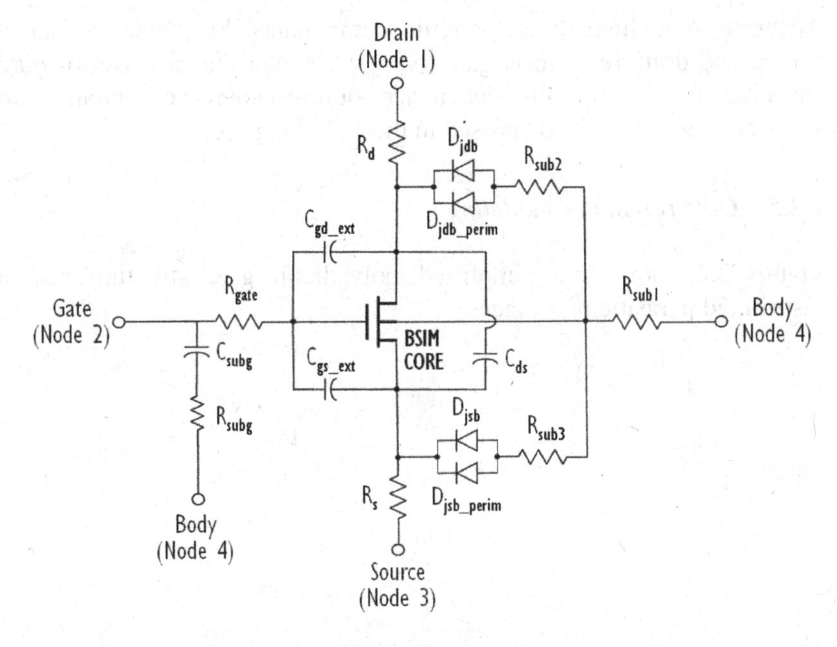

Figure 2.1 RF equivalent sub-circuit model [4].

The resistance R_{gate} represents the effective lumped gate resistance that consists of both the electrode resistance and the distributed channel resistance [48]. The resistances R_s and R_d represent the effective source and drain resistance that consists of the metal line, via and contact resistances.

The capacitance C_{gs_ext} and C_{gd_ext} represent the effective gate to source and gate to drain and consist of both the overlap and fringing capacitance between the terminals. C_{ds} represents the drain to source fringing capacitance between the metal lines that connect to the source and drain diffusions. As the internal junction capacitances of the core model are turned off, the external diodes D_{jdb}, D_{jdb_perim}, D_{jsb} and D_{jsb_perim} are added as the junction capacitances to connect the substrate resistance network. D_{jdb} stands for the area intensive diode while the D_{jdb_perim} stands for the perimeter intensive diode and the definition is the same for the source to body junction diodes. The parameters R_{sub1}, R_{sub2}, and R_{sub3} represent the substrate network resistances. Finally, C_{subg} and R_{subg} are the gate-to-substrate capacitance and resistance over the shallow trench isolation (STI) region.

Accurate modeling of sub-circuit components like gate resistance, source-and-drain resistance, gate-to-substrate capacitance, gate-to-source capacitance, gate-to-drain capacitance, drain-to-source capacitance, and substrate resistance are discussed in the following sections.

2.2.5.2 Gate resistance modeling

Figure 2.2 shows the simplified polysilicon gate structure and its distributed parasitic resistances.

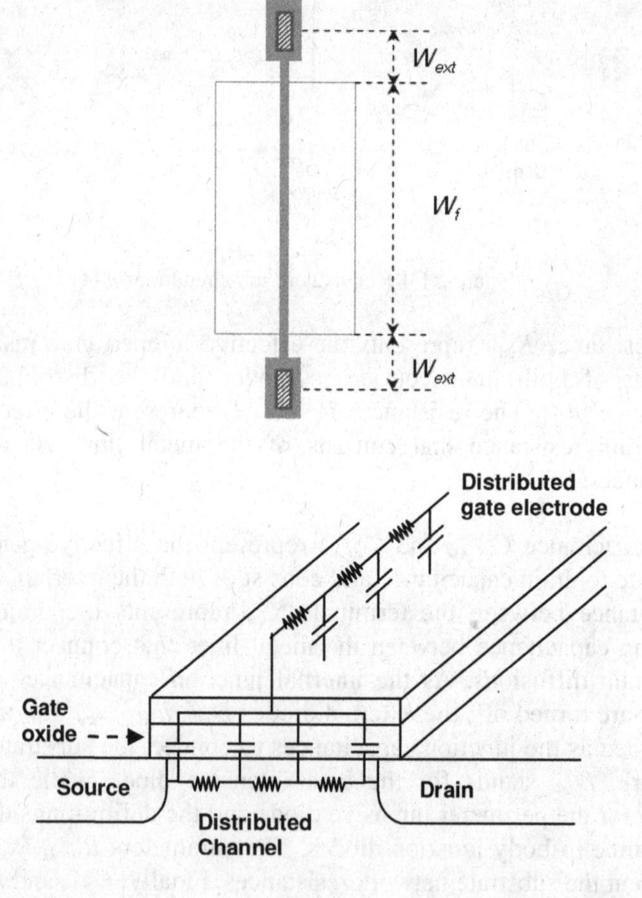

Figure 2.2 Simplified polysilicon gate structure and its distributed parasitic resistances.

The total resistance is given by

$$R_{gate} = R_{g,poly\_W_f} + R_{g,ch} + R_{g,poly\_W_{ext}} \tag{2.22}$$

In [18], the distributed effect of the gate electrode has been studied and the following equations have been derived to calculate the distributed gate electrode resistance.

$$R_{g,poly\_W_f} = \frac{\rho_{poly} \cdot W_f / L_g}{3 \cdot N_f} \tag{2.23}$$

$$R_{g,poly\_W_f} = \frac{\rho_{poly} \cdot W_f / L_g}{12 \cdot N_f} \tag{2.24}$$

In the single-contacted gate (2.23) and the double-contacted gate (2.24), the variable N_f is the number of fingers, ρ_{poly} is the gate sheet resistance, and L_g and W_f are the channel length and unit width of a single finger, respectively. The factors of 1/3 and 1/12 are used in (2.23) and (2.24) to account for the distributed gate resistance effect and the different gate connection configurations at the ends of the gate structure.

In (2.25), the polysilicon gate extension W_{ext} also contributes to the total gate resistance. Note that W_{ext} is divided by 2 in (2.25), since the polysilicon gate extension regions at both ends are connected in parallel as shown in Figure 2.2.

$$R_{g,poly\_W_{ext}} = \frac{\rho_{poly} \cdot \left(W_{ext} / 2 \right)}{N_f \cdot L_g} \qquad R_{g,ch} = \frac{x_1 \cdot \left(L_g \right)}{N_f \cdot W_f} \tag{2.25}$$

where the variable x_1 is the factor of the channel sheet resistance.

It is observed that R_{gate} is inversely proportional to N_f. The dependence of R_{gate} on N_f and W_f can be explained by considering the equations (2.22) to (2.25). The 3 physical effects on the gate resistance are inversely proportional to N_f. As shown in (2.24), the resistance $R_{g,poly\_Wf}$ is directly proportional to W_f, but in (2.25), the resistance $R_{g,ch}$ is inversely proportional to W_f.

2.2.5.3 Source and drain resistance modeling

The resistances R_s and R_d shown in Figure 2.1 are the effective resistance, which consists of the metal line, via, and contact resistances shown in the layout in Figure 2.3. It is assumed that the source and drain resistance in BSIM3v3 only models the active region of the parasitic resistances. Based on the layout below, the following equations can be derived to represent R_s and R_d [4]:

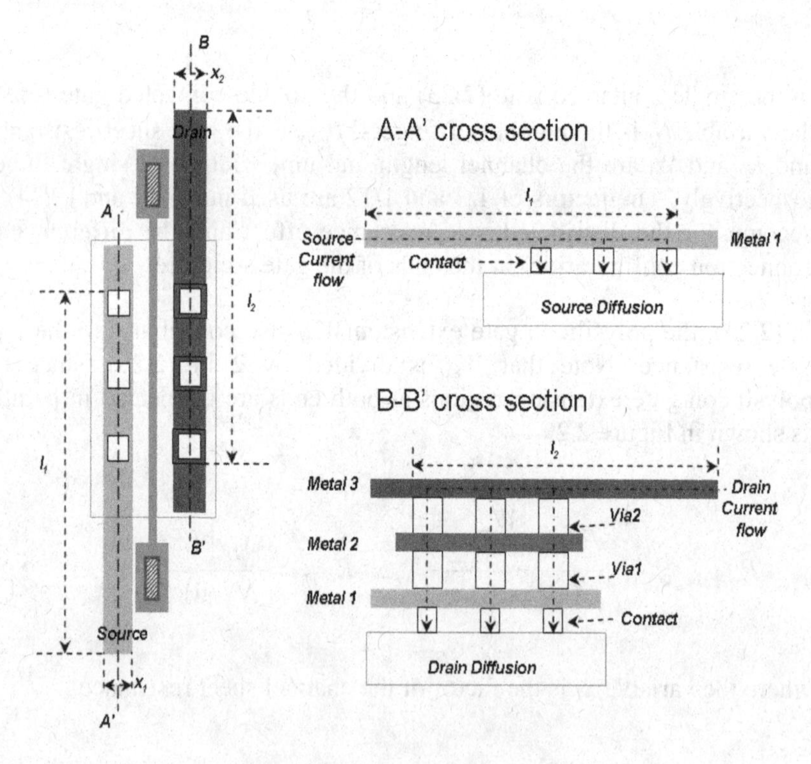

Figure 2.3 Source and drain metal structure.

$$R_s = \frac{\left(\frac{\rho_{m1} \cdot l_1}{x_1} + \frac{R_{con}}{n_{con}} \right)}{n_{diff,source}} \tag{2.26}$$

$$R_d = \frac{\left(\frac{\rho_{m3} \cdot l_2}{x_2} + \frac{(R_{con} + R_{via1} + R_{via2})}{n_{con}} \right)}{n_{diff,drain}} \tag{2.27}$$

The variables ρ_{m1} and ρ_{m3} represent the sheet resistance for metal layers 1 and 3; variables R_{con}, R_{via1}, and R_{via2} represent the contact, via_1, and via_2 resistance; and n_{con}, $n_{diff,source}$, and $n_{diff,drain}$ are respectively the number of contacts and the source and drain diffusions of the transistor.

It is observed that both resistances are inversely proportional to N_f. The $n_{diff,source}$ and $n_{diff,drain}$ are proportional to N_f. Hence, R_s and R_d resistance show inverse proportionality with N_f. The variables l_1, l_2 and n_{con} are proportional to the W_f of the transistor.

2.2.5.4 Gate to substrate capacitance and resistance modeling

The components C_{subg} and R_{subg} shown in Figure 2.1 are the gate to substrate capacitance and resistance over the STI region, respectively. They are shown in the cross-sectional structure in Figure 2.4. The dotted enclosed region is the gate area that is on top of the STI region generating the parasitic components C_{subg} and R_{subg}. Based on the layout geometry, the following equations are formulated [4]:

$$R_{subg} = \frac{R_{substrate,STI}}{N_f} \tag{2.28}$$

$$C_{subg} = C_{M1,STI} \cdot a_{M1} + C_{M2,STI} \cdot a_{M2} \tag{2.29}$$

The variables $C_{M1,STI}$ and $C_{M2,STI}$ are the parasitic capacitance per unit area of metals 1 and 2 over the STI region while the variables a_{M1} and a_{M2} are the area of the dotted enclosed region of metal 1 and metal 2 as shown in Figure 2.4.

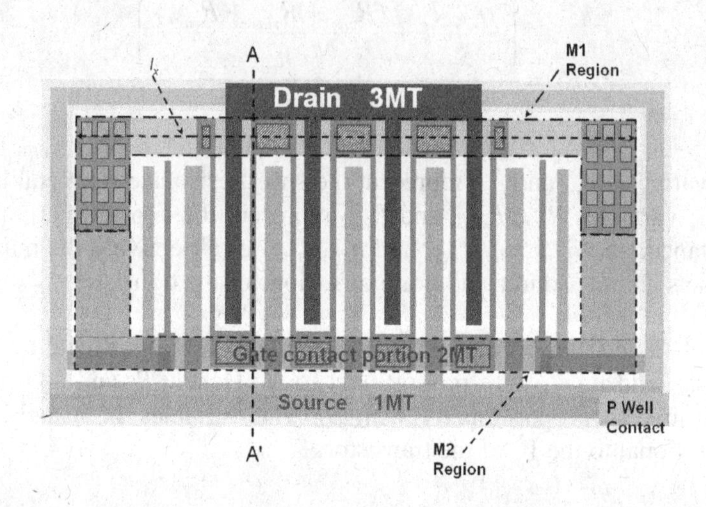

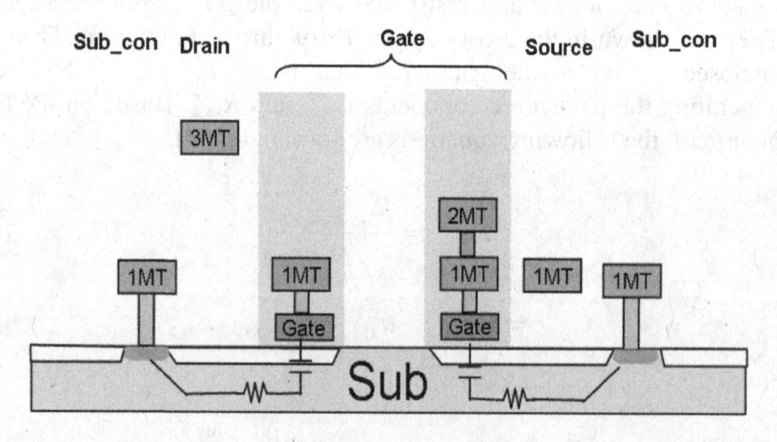

Figure 2.4 Gate to substrate capacitance and resistance structure.

Since the dielectric thickness between metal 2 and the substrate is higher than that of metal 1, it is expected that the extracted $C_{M1,STI}$ will be higher than that of $C_{M2,STI}$. Since some areas are under enclosed metals 1 and 2 (i.e., overlapped with the poly-silicon gate), the proposed Eqn. (2.16) may over-estimate C_{subg} slightly such that a small capacitance may be required to be subtracted.

C_{subg} and R_{subg} are extracted respectively using *Seneca* and *Substrate Storm* [49] which simulate the layout structure in Figure 2.4. Based on the extracted results of R_{subg}, it is only dependent on N_f and formulated as shown in (2.15). The extracted $R_{substrate,STI}$ is the substrate parasitic resistance under the STI region.

2.2.5.5 Gate to source and gate to drain capacitance modeling

The capacitances C_{gs_ext} and C_{gd_ext} in Figure 2.1 represent the overlap and fringing capacitances between the gate-to-source and gate-to-drain terminals shown in Figure 2.5. The equations are formulated as [4]

$$C_{gs\_ext} = C_{M2-M1,gs\_overlap} \cdot n_{diff,source} + C_{M1-Poly,gs\_fringing} \cdot N_f \cdot W_f \quad (2.30)$$

$$C_{gd\_ext} = C_{M3-M1,gd\_overlap} \cdot n_{diff,drain} + C_{M1-Poly,gd\_fringing} \cdot N_f \cdot W_f \quad (2.31)$$

$C_{M2-M1,gs\_overlap}$ is the overlap capacitance (*fF*) between the gate (metal 2) and source (metal 1) metal lines, whereas $C_{M1-Poly,gs\_fringing}$ is the fringing capacitance per unit width (*fF/μm*) between the gate structure (poly-silicon) and the source (metal 1) metal lines. $C_{M3-M1,gd\_overlap}$ is the overlap capacitance (*fF*) between the gate (metal 1) and drain (metal 3) metal lines, whereas $C_{M1-Poly,gd\_fringing}$ is the fringing capacitance per unit width (*fF/μm*) between the gate structure (poly-silicon) and the drain (metal 1) metal lines.

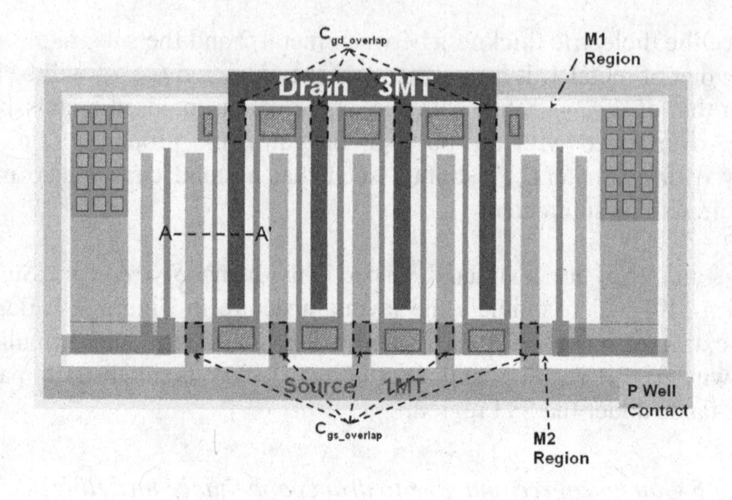

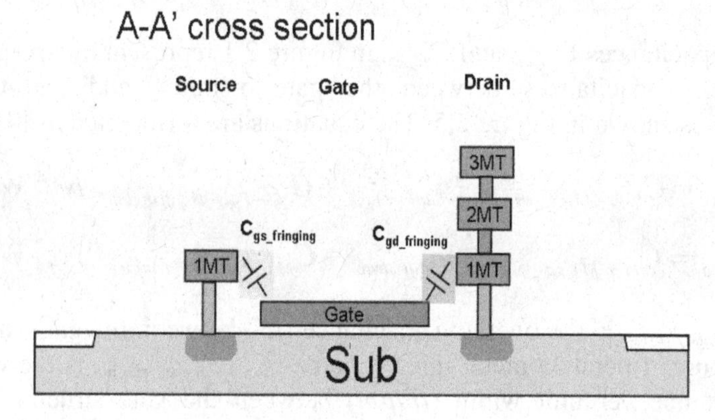

Figure 2.5 Gate to source and gate to drain capacitance structure.

It is observed that both capacitances are proportional to N_f and W_f, and the extracted C_{gs_ext} capacitance is slightly larger.

2.2.5.6 Drain to source capacitance modeling

C_{ds} is defined as the fringing capacitance between the metal lines that connect the source and drain diffusions. The location of the fringing capacitance is indicated in Figure 2.6. Based on the layout structure, it is predicted that the fringing capacitance is proportional to N_f and W_f of the transistor. Hence, the following equation is formulated for C_{ds} [4]:

$$C_{ds} = C_{ds\_fringing} \cdot N_f \cdot W_f \qquad (2.32)$$

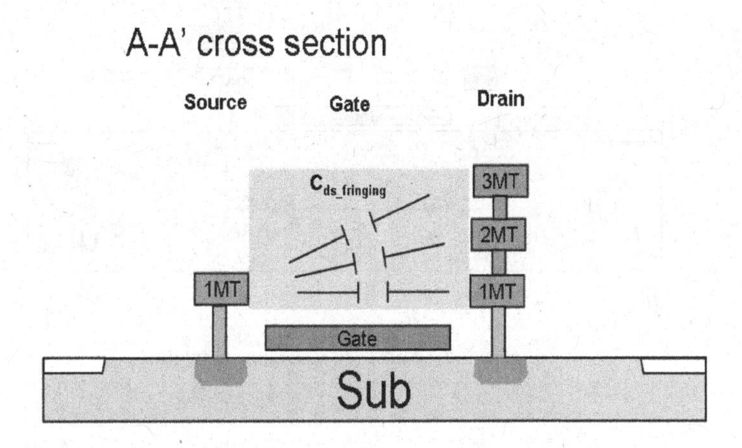

Figure 2.6 Drain to source capacitance structure.

$C_{ds\_fringing}$ is the fringing capacitance per unit width ($fF/\mu m$) between the metal lines of the source and drain metal structure. Observations show that as N_f and W_f increases, the amount of fringing capacitance between the source and drain metal increases. C_{ds} has an increasing trend with N_f and W_f.

2.2.5.7 Substrate resistance modeling

In the cross-sectional view A-A' in Figure 2.5, the substrate resistances network is added into the structure to indicate the location of the parasitics (Figure 2.7) [4]. C_{jsb} and C_{jdb} are the junction capacitances that are replaced by the junction diodes shown in Figure 2.1. R_{sub2} and R_{sub3} represent the substrate resistances under the channel, while R_{sub1} connects the intrinsic bulk node to the body terminal. Based on the layout structure, it is predicted that R_{sub2} and R_{sub3} are proportional to $L_g/(N_f{*}W_f)$ while R_{sub1} is inversely proportional $N_f{*}W_f$. Hence, the following equations (2.33) and (2.34) are formulated for the substrate resistances. Note that the layout of the body contact structure is a ring that encompasses the active region of the transistor.

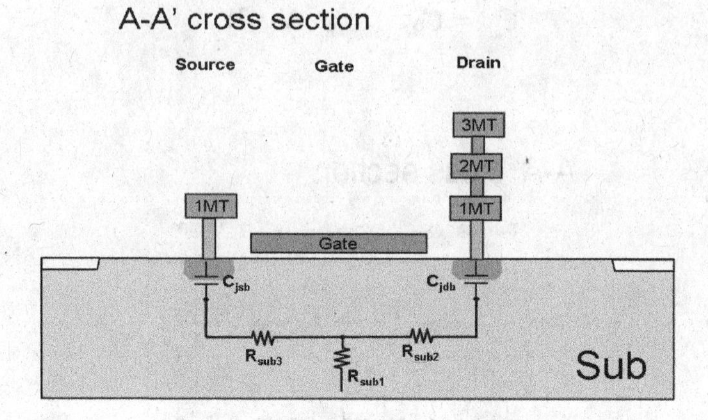

Figure 2.27 Substrate resistance network.

$$R_{sub1} = \frac{\rho_{substrate}}{N_f \cdot (W_f + 2 \cdot XJ)} \tag{2.33}$$

$$R_{sub2} = R_{sub3} = \frac{\rho_{substrate-sheet}}{N_f \cdot (W_f + 2 \cdot XJ)} \cdot \left(\frac{L_g}{2}\right) \tag{2.34}$$

where the variable $\rho_{substrate}$ is the substrate resistivity (*Ω-μm*) while $\rho_{substrate\text{-}sheet}$ is the substrate sheet resistance (*Ω/number of squares*) under the active region. The parameter *XJ* represents the source and drain junction depth. Its value can be found in the BSIM3v3 model parameters. Substrate resistances are inversely proportional to N_f and W_f. This is mainly due to the substrate resistance networks that are all connected in parallel with N_f fingers and the substrate resistance has a large W_f path whereby the signal flows through the substrate region.

2.3 On-chip Inductors

Inductors are passive electronic components that store energy in the form of magnetic flux. In its simplest form, an inductor consists of a wire loop or coil. Its inductance is dependent on the total conductor length of the coil and also the magnetic properties of the material around which the coil is wound. Inductors when connected in series or parallel configurations can provide discrimination against unwanted signals and they are essentially used in matching networks, resonators in voltage-controlled oscillators (VCOs), and degenerators in low-noise amplifiers (LNAs). Technological progress in microelectronics and microwave integrated circuits has seen thin-film microelectronic inductors used to an increasing extent. Technical literature for such inductor design has been based more or less on theories and derivations from [50]-[52].

2.3.1 Spiral Inductors on Silicon

Discrete inductors are usually in the form of solenoid coils because of the large mutual coupling between turns and the ease of inserting high-permeability (μ) material inside the coil to increase the inductance, L and quality factor Q. L is a measure of the amount of electromotive force generated for a unit change in current; whereas Q is the ratio of the inductive reactance ωL to its resistance R at a given frequency. In silicon technology, conventional 3-dimensional coils are difficult to realize as there are limitations to the maximum number of metal layers in the existing process flows.

The inter-metal dielectric between metal layers results in huge overlap capacitances and thus lowers the maximum usable frequency considerably. The quality factor of such coils also suffers due to large via resistances. Spiral planar inductors are therefore more compatible with the back-end metallization schemes of present silicon technologies.

2.3.1.1 Figure of merits

The important parameters that define the performance of an inductor are inductance value, quality factor and self-resonant frequency. Quantitatively, the impedance of a real inductor increases with frequency until it reaches self-resonance whereby the parasitic capacitances resonate with the inductor. Beyond its self-resonant frequency, the inductor behaves like a capacitor and its impedance decreases with frequency.

2.3.1.1.1 Inductance L

According to Greenhouse [53], the overall inductance L_T of a monolithic inductor is given by the sum of all the self-inductance of individual segments L_0, plus the sum of all positive mutual inductances between adjacent segments M_+, minus the sum of all negative inductances between segments on opposite sides of the spiral M_-:

$$L_T = L_0 + M_+ - M_- \tag{2.35}$$

The mutual inductance of two metal segments is positive if the currents are flowing in the same direction and negative if the currents are in the opposite direction. The exact self-inductance for a straight conductor is [54]

$$L = 0.002L\left[\ln\left(2 \quad / GMD\right) - 1.25 + \left(AMD / \quad\right) + \left(\mu / 4\right)T\right] \tag{2.36}$$

L is the inductance in microhenries, is the conductor length in centimeters. GMD and AMD represent the geometric and arithmetic mean distances of the conductor cross section, respectively. μ is the conductor permeability, and T is a frequency correction parameter.

Based on [55], the geometric mean distance (GMD) of a conductor cross section is the distance between two segments normal to the cross section, whose mutual inductance is equal to the self-inductance of the conductor.

The arithmetic mean distance (AMD) is the average of all the distances between the points of one conductor and the points of another.

For circular cross section, substituting associated GMD and AMD values [50] into (2.36) gives

$$\begin{aligned} L &= 0.002 \ \left[\ln(2 \ /0.7788r) - 1.25 + r/ \ + (\mu/4)T\right] \\ &= 0.002 \ \left[\ln(2 \ /r) - \ln 0.7788r - 1.25 + r/ \ + (\mu/4)T\right] \\ &= 0.002 \ \left[\ln(2 \ /r) - 1 + r/ \ + (\mu/4)T\right] \end{aligned} \tag{2.37}$$

where r is the radius. For near-direct-current condition, $T = 1$. The equation is thus

For thin-film inductors with rectangular cross sections,

$$\begin{aligned} L &= 0.002 \ \{\ln[2 \ /0.2232(a+b)] - 1.25 + [(a+b)/3 \ ] + (\mu/4)T\} \\ &= 0.002 \ \{\ln[2 \ /(a+b)] - \ln 0.2232 - 1.25 + [(a+b)/3 \ ] + (\mu/4)T\} \\ &= 0.002 \ \{\ln[2 \ /(a+b)] + 0.25049 + [(a+b)/3 \ ] + (\mu/4)T\} \end{aligned} \tag{2.38}$$

where a and b are the rectangular dimensions of the cross section. For the near-direct case in which magnetic permeability is 1, Eqn. (2.38) becomes

$$L = 0.002 \ \{\ln[2 \ /(a+b)] + 0.50049 + [(a+b)/3 \ ]\} \tag{2.39}$$

In general,

$$M = 2 \ Q \tag{2.40}$$

M is the mutual inductance in nanohenries, is the conductor length in centimeters, and Q is the mutual-inductance parameter from

$$Q = \ln\left\{(\ /GMD) + \left[1 + (\ ^2/GMD^2)\right]^{1/2}\right\} - \left[1 + (GMD^2/ \ ^2)\right]^{1/2} + (GMD/ \ ) \tag{2.41}$$

where

$$\begin{aligned} \ln GMD = \ln d - \{&\left[1/12(d/w)^2\right] + \left[1/60(d/w)^4\right] + \left[1/168(d/w)^6\right] + \\ &\left[1/360(d/w)^8\right] + \left[1/660(d/w)^{10}\right] + \ldots\} \end{aligned} \tag{2.42}$$

where w is the track width and d is the distance between two opposite segments.

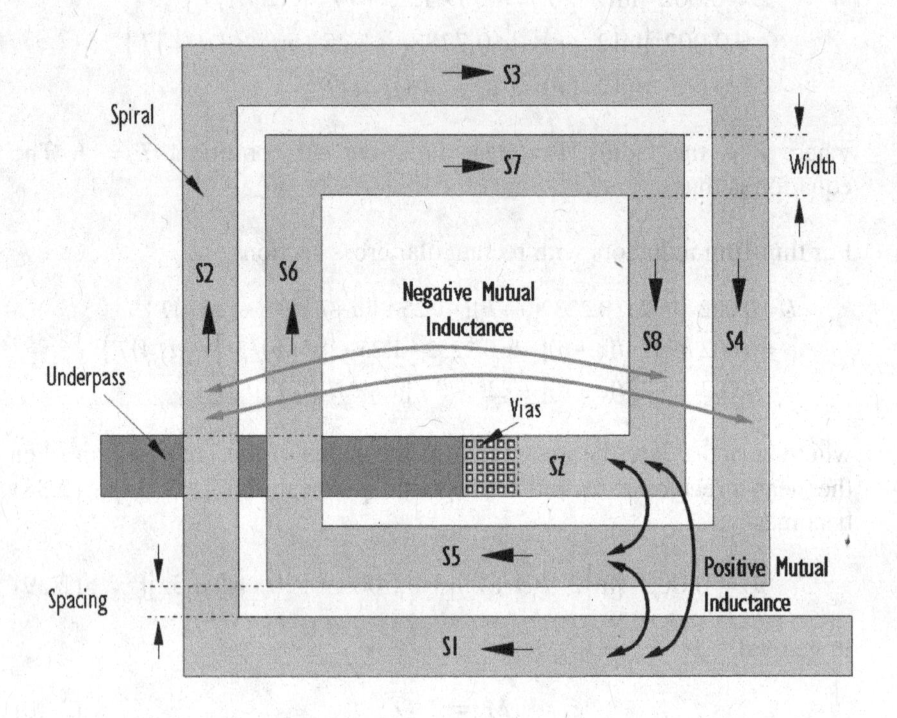

Figure 2.8 Positive and negative mutual inductance components in a rectangular inductor.

For a rectangular planar inductor with two complete turns and eight segments (Figure 2.8), its total inductance is equal to the sum of the self-inductances, L of each segment and all the mutual inductances, M between opposite segments:

$$L_{total} = \sum L + \sum M \tag{2.43}$$

$$\begin{aligned} L_{total} &= L_1 + L_2 + L_3 + L_4 + L_5 + L_6 + L_7 + L_8 \\ &+ 2\left[\left(M_{1,5} + M_{2,6} + M_{3,7} + M_{4,8}\right)\right. \\ &\left.- \left(M_{1,7} + M_{1,3} + M_{5,7} + M_{5,3} + M_{2,8} + M_{2,4} + M_{6,8} + M_{6,4}\right)\right] \end{aligned} \tag{2.44}$$

Inductive characteristics of spiral inductors can be improved if there is no winding of metal conductors in the centre of the spiral. This is so because negative mutual coupling effect reduces when distance between the four segments (S6, S7, S8 and SZ in Figure 2.8) with opposite current flow increases. However, the core diameter is expected to become larger which inevitably increases the overall size of the inductor.

The width of these metal segments should be large so that resistances of the metal segments are kept low. Nonetheless, having a large width reduces the inductance per unit conductor length and increases the parasitic shunt capacitances of the inductor, which in turn degrades the resonant frequency as well as quality factor of the inductor. In addition to the above findings, metal spacing between conductors should be kept as close as possible, i.e., the minimum design rule for the process. This will save on the silicon area occupied by the inductor and promote constructive mutual coupling, thereby increasing the overall inductance of the inductor. Such preliminary analyses reveal some ideas on how the physical layout of spiral inductors can be optimized. Even so, all the physical design parameters are inter-related and hence, to understand the effects of each design parameters would require in-depth studies, especially for inductors fabricated using high resistive metallization on lossy silicon substrates.

2.3.1.1.2 Quality factor, Q

The efficiency of an inductor is determined by its quality factor or *Q* factor. Basic physical definitions for *Q* are defined as either of the following:

$$2\pi \times \frac{\text{Energy Stored}}{\text{Energy Dissipated per Cycle}} \tag{2.45}$$

$$\omega \times \frac{\text{Energy Stored}}{\text{Average Power Dissipated}} \tag{2.46}$$

Quality factor is dimensionless and proportional to the ratio of energy stored to energy dissipated per unit time cycle. The quality factor expressions shown in Equations (2.45) and (2.46) are fundamental definitions which can be applied in different systems. They do not specify the type or form in which the energy is stored or dissipated. For electrical systems, both equations also define the quality factor for an LC tank. The subtle distinction in quality factor between an inductor and an LC tank lies in the forms of energy storage.

For an LC tank, the Q factor serves to provide an indication of how much energy is lost as it is being transferred between the capacitor and the inductor. The energy stored in an LC tank is actually the sum of the average magnetic and electrical energies; and for a lossless LC tank, this energy is constant because it oscillates between magnetic and electrical forms. As such, it can be equal to the peak magnetic energy, or the peak electrical energy [56]. The rate of this oscillation is known as the resonant frequency of the tank. Since there is no energy dissipation, the Q factor is infinite for a lossless LC tank.

On the contrary, an inductor is designed primarily to store magnetic energy and hence only energy stored in the magnetic field is of interest. All other energies stored in the form of electric field or energies dissipated due to parasitic resistances are counterproductive. Therefore, the quality factor of an inductor is proportional to the net magnetic energy stored, which is the difference between the peak magnetic and electrical energies. Also, the inductor is said to be at self-resonance, when its peak magnetic and electrical energies are equal. Hence, at this self-resonant frequency, the inductor's Q becomes zero and no net magnetic energy outputs from the inductor to any external circuit. Above its self-resonant frequency, the inductor behaves like a capacitor.

To demonstrate the subtle distinction between the inductor and the LC tank, consider a simple parallel RLC circuit to be first modeled as a real inductor with substrate parasitics, R and C, and subsequently as an LC tank. Figure 2.9 shows the circuit of a parallel RLC tank.

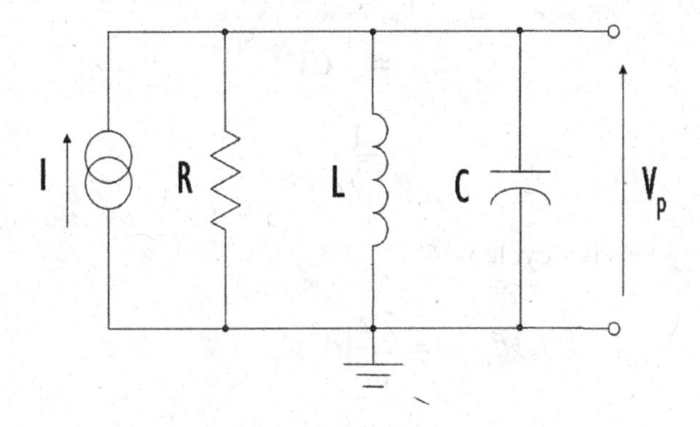

Figure 2.9 Parallel RLC circuit.

The following shows the expressions for the electrical and magnetic energies which are required to derive the quality factor for both cases, as an inductor and as an LC tank. The peak magnetic energy of the circuit shown in Figure 2.9 is given by

$$
\begin{aligned}
\xi_{PM} &= \frac{1}{2} L I^2 \\
&= \frac{1}{2} L \left(\frac{V_p}{X_L} \right)^2 \\
&= \frac{1}{2} L \left(\frac{V_p^{\,2}}{\omega^2 L^2} \right) \\
&= \frac{V_p^{\,2}}{2\omega^2 L}
\end{aligned}
\qquad (2.47)
$$

The peak electrical energy is given by

$$\xi_{PE} = \frac{1}{2}CV^2 = \frac{1}{2}CV_p^2 \tag{2.48}$$

The energy loss per cycle is

$$\xi_{LOSS} = \frac{2\pi}{\omega}|I|^2 R = \frac{2\pi}{\omega}\left(\frac{V_P}{\sqrt{2}R}\right)^2 R = \frac{\pi}{\omega} \times \frac{V_P^2}{R} \tag{2.49}$$

The average magnetic energy is given by

$$\xi_{AM} = \frac{V_p^2}{4\omega^2 L} \tag{2.50}$$

The average electrical energy is given by

$$\xi_{AE} = \frac{1}{4}CV_p^2 \tag{2.51}$$

The resonant frequency is given by

$$f_o = \frac{1}{2\pi\sqrt{LC}} \tag{2.52}$$

where V_p represents the peak voltage across the circuit terminals of Figure 2.9. Using the expressions from (2.47) to (2.52), the Q factor for the inductor and the LC tank are defined as follow:

$$Q_{Inductor} = 2\pi \cdot \frac{Peak\ Magnetic\ Energy - Peak\ Electric\ Energy}{Energy\ Loss\ Per\ Cycle}$$
$$= \frac{R}{\omega L}\left[1-\left(\frac{\omega}{\omega_o}\right)^2\right] \tag{2.53}$$

In a spiral inductor built on silicon, C can be regarded as representing the parasitic capacitances of the substrate. R on the other hand represents the resistance of the substrate underneath the inductor. $Q_{Inductor}$ is equal to zero when the inductor is operating at its resonant frequency (i.e., $\omega = \omega_o$). This self-resonance phenomenon occurs due to the undesirable energy dissipation or energy loss into the conductive silicon substrate.

$$Q_{LC\ Tank} = 2\pi \cdot \left.\frac{Average\ Magnetic\ Energy + Average\ Electric\ Energy}{Energy\ Loss\ Per\ Cycle}\right|_{\omega=\omega_o}$$
$$= 2\pi \cdot \left.\frac{Peak\ Magnetic\ Energy}{Energy\ Loss\ Per\ Cycle}\right|_{\omega=\omega_o} \tag{2.54}$$
$$= \frac{R}{\omega_o L}$$
$$or \quad = 2\pi \cdot \left.\frac{Peak\ Electric\ Energy}{Energy\ Loss\ Per\ Cycle}\right|_{\omega=\omega_o}$$
$$= \omega_o RC$$

The Q factor of the LC tank circuit is very important in electronic communications as it determines the 3-dB bandwidth of resonant circuits. This bandwidth of the LC tank defines the amount of signal that can be transmitted and the amount of noise that can be attenuated by a circuit.

Both definitions (2.53) and (2.54) are important and their applications are actually determined by the intended function of a circuit. However, to evaluate the Q factor of an on-chip inductor as a single element, the definition in (2.53) will be used. In actual device measurements, both L and Q are extracted from measured S-parameters after employing proper test structure de-embedding techniques.

2.3.2 Advantages of Silicon-based Spiral Inductors

In reality, cost is the key concern in bringing high frequency personal communication equipment into the consumer market. Silicon's ever-improving capabilities, driven by rapid microprocessor development, are making itself more cost efficient than GaAs technology for RF applications up to 5GHz. Besides affordable manufacturing costs, other advantages included the following: they are good conductors of heat; they have a high level of integration through multi-level interconnect metallization; there is availability of CMOS, BJT, BiCMOS and SiGe transistors; they also have excellent planarity for all existing bonding technologies and possibility of system-on-chip (SOC) solutions.

Compared to GaAs or quartz substrates, silicon does not exhibit any outstanding microwave properties. Silicon substrates for CMOS applications are predominantly lossy and hence degrade the Q factor at high frequencies. This situation is expected to be even more critical as the CMOS technology node scales down further. In advanced nodes, silicon substrates with increasingly higher conductivity would have to be used to eliminate latch-up occurrences under tighter design rules.

2.3.3 Identifying Loss Mechanisms in Silicon-based Spiral Inductors

Loss mechanisms in silicon-based integrated inductors can be classified into four categories, namely, metallization resistive loss, substrate capacitive and resistive losses as well as formation of substrate eddy current. These are the main reasons why performance of silicon-based inductors is sometimes not acceptable and accurate predictions of quality factor are often difficult to make, even with reliable electromagnetic simulation tools.

2.3.3.1 Metallization resistive loss

At low frequencies, the current flow within a metal conductor trace is uniform and its series resistance R can be easily determined by

$$R = \rho \frac{\ell}{A} \tag{2.55}$$

where ρ is the resistivity of the metal used, refers to the length and A the cross-sectional area of the metal conductor. However, at high frequencies, the induced electromotive force (EMF) causes a non-uniform current distribution in the inductor. This non-uniform current flows along the conductor surface, effectively reducing the cross-sectional area of the conductor. As a result, there is an increase in the series resistance, which causes a net reduction in the quality factor of the inductor. This phenomenon is known as the *skin effect* and is directly proportional to the operating frequency.

Skin depth, δ is defined [57] as,

$$\begin{aligned}\delta &= \sqrt{\frac{2}{\omega\mu\sigma}} \\ &= \frac{1}{\sqrt{\pi f\mu\sigma}}\end{aligned} \tag{2.56}$$

where f refers to the frequency, σ is the conductivity and μ is the permeability of the metal conductor. Since skin depth δ is inversely proportional to frequency, at higher frequencies, δ decreases and this suggests that there is lesser low resistive path within the conductor for current to flow easily. Hence, resistance of the conductor increases with operating frequency.

From Eqn. (2.56), the skin resistance, R_{skin} (Ω/m^2), can be defined as,

$$\begin{aligned}R_{skin} &= \frac{1}{\sigma\delta} \\ &= \sqrt{\frac{\pi f\mu}{\sigma}}\end{aligned} \tag{2.57}$$

Increasing the metal thickness of the inductor coil can minimize resistive losses, thereby improving the inductor's quality factor. This is especially so for CMOS technology because the metal thickness is typically between 0.25μm to 0.7μm.

Many experimental results have shown that the Q factor improves significantly when the metal thickness increases from 0.25μm to 1.0μm. This is due to massive reduction in the series resistance of the metal coil at low frequencies.

However, because of skin effect, the Q factor will only increase at a much slower rate when the metal thickness continues to increase any further. According to [58], the skin depth of aluminum is about 2.5μm at 2GHz. This suggests that when the metal thickness is thinner than 2.5μm, current still flows uniformly in the metal coil. When the metal conductor thickness is greater than the skin depth (i.e., 2.5μm), current will only flow along the surface of the metal at a depth of only 2.5μm. In addition, it also shows that the series resistance R_s does not decrease with any further increase in the metal thickness. In general, to avoid skin effect at any desired operating frequencies, the inductor's metal thickness should be as large as possible so that high quality factor can be achieved.

2.3.3.2 Substrate capacitive and resistive loss

The conventional silicon-based spiral inductor is usually fabricated using the top and second highest metal layers. It sits on top of the inter-metal dielectric layers (which are made up of silicon dioxide) and the silicon substrate. The layers of silicon oxide and the substrate itself contribute to unwanted capacitive coupling, which degrades the self-resonant frequency of the inductor. The self-resonant frequency f_o [59] of an inductor is defined as the frequency at which parallel resonance is achieved between the device inductance and its parasitic capacitances and resistances. Here, the parasitic capacitances refer to the oxide capacitance between the inductor and silicon substrate and the capacitance of the silicon substrate. The parasitic resistance, on the other hand, refers to the resistance of the silicon substrate. Ideally, the substrate and the oxide layer should not be capacitive and should have infinite resistivity in order to minimize unwanted substrate coupling. Inductors built on such ideal substrate will have high quality factor and high self-resonant frequency.

2.3.3.3 Substrate eddy current

Inductors experience magnetic losses when they are built flatly on a highly conductive silicon substrate. According to Faraday's law, an image current or eddy current is induced in the substrate underneath the spiral coil when a current is flowing in the inductor. Since the silicon substrate has low resistivity, this image current can flow easily. In compliance with Lenz's law, the direction of flow for this induced current is opposite to that of the inductor. This generates an opposing parasitic magnetic field in the substrate, which interacts with the magnetic field of the inductor and results in a degradation of the inductor's overall useful inductance.

2.3.4 Q-Factor Enhancement Techniques

A number of researchers have come up with techniques to improve the Q factor of the spiral inductor with many reported breakthroughs. The following sections will consolidate some of these methods and techniques that are found to have improved the performances of the spiral inductors.

2.3.4.1 Q-factor enhancement using processing technologies

Kamogawa [60] presented a high-Q inductor fabricated on a conductive Si substrate using monolithic microwave integrated circuit (MMIC) technology. Conductor resistive loss is minimized by the implementation of gold (Au) metallization while substrate losses and the parasitic capacitance are eliminated by the special dielectric stack and ground plane placed between the spiral and the substrate. The special polyimide layer with low dielectric constant further reduces the shunt capacitance to the ground plane, allowing the inductor to have a high resonant frequency.

Another publication by Burghartz, Soyuer, and Jenkins [61][62] presented a high-Q factor, three-layer metal inductor, fabricated using the multi-level interconnect technology available in the CMOS process. The Q factor of the inductor improves significantly due to a huge reduction in series resistance of the spiral coil. Nevertheless, the whole inductor structure is brought closer to the silicon substrate and inevitably increases the parasitic capacitance. This not only degrades the inductor's Q factor at higher frequencies but also causes a reduction in its self-resonant frequency. Such techniques work well for small inductors but may not be applicable to large-inductance inductors since they require much bigger area, suggesting that their parasitic capacitance would be too huge to allow for high Q-factor and self-resonant frequency.

In 2006, Yang et al. [63] examined the performance of spiral inductors fabricated using silicon-on-insulator (SOI) technology. They reported that spiral inductors on high resistive SOI (2kΩ-cm) substrates out-performed those designed on conventional bulk silicon (20Ω-cm). The 2.3nH spiral inductor on high-resistive SOI substrate has a maximum Q factor value that is 50% larger compared to the spiral inductor constructed on conventional bulk silicon substrate. In the same year, Tekin [64] used high Q-factor inductors in SOI CMOS technology to achieve a low power 700MHz VCO with a phase noise performance of –121 dBc/Hz at 600MHz offset. They also testified that the high resistive substrate allows 2.5 to 8.5-turn square spiral inductors to have a high Q factor.

Malba [65] also reported successful implementation of a monolithic 3-dimensional copper coil inductor. Thick copper traces of the coil are fabricated using electroplating process. Alumina is used as the core material because of its negligible loss tangent at high frequencies [66]. Compared to conventional spiral inductors, this 3-dimensional copper coil inductor achieves superior performance with Q values up to 30 at 1GHz.

Another team from the Nokia research centre [67] in Finland has also developed a novel 3-dimensional toroidal inductor using their polymer replication technique. This inductor was put together on silicon substrate, with gold as primary material for the coil. The characteristics of the toroidal inductor were excellent with a very high Q factor of 50 at 3GHz.

Mernyei et al. [68] reported a new technique to improve the Q factor of spiral inductors by reducing the substrate magnetic loss, which is primarily caused by induced eddy currents from the inductor's magnetic field. This is achieved by inserting n^+ regions (narrow strips) in the top p^+ layer perpendicular to the eddy-current flow. P-N-P junctions are thus created and can be used to eliminate the eddy-current closed loop paths in the top heavily doped p^+ layer.

In 1998, P. Yue and S. Wong [69] presented a shielded inductor which can improve the inductor's Q factor. The implementation of such inductors also saw reduction in substrate coupling between two adjacent inductors by as much as 25dB at 1-2GHz. These shielded inductors are fabricated by inserting a patterned polysilicon ground shield between the spiral inductor and the silicon substrate. This ground plane helps to terminate the electric field of the inductor before it penetrates into the substrate, thereby eliminating unwanted substrate losses. The ground shield is patterned so that it hinders the formation of closed loop current paths in the substrate and prevents negative mutual coupling between the induced substrate current and the current flowing in the inductor such that the inductor's magnetic field and its overall inductance are not reduced.

2.3.4.2 Q-factor enhancement using active inductors

Yang et al. [70] designed a CMOS broadband amplifier with high Q-factor active inductors. Adopting common-gate configuration and high Q-factor active inductors, the broadband amplifier has achieved high power gain, wider bandwidths, and better matching characteristics. In the absence of passive inductors, the chip area is significantly reduced although circuit complexity increases.

2.3.4.3 Q-factor enhancement using coupled spiral coils

A novel tunable inductor is presented by Pehlke in 1997 [71]. The inductor architecture consists of coupled RF and drive coils which employs phase shifting of the mutual components, demonstrating inductance tuning and significant decrease in resistive losses. Extremely high Q factor of about 2000 can be achieved using this elaborate technique.

2.3.4.4 Q-factor enhancement using layout optimization

Lopez-Villegas et al. [72] presented a method to improve the Q factor of RF integrated inductors by optimizing the inductor's layout. It is applied to the design of square spiral inductors using a silicon-based multi-chip module (MCM) technology and silicon micro-machining post processing. The layout is optimized by using a variable width spiral coil inductor to achieve high Q. This optimized inductor produced the best performance in a wide range of frequencies around 3.5GHz.

2.3.4.5 Q-factor enhancement using inductor device model

Post [73] proposed optimizing the physical layout of spiral inductors through analyses of the physical model. This lumped element empirical model for silicon-based spiral inductors was developed by Yue et al. [74]. The simplicity of the physical model enables a computational procedure for efficient layout selection, optimizing both Q factor and self-resonant frequency. The best width and number of turns is found to be about 10 μm and 6 turns respectively. Sia [75][76] proposed a scalable lumped-element RF sub-circuit model for symmetrical inductors (Fig. 2.10). In the sub-circuit model, L_S and R_S account for the self-inductance and resistive loss of the spiral coil respectively, while L_{SK} and R_{SK} model the skin effects of the metallization at giga-hertz frequencies. C_S portrays the capacitive coupling between the input and output ports. Substrate loss of the symmetrical inductor is modeled by the parasitic oxide capacitance between the silicon substrate and inductor C_{OX}, and the capacitive C_{SUB} and resistive R_{SUB} losses of the silicon substrate respectively. Since the layout is symmetrical, the extracted model parameters such as R_S, L_S, R_{SK}, L_{SK}, C_{OX}, R_{SUB}, and C_{SUB} are identical. A double-π model instead of a simple single-π model is used for the symmetrical inductors because the total conductor lengths of these inductors are typically very long. As operating frequency increases, these metallizations behave like transmission lines with distributed characteristics, thus a simple lumped element π model is insufficient to describe its behavior. The extraction strategy to obtain values of each element in the symmetrical inductor model is outlined in [5]. In this strategy, the open de-embedded Y-parameters for the symmetrical inductors are manipulated to derive parameters such as inductance L, Q factor, and series resistance R.

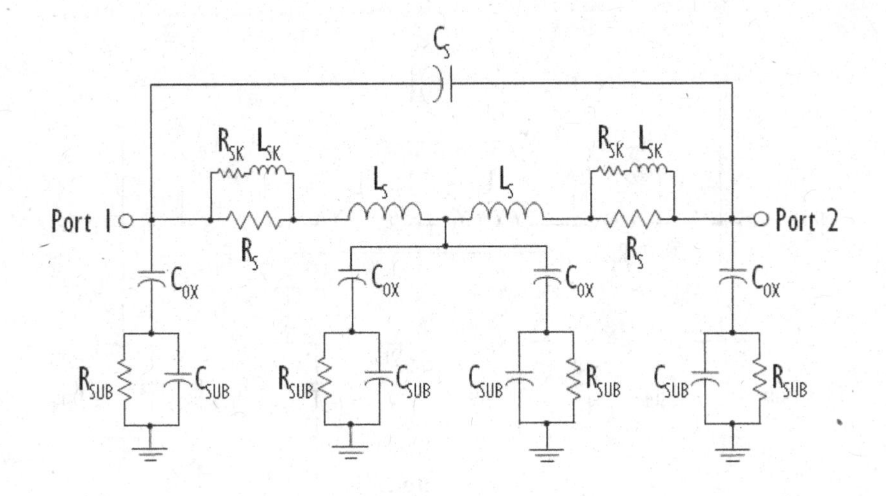

Figure 2.10 SPICE-Compatible RF sub-circuit model for Symmetrical Spiral Inductors [5].

Sia [5] also proposed another sub-circuit model for the differential spiral inductor [77]-[81]. Figure 2.11 depicts the proposed scalable lumped-element RF sub-circuit model for the differential inductors in this paper. L_S and R_S emulate the self-inductance and resistive loss of both spiral coils respectively while L_{SK} and R_{SK} model the skin effects of the metallization at giga-hertz frequencies.

M_1 describes the inductive mutual coupling effects between the 2 coils. C_S accounts for the capacitive coupling between coil 1 and coil 2. R_1 models the resistive loss of the centre-tap underpass. Substrate loss of the differential inductor is taken care by C_{OX}, C_{SUB} and R_{SUB} which describe the parasitic oxide capacitance between the silicon substrate and inductor, the capacitive and resistive losses of the silicon substrate respectively.

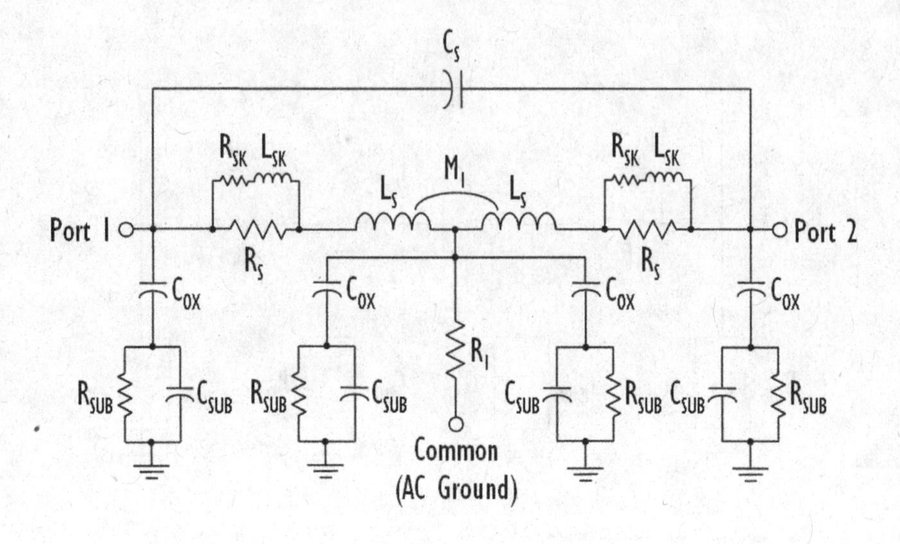

Figure 2.11 SPICE-compatible RF sub-circuit model for differential inductor [5].

2.3.4.6 Figure of merits for differential spiral inductors

From [82][83], the figure of merits, differential inductance L_{DIFF}, and differential quality factor Q_{DIFF} of the differential inductor are expressed in (2.58) to (2.61). The differential one-port S-parameter is first obtained from the de-embedded S-parameters as follows:

$$S_{DIFF} = \frac{S_{11} + S_{22} - S_{12} - S_{21}}{2} \tag{2.58}$$

With Z_o as the differential system impedance, the differential Z-parameter is derived as:

$$Z_{DIFF} = \frac{2Z_o\left(1 + S_{DIFF}\right)}{1 - S_{DIFF}} \tag{2.59}$$

Differential inductance and quality factor are evaluated from the differential Z-parameter as:

$$L_{DIFF} = \frac{Imag(Z_{DIFF})}{2 \times \pi \times Frequency} \tag{2.60}$$

$$Q_{DIFF} = \frac{Imag(Z_{DIFF})}{Real(Z_{DIFF})} \tag{2.61}$$

The inductance related to single-ended excitation (at Port 1) to decouple the inductive mutual coupling effects between the two spiral coils can be evaluated as:

$$L = \frac{Imag\left[Z_o \times \left(\frac{1+S_{11}}{1-S_{11}}\right)\right]}{2 \times \pi \times Frequency} \tag{2.62}$$

Differentially driven symmetrical spiral inductors (Fig. 2.12) are extensively used in matching networks and LC tanks for differential circuits such as low noise amplifiers and voltage controlled oscillators. By inter-winding two coils together, a differential inductor consumes smaller silicon area, and in doing so reduces the overall chip size. They also exhibit a higher Q factor over a broader range of frequencies, making them essential device components for radio frequency integrated circuit (RFIC) design.

Despite these advantages, such inductors are not readily available in silicon-verified device libraries but only offered by established semiconductor foundries. Reliable techniques to optimize the physical design of differential inductors are also not well established in the literature.

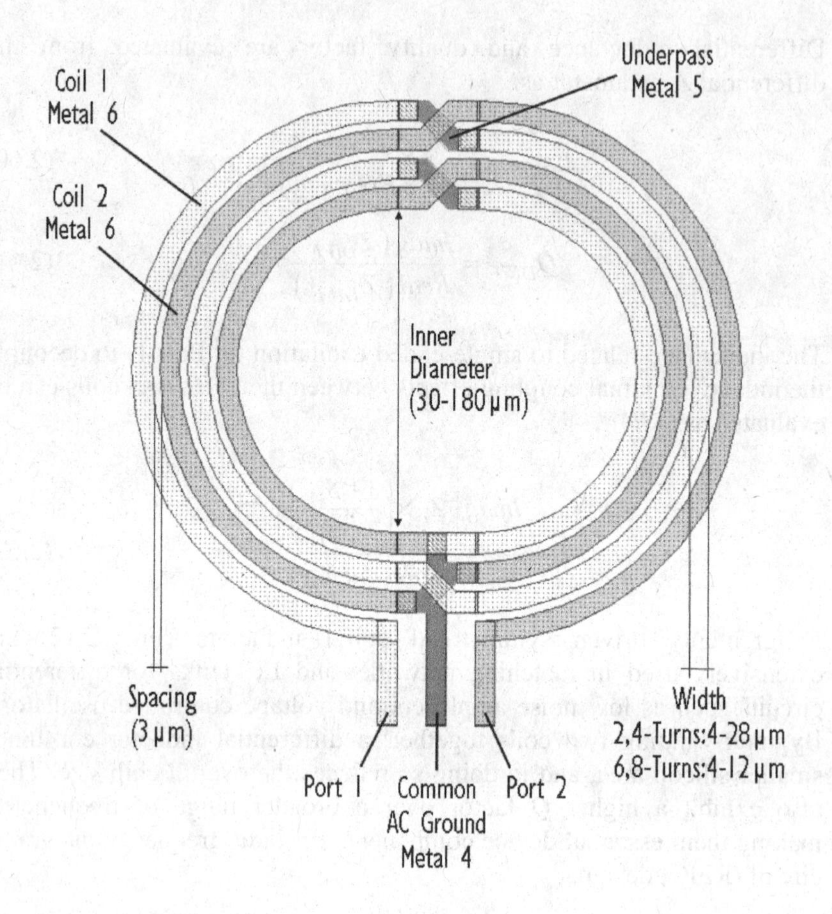

Figure 2.12 Layout scheme of differential inductor in 0.18 μm RFCMOS process [5].

2.4 Baluns/Transformers

RF transformers are widely used in electronic circuits for impedance matching to achieve maximum power transfer for suppressing undesired signal reflection, and as baluns for conversion between differential and single-ended signals. Depending on the circuit application and frequency of operation, transformer design requirements can be quite different.

For some LNA and VCO designs [7][84]-[86], the transformer is optimized for voltage transfer and magnetic coupling. For a balun circuit, single-ended to differential transformation and impedance transformation are emphasized.

2.4.1 The Ideal Transformer

In a typical transformer, the signal current goes through the primary winding and generates a magnetic field which induces a voltage across the secondary winding. Figure 2.13 shows a circuit model of an ideal transformer [7].

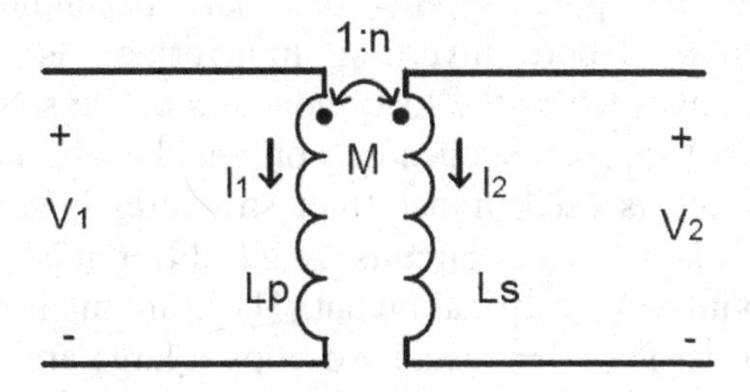

Figure 2.13 An ideal transformer [74].

The magnetic flux produced by the time-varying current I_1 flowing into the primary winding, induces a time-varying current of I_2 in the secondary winding. The terminal voltages and currents of this ideal transformer are related as follows:

$$\begin{bmatrix} V_1 \\ V_2 \end{bmatrix} = \begin{bmatrix} j\omega L_p & j\omega M \\ j\omega M & j\omega L_s \end{bmatrix} \begin{bmatrix} I_1 \\ I_2 \end{bmatrix} \tag{2.63}$$

where L_p and L_s are the self-inductance of the primary and secondary. M is the mutual inductance between the primary and secondary. The magnetic coupling coefficient k is given by

$$k = \frac{M}{\sqrt{L_p L_s}} \tag{2.64}$$

For an ideal transformer, $k = 1$. For most in-chip transformers, k lies between 0.3 and 0.9. Another electrical parameter of the transformer is the turn ratio n. It is related to the current and voltage of the primary and secondary coil by the relationship:

$$n = \sqrt{\frac{L_s}{L_p}} = \frac{V_2}{V_1} = \frac{I_1}{I_2} \tag{2.65}$$

2.4.2 Transformer Types

Monolithic transformers have been widely used in RF circuits. A transformer can be 2 ports, 3 ports, or 4 ports depending on the design requirement. The 2-port inverting transformer is used in many differential circuits such as the load inductors or the source degenerated inductors in LNA circuits. Depending on whether the lateral or vertical magnetic coupling is used, transformer structures have two categories: planar and stacked. Some circuits (e.g., differential circuits) prefer transformers with a symmetrical layout. There are many variations of the planar transformer [7]. They usually occupy a large area. In a multilayer metal process, several top metal layers can be strapped together to reduce resistance. The tradeoffs are a larger terminal-to-substrate capacitance and a smaller self-resonant frequency. Examples of the planar transformer (Fig. 2.14) are the tapped 2-port transformer, the Shibata transformer, the Frlan transformer, the symmetric, and the step-up. Figure 2.14(d) shows the layout that is suitable for 3 or 4-port applications with center taps at about half the physical length.

A stacked transformer using three metal layers is shown in Figure 2.15. As both vertical and lateral magnetic coupling is utilized, self-inductance and coupling coefficient ($k \approx 0.9$) are high. The method to calculate the effective capacitance can be found in [87]. The capacitance is inversely proportional to the square of the number of layers.

Another concern is that in modern multi-level processes, different metal layers usually have different thickness with the top metal layer usually thicker than the lower layers.

Figure 2.15 shows the symmetrical layout for the multilayer stacked transformer. Other variations may combine interleaved and stacked features to achieve symmetry as in [88]. In short, different transformers are used for different applications different characteristics. Table 2.1 compares the performances between the various types of transformer structures.

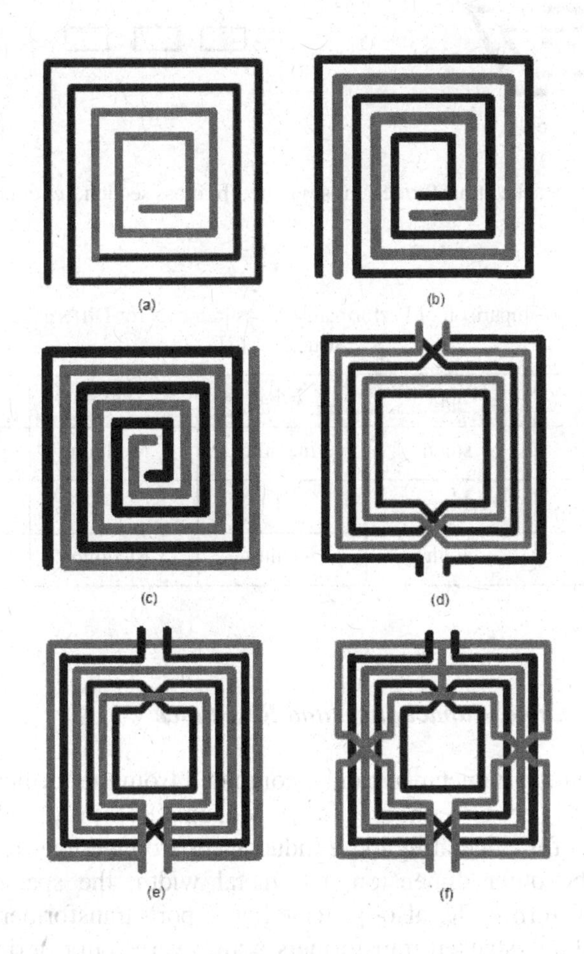

Figure 2.14 Planar transformer structures [7]: (a) tapped; (b) parallel (Shibata); (c) intertwined (Frlan); (d) symmetric (3:2); (e) step-up; (f) step-up variation.

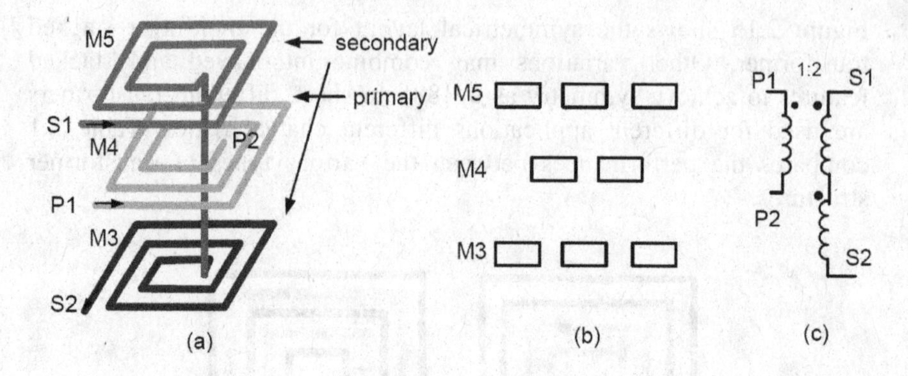

Figure 2.15 Stacked transformer: (a) structure; (b) cross section; (c) circuit model [7].

Table 2.1 Comparison of Performance Measurements for Different Transformer Structures [7].

	Tapped	Frlan	Step-up	Stacked
k	small	medium	Medium	high
n	arbitrary	1	1~4	arbitrary
Q	high	medium	Medium	small

2.4.3 Inductance, Capacitance, and Resistance

Self and mutual inductance can be computed from the formulae given by Grover [54] or Greenhouse [53]. Mohan [88] proposed simplified expressions for calculating these inductances from geometric parameters, such as the outer dimension, the metal width, the spacing, and the number of turns. He also provided a 2-port transformer model for interleaved and stacked transformers with a very rough series resistance approximation. The relationship for tapped and interleaved transformers is given by the expression

$$M = \frac{(L_T - L_p - L_s)}{2} \tag{2.66}$$

where L_T is the total inductance of a single spiral with all the segments of both primary and secondary traces; L_p and L_s are the self-inductances of the primary and secondary; M is the mutual inductance. For different trace widths in the primary W_p and secondary W_s coils, the weighted average width W_{avg} can be used for L_T calculation:

$$W_{avg} = \frac{(W_p N_p + W_s N_s)}{(N_p + N_s)} \tag{2.67}$$

where N_p and N_s are the number of turns for primary and secondary.

Capacitances C_p and C_s can be approximated using the equation:

$$C_i = \frac{\varepsilon_{ox}}{t_{ox}} l_i W_i \quad \text{where} \quad i = p \text{ or } s \tag{2.68}$$

where l_i and W_i are the length and width of the primary or secondary trace, and t_{ox} is the thickness of the oxide from the bottom of the metal layer to the top surface of the PGS. The shield parasitic resistance is given by

$$R_b = \left[\frac{D_{out}}{2W_{pgs}} + \frac{1}{12}\left(\frac{D_{out}}{2W_{pgs}}\right)\right] Rsh_{M1} + \frac{1}{4} Rsh_{poly} \tag{2.69}$$

where D_{out} is the outer dimension of the transistor (Fig. 2.16) and W_{pgs} is the width of the PGS strip. Rsh_{M1} and Rsh_{poly} are the sheet resistances of metal 1 and silicided poly, respectively.

The total resistance including all the losses can be expressed as

$$R_{total_p} = R_{dc} + R_{rfp} = R_{dc}\left(1 + r_{rf}\frac{(f/f_0)^2}{1+(f/f_0)^2}\right) \tag{2.70}$$

where R_{rfp} is the increased resistance due to the proximity effect; r_{rf} is a transformer geometry and technology related coefficient, and f_0 is a frequency factor given by

$$f_0 = \frac{2R_{sh}(1+3q)}{\mu_0 W(1-q^2)} \tag{2.71}$$

where μ_0 is the permeability of free space; R_{sh} is the sheet resistance of the metal; W is the width of the metal trace, and q is the ratio of W_i/W.

2.4.4 Coupling Coefficient k, Turn Ratio n, and Quality Factor Q

For voltage or current coupling such as the transformer interstage coupling and feedback circuit, high coupling coefficient k and specific or high turn ratio n, or both, are desired [84][86]. To increase k, the spacing between the primary and secondary should be as small as possible, constrained by the coupling capacitance and self-resonant frequency.

For a simple transformer, the quality factor Q_p and Q_s of the primary and secondary windings are defined as

$$Q_p = \frac{\omega L_p}{R_p} \qquad Q_s = \frac{\omega L_s}{R_s} \tag{2.72}$$

where R_p and R_s are the series resistances of the primary and secondary coils respectively.

2.4.5 Patterned Ground Shield

There are several ways to improve Q. Some improvements can be made through the fabrication process, such as using thick copper top layers or strapping multiple levels of metal layers to reduce ohmic losses. We can also use a thick oxide layer and/or highly resistive substrate to reduce substrate loss.

Another way is to shield the transformer from the silicon substrate. We use a patterned ground shield (PGS) [89]-[91]. The ground shield prohibits the electric field from entering the substrate. Due to the close proximity of the device and shield, using solid metallization would allow image eddy currents to flow, which would produce an opposing magnetic field. This would reduce the device's magnetic energy and thus increase the device's insertion loss. To minimize the parasitic capacitance between the transformer and the PGS, the transformer can be formed with a bottom metal and silicided polysilicon. The poly fingers are constructed along the directions perpendicular to the direction of eddy-current flows. By doing so, we slice the path of the eddy current. The X shape in the PGS enables lower shield resistance because the capacitive current flows the smallest distance to reach the ground.

2.4.6 Designing the Transformer

According to [7], for the same width W and spacing s, there is a large improvement in k as N increases from 1 to 2 because of the coupling between adjacent lines. However, further increase of N does not improve k considerably. For the same N, k decreases when W and s increase. However, wider metal traces usually have a better Q. For some applications such as the output baluns for power amplifiers, the large DC current requires wide traces for reliability. There must be an optimum metal width to balance k and Q for the maximum $k^2Q_pQ_s$. For power transfer and impedance transformation applications, such as baluns, the insertion loss is used as a metric to measure the performance of a transformer. The minimum insertion loss (*ILm*) is given by [92]

$$ILm\,(dB) = -10\log_{10}\left(1+2\left(x-\sqrt{x^2+x}\right)\right) \tag{2.73}$$

To obtain *ILm*, the product of $k^2Q_pQ_s$ needs to be maximized.

The wide traces can be split into multiple parallel segments and interleaved (Fig. 2.16) to improve *k*. Another advantage is that the proximity effect is also mediated. Since the effective width of the primary or secondary winding is enlarged, the self-inductances L_p and L_s are reduced while the resistance remains about the same. Q_p and Q_s will be reduced at low frequencies. The coupling capacitance between the primary and secondary windings increases. These contradictory effects need to be balanced in order to achieve an optimum *ILm*.

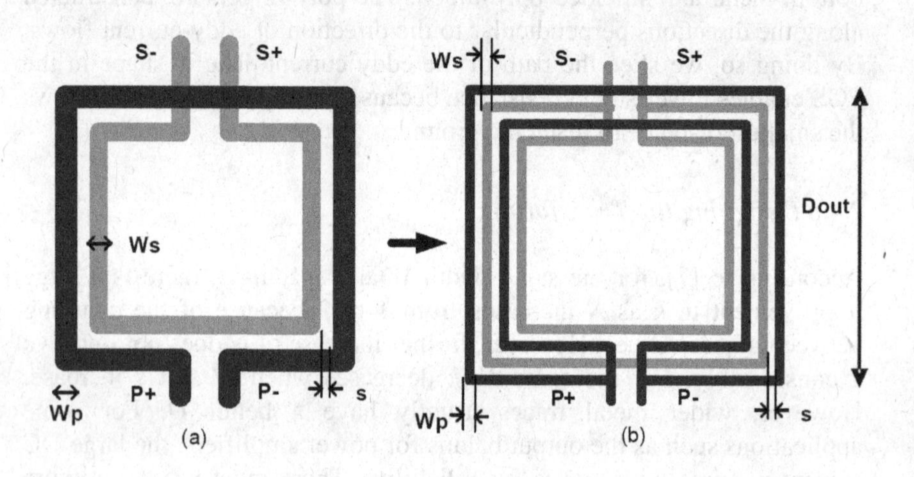

Figure 2.16 (a) A typical single-turn transformer; (b) a transformer with interleaving coils [7].

The *k* factor represents the strength of the magnetic coupling between the primary and secondary windings. The S-parameters and minimum insertion loss are used to validate the model. The *Q* factor, the coil inductance, and the coupling coefficient are used to demonstrate the scalability of the model.

The series resistance is composed of dc and ac resistances. Due to eddy currents, the ac resistance is frequency dependent, which makes it difficult to predict using physics-based expressions. The eddy-current effect occurs when a conductor is subject to time-varying magnetic fields. This current is composed of skin and proximity effects. The skin effect increases proportionally to the square root of frequency, while the proximity effect depends on the geometry and orientation of the conductor. Eddy currents reduce the net resistance flow in the conductor, and hence increase the ac resistance.

The Q factor and bandwidth can be improved by increasing the oxide and metal thickness. The Q factor initially rises with frequency as the reactive component of the impedance increases, peaks, and then decreases due to increasing energy dissipation at high frequencies. The width and thickness of the metallization added to the resistivity of the substrate, thus limiting the transformer quality factor as energy is dissipated by finite resistivity of metallization (skin and proximity effect) as well as in the conductive substrate. Inductance and k increase as the outer diameter increases for a fixed number of turns and pitch. A transformer with a wider area leads to high parasitic capacitance and thus a lower resonance frequency. The coupling coefficient and the inductance increase as the number of turns increases. k is higher for a symmetrical transformer with an important number of turns because the magnetic edge coupling is more important. The resonance frequency and working frequency decrease as the area increases.

2.5 RF Interconnects

Interconnects [5][93] exist to serve the essential task of providing electrical connections for devices and they can be classified into different levels. As an example, at the lowest level, on-chip backend metallization interconnects provide electrical linkage between semiconductor devices such as capacitors, resistors, MOS and bipolar junction transistors. In the higher levels, interconnects are backplanes or gateways providing infrastructure for high speed, large bandwidth data transfer between networks.

Sizes of interconnect increase as the interconnect level escalates and interconnects with the smallest dimensions can be found at the lowest level, in this case, on a semiconductor chip. At various levels in the hierarchy, due to different requirements and operating frequencies, interconnects are designed and modeled differently. Modeling techniques [94] such as electromagnetic (EM) simulation using HFSS (High-Frequency Structure Simulation), circuit modeling using partial element equivalent circuits (PEEC), or passive device physical modeling, can be applied at different levels of the hierarchy.

Further explorations of silicon technologies for a wider range of possible RF applications reveal that interconnects, just like inductors, are fast becoming technological bottle-necks limiting any potential utilization. This is because CMOS on-chip interconnects are fabricated using resistive aluminum or copper metallization, isolated by silicon oxide on lossy silicon substrates. Such interconnects have detrimental effects on circuits. As frequency increases, undesirable parasitic responses, both inductive and resistive in nature, as well as capacitive coupling to the substrate, become even more pronounced, therefore leading to significant deteriorations in circuit performance as well as shifts in circuit frequency response.

2.5.1 Transmission Line Concept

The most important difference between transmission line theory and circuit theory is electrical size. Circuit analysis assumes that the physical length of an electrical connection is much smaller than the electrical wavelength of the propagating signal, while transmission lines are much shorter, just fractions of a wavelength, or many wavelengths [95]. Hence, for transmission line structures with physical dimensions which are considerably smaller than the wavelengths of the signals being transmitted, they may be acceptably described with line voltages and currents. And if the signal level is fairly constant along the entire length of the interconnect, it can be treated as a lumped element and not a transmission line.

When the signal is sinusoidal, it is generally agreed in [96] that the signal does not change much in time over an interval equal to one twentieth of the period T of the signal. When the interconnect line is less than $1/20^{th}$ of the wavelength λ of the signal, it is regarded as safe to use a lumped circuit model for the interconnect. For lengths greater than $\lambda/10$, it is best to use the transmission line model.

High frequency response of an interconnect line depends on the source resistance and termination resistance looking in and out of the interconnect structure [97]. When the source resistance is much larger than the interconnect line impedance, only a small portion of the applied voltage will be placed on the line initially. The voltage will slowly increase, in an exponential fashion until reaching steady state over time. In such cases, the interconnect line effect is negligible and the line could be approximated by a lumped *RLC* physical model. If the source resistance is reduced and the line's response begins to include more and more ringing over time, this demonstrates the delay through the transmission line. In such cases, transmission line is a better approximation and these kind of structures can be approximated by cascading several interconnect lump models to achieve reasonable accuracy.

2.5.1.1 Transmission line constants

In a uniform transmission line, the line parameters R′, L′, G′ and C′ are referred to as resistance, inductance, conductance and capacitance per unit length respectively. In most radio frequency transmission lines, the effects due to L′ and C′ tend to dominate because of the relatively high inductive reactance and capacitance. They are generally referred to as "loss-free" or "lossless" lines although in practice some information about R′ or G′ may be necessary to determine the actual power loss. The lossless concept offers relatively good approximation. The propagation of a wave along the line is characterized by the propagation coefficient γ

$$\gamma = \sqrt{(R' + j\omega L')(G' + j\omega C')} \quad \text{or} \tag{2.74}$$

$$\gamma = \alpha + j\beta \tag{2.75}$$

where, α = attenuation coefficient, in Nepers per meter.

β = phase change coefficient, in degrees, or radians, per meter.

ω = angular frequencies in radians per second

At sufficiently high radio frequencies, Eqns. (2.74) and (2.75) yield

$$\beta = \omega\sqrt{L'C'} \tag{2.76}$$

β is called the "wave number". From the relationship of propagation velocity, the signal velocity propagation can be written as

$$v_p = \omega / \beta \text{ (m/sec)} \tag{2.77}$$

Using Eqn. (2.76), v_p can be expressed as

$$v_p = 1/\sqrt{L'C'} \text{ (m/sec)} \tag{2.78}$$

The velocity of propagation is also given in terms of the absolute permeability μ and permittivity ε of the medium through which the wave passes,

$$v_p = 1/\sqrt{\mu\varepsilon} = c/\sqrt{\mu_r \varepsilon_r} \text{ (m/sec)} \tag{2.79}$$

where c = 2.99793 x 10^8 m/sec, the velocity of light in free space; μ_r = relative permeability of the medium through which the wave passes; ε_r = relative permittivity of the medium through which the wave passes. From Eqn. (2.79), it is useful to note that the wave propagation speed is dependent on the dielectric medium.

2.5.1.2 Transmission line impedances

The characteristic impedance Z_0 of a transmission line is generally given by

$$Z_0 = \sqrt{\frac{R' + j\omega L'}{G' + j\omega C'}} \tag{2.80}$$

At high radio frequencies this simplifies to

$$Z_0 = \sqrt{\frac{L'}{C'}} \tag{2.81}$$

Changing Z_0 requires a change in the physical dimensions of the transmission line. The impedance looking into a transmission line varies with the distance progressed along the line. By setting down the distance limits d into expressions for the voltage and current at any point along the line, for a specific impedance Z_L, the following expression for the input impedance of the line is given by

$$Z_{in} = Z_0 \left(\frac{Z_L \cosh \gamma d + Z_0 \sinh \gamma d}{Z_0 \cosh \gamma d + Z_L \sinh \gamma d} \right) \tag{2.82}$$

Eqn. (2.82) is a general expression for Z_{in}. For lossless lines, it simplifies to

$$Z_{in} = Z_0 \left(\frac{Z_L + jZ_0 \tan \beta d}{Z_0 + jZ_L \tan \beta d} \right) \tag{2.83}$$

2.5.1.3 Reflection and voltage standing wave ratio

In all cases, unless the line is in completely matched condition, the load termination will reflect some of the energy originally injected into the transmission line. When a signal is reflected back to the launching point, interference between the incident and reflected waves, traveling at the same velocity but in opposite directions, causes a standing-wave field pattern to be formed. The voltage reflection coefficient Γ, ratio of reflected to incident voltage at the load, is given by

$$\Gamma = \frac{Z_L - Z_0}{Z_L + Z_0} \tag{2.84}$$

The ratio of maximum to minimum amplitude of the standing wave is called the Voltage Standing Wave Ratio (or VSWR) and is given by

$$\mathrm{VSWR} = \frac{1 + |\Gamma|}{1 - |\Gamma|} \tag{2.85}$$

For matched condition, $\Gamma = 0$ and $VSWR = 1$. The reflection coefficients and standing wave ratios for short-circuit and open-circuit terminated conditions are –1 and +1 respectively. In both cases mentioned above, the VSWR is infinite.

2.5.1.4 Frequency-dependent charge distribution

On-chip interconnects are essentially planar structures with cross-sectional view as shown in Figure 2.17. Such lines have frequency dependent behaviors. Figure 2.18 shows the charge distribution of the interconnect structure under different operating conditions. When a positive DC voltage is applied on the top conductor, positive charges on the top conductor are generally arranged in a fairly uniform distribution as illustrated in Figure 2.18(a). The bottom conductor, which is the ground plane, has balanced negative charges, so that the electric field lines begin on the positive charges and terminate on the negative charges. The negative charges on the ground plane are uniformly distributed over the entire ground plane.

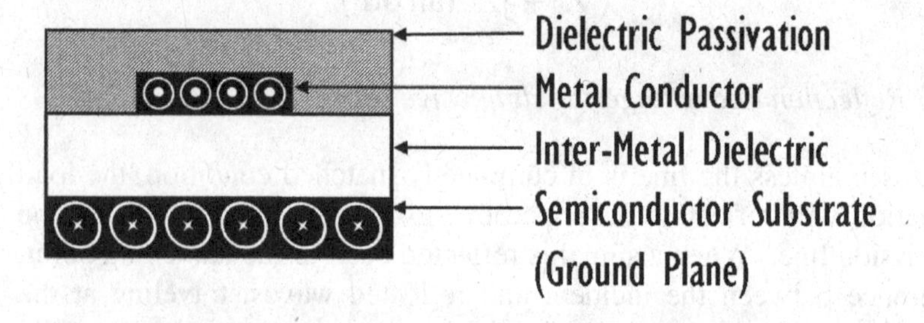

Fig. 2.17 Cross-sectional view of on-chip interconnects.

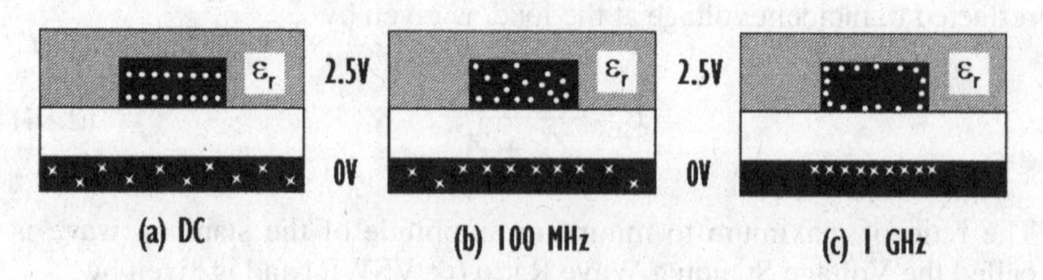

Fig. 2.18 Cross-sectional view of the charge distribution in an interconnect at different frequencies [5]. (The dots and crosses indicate charge concentration of different polarity).

As frequency of the signal propagating in the interconnect increases, the charges do not have sufficient time to rearrange and so they are not distributed homogeneously. Such effect is observed at 100MHz, as shown in Figure 2.18(b). This phenomenon becomes more prominent when the frequency of the signal is at 1GHz as depicted in Figure 2.18(c). The electromagnetic fields, being time varying, are not able to penetrate the conductor as before due to skin effects. As such, charges in the top conductor are only distributed with respect to the depth of penetration into the conductor. Consequently, current flow is mostly concentrated near the surface of the conductor and the effective cross-sectional area of the conductor for current to flow is reduced. Hence, resistance of the conductor increases significantly. Redistribution of the charges results in a change of the conductor's inductance with frequency. At low frequencies, for the same current, magnetic energy is stored inside as well as outside the conductors. At intermediate frequencies, the inductance reduces with frequency and only at sufficiently high frequencies, the magnetic field becomes confined to the region outside the conductors and the inductance remains approximately constant. As frequency continues to increase, parasitics become more pronounced and the inductance starts to increase. Nonetheless, the most significant frequency dependent effect is the resistance of the interconnect, which increases approximately with the square root of the frequency.

2.5.1.5 Effects of dielectric on interconnects

On-chip interconnects are usually constructed within insulating materials such as silicon dioxide. The presence of such insulating material between conductors alters the electrical characteristics of the interconnect. Application of an electric field across the dielectric moves the centers of positive and negative charges, changing the amount of energy stored in the electric field. The extra energy storage property is described by the relative permittivity ε_r, which is a ratio of the material's permittivity to that of free space.

Relative permittivity of materials used in on-chip interconnects varies from 3.5 to 3.9 for doped silicon dioxide (e.g., boro-phospho-silicate glass (BPSG), fluorine-doped silicate glass (FSG), etc.) to 11.9 for silicon. When interconnect is enclosed by more than one medium i.e. a non-homogeneous transmission line, the effective permittivity ε_{eff}, has to be used. ε_{eff} will change with frequency as the proportion of energy stored in the different regions changes. This effect is called *dispersion* and can cause an input pulse to spread out as the different frequency components of the pulse travel at different speeds.

A similar effect on energy storage in the magnetic field occurs for a few materials. The magnetic properties of materials are due to the magnetic dipole moments resulting from alignment of electron spins. In most materials, the electron spins occur in pairs with opposite polarity and the net outcome is zero magnetic moment.

However, when the net outcome is not zero, this remaining net magnetic moment will align with a magnetic field and provides a mechanism for additional storage of magnetic energy. The relative permeability μ_r describes this effect. Nearly all materials used with interconnects have $\mu_r = 1$.

2.5.2 Existing Methodologies to Tackle Post Layout Parasitics

Silicon-based RF designs are now conceptualized on CAD tools that utilize feature-rich process design kits (PDKs) with accurate and scalable device models developed based on extensive and reliable on-wafer RF device characterization. Despite having accurate device models that are capable of predicting device degradation effects at radio frequencies, a typical RFIC still requires several design iterations before it can comply with the design specifications. These design iterations can be avoided if metallization interconnects, which become parts of the circuit when they are used to provide electrical connections between devices, are carefully considered during post-layout circuit simulations.

Sia [5] proposed the scalable double-π RF sub-circuit model for RF interconnects is shown in Figure 2.19. L_S and R_S describe the parasitic self-inductance and resistive loss on the metallization respectively. L_{SK} and R_{SK} model the skin effects of the metallization at radio frequencies. Substrate loss for interconnect is modeled by the RC network that consists of C_{OX}, C_{SUB} and R_{SUB}. These 3 elements describe the oxide capacitance between the silicon substrate and metallization, the capacitive and resistive losses in the silicon substrate respectively.

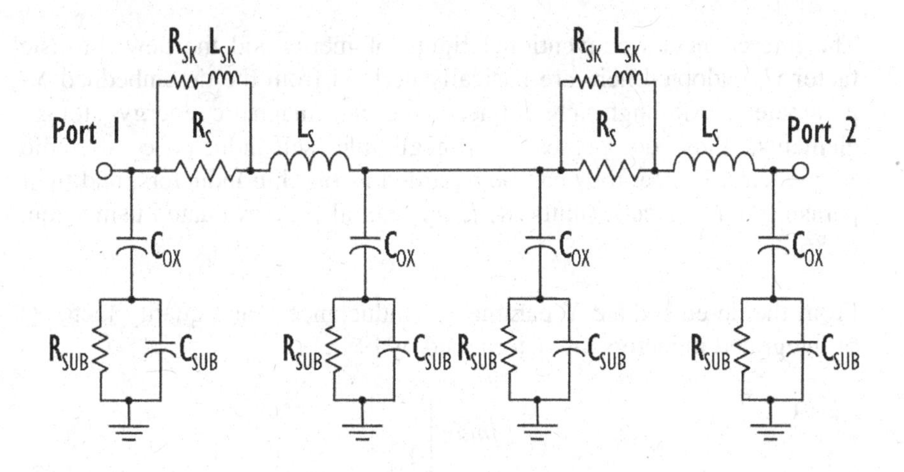

Figure 2.19 Double-π RF sub-circuit model for interconnects [5].

2.5.3 Proposed Figure of Merit for RF Interconnects

Although the proposed double-π model is complicated with numerous RLC elements, this model is symmetrical and values for the model parameters are identical for the two π networks. Furthermore, for RF applications, a more suitable figure of merit called the *intrinsic factor* I_F for describing interconnects is proposed by Sia [5]:

$$I_F = \frac{1}{Imag\left[\frac{1}{Y_{11}}\right] \times Real\left[\frac{1}{Y_{11}}\right]} \tag{2.86}$$

I_F will conveniently guide engineers to choose interconnects with the smallest parasitics for device routing where, from the de-embedded *Y* parameters, *L*, *Q*, and *R* are conventionally determined by methods in [98].

The interconnect's conventional figure of merits and the new intrinsic factor (I_F) adopted here are basically derived from the de-embedded Y-parameters. Although interconnects are not magnetic energy storage elements, they do possess non-negligible self-inductance at radio frequencies. Hence, they can be regarded as on-chip inductors, and their parasitic inductance *L* (units in *Henry*) can also be evaluated using Eqn. (2.86).

From the de-embedded Y parameters, inductance *L* and quality factor *Q* of integrated inductors are determined by [98],

$$L = \frac{Imag\left[\frac{1}{Y_{11}}\right]}{2 \times \pi \times Frequency} \tag{2.87}$$

$$Q = -\frac{Imag\left[Y_{11}\right]}{Real\left[Y_{11}\right]} \tag{2.88}$$

And series resistance *R* for spiral inductors can be expressed as follows [99],

$$R = Real\left[-\frac{1}{Y_{12}}\right] \tag{2.89}$$

The *Q* shown in (2.88) for an inductor can also be used to correlate how interconnects could influence the circuit performance. *L* and *Q* are both extracted from Y_{11} and not Y_{12} parameter, since it is important to include and consider the effects of the lossy silicon substrate when evaluating the performance of interconnects. The Q factor is defined as a ratio of *imaginary* $[Y_{11}]$ over *real* $[Y_{11}]$. This suggests that at a fixed resistive loss, when the inductance increases, the Q factor increases. However, for the case of interconnects, desirably low loss metallization should possess negligible parasitic inductance and resistance.

If we refer to Eqn. (2.86), interconnects with large intrinsic factors insinuate that they have small resistive and inductive parasitics. Parasitic series resistance R associated with the metallization is extracted from Y_{12} (2.89) to show the skin effects of metallization at radio frequencies.

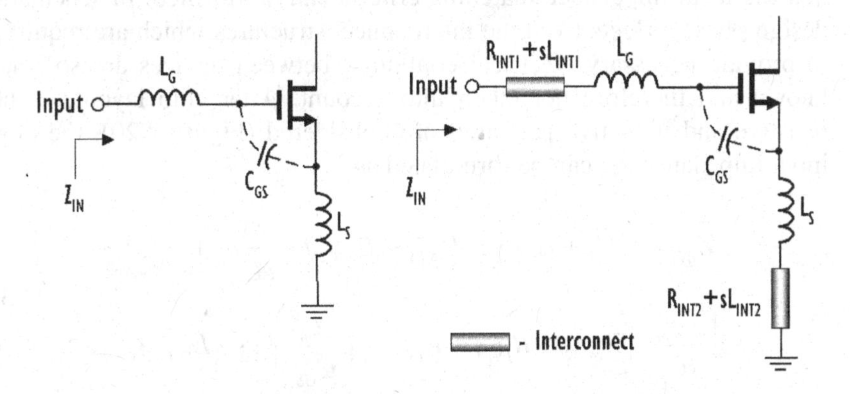

Figure 2.20 Inductive source degeneration impedance matching without (left) and with (right) consideration of interconnect metallization [5].

The input impedance Z_{IN} can be expressed as

$$Z_{IN} = s(L_G + L_S) + \frac{1}{sC_{GS}} + \left(\frac{g_m}{C_{GS}}\right)L_S \tag{2.90}$$

where g_m is the transconductance of the nMOSFET; C_{GS} is the gate-source capacitance of the nMOSFET; and $s = j*2\pi*f$.

For maximum power transfer to occur at the operating frequency, *real* $[Z_{IN}]$ have to match the source resistance R_{SOURCE}, which is typically 50 ohm:

$$\left(\frac{g_m}{C_{GS}}\right)L_S = R_{SOURCE} \tag{2.91}$$

and the imaginary impedances must satisfy the following condition:

$$(L_G + L_S) = \frac{1}{C_{GS}} \tag{2.92}$$

From Eqns. (2.91) and (2.92), values of L_S and L_G are first determined so that the ideal impedance matching criteria can be fulfilled. In schematic design phase, effects from the interconnect structures which are required to provide necessary electrical continuity between devices are still not known and therefore not taken into account. After chip layout, if the resistive and inductive parasitics are considered (Figure 2.20), the new input impedance Z_{IN} can be formulated as

$$Z_{IN} = s(L_G + L_S) + s(L_{INT1} + L_{INT2}) + \frac{1}{sC_{GS}} + \left(\frac{g_m}{sC_{GS}}\right)R_{INT2} + R_{INT1} + R_{INT2} + \left(\frac{g_m}{C_{GS}}\right)(L_S + L_{INT2}) \tag{2.93}$$

For impedance matching at the operating frequency, *real* [Z_{IN}] will have to match to the source resistance

$$R_{INT1} + R_{INT2} + \left(\frac{g_m}{C_{GS}}\right)(L_S + L_{INT2}) = R_{SOURCE} \tag{2.94}$$

and the imaginary impedance will have to satisfy the new condition:

$$(L_G + L_S) + (L_{INT1} + L_{INT2}) = \frac{1}{C_{GS}} + \left(\frac{g_m}{C_{GS}}\right)R_{INT2} \tag{2.95}$$

To achieve maximum power transfer, *real* [Z_{IN}] must be matched to the source resistance but evaluating Eqns. (2.91) and (2.94) reveal that additional components of ($R_{INT1} + R_{INT2}$) and $L_{INT2}*(g_m/C_{GS})$ have impeded the performance of the original matching inductor L_S. Comparing Eqns. (2.92) and (2.95), the resonant frequency of the impedance matching network differs by the extra components of (L_{INT1} + L_{INT2}) and $R_{INT2}*(g_m/C_{GS})$. These parasitic components from the interconnects contribute to unfavorable shifts in the frequency response of the circuit as opposed to when it was previously optimized with L_S and L_G which excluded effects from the interconnects. Therefore, without prior knowledge of the parasitic resistances and inductances introduced by the interconnects, operating frequency in this narrow-band RFCMOS design is altered and power is not transferred efficiently into the nMOSFET. If both interconnects in Figure 2.20 are identical, i.e., $R_{INT1} = R_{INT2} = R_{INT}$ and $L_{INT1} = L_{INT2} = L_{INT}$, Eqns. (2.94) and (2.95) simplify to

$$2R_{INT} + \left(\frac{g_m}{C_{GS}}\right)(L_S + L_{INT}) = R_{SOURCE} \tag{2.96}$$

$$(L_G + L_S) + (2L_{INT}) = \frac{1}{C_{GS}} + \left(\frac{g_m}{C_{GS}}\right)R_{INT} \tag{2.97}$$

Using interconnects with large I_F will have less significant impact on circuit performances. In this example, for the interconnects shown in Figure 2.20, without having any shunt parasitic capacitances, from Eqn. (2.86), I_F can be extracted from Y_{12} parameters and therefore written as

$$\begin{aligned} I_F &= \frac{1}{Imag\left[\frac{1}{Y_{12}}\right] \times Real\left[\frac{1}{Y_{12}}\right]} \\ &= \frac{1}{2*\pi*Frequency \times L_{INT} \times R_{INT}} \end{aligned} \tag{2.98}$$

In Eqn. (2.98), when I_F is very large, it suggests that both R_{INT} and L_{INT} are negligibly small. Hence, Eqns. (2.96) and (2.97) tends to the original matching conditions described in Eqns. (2.91) and (2.92):

$$2R_{INT} + \left(\frac{g_m}{C_{GS}}\right)(L_S + L_{INT}) \rightarrow \left(\frac{g_m}{C_{GS}}\right)L_S \text{ and} \quad (2.99)$$

$$(L_G + L_S) + (2L_{INT}) = \frac{1}{C_{GS}} + \left(\frac{g_m}{C_{GS}}\right)R_{INT} \rightarrow (L_G + L_S) = \frac{1}{C_{GS}} \quad (2.100)$$

Therefore, choosing high I_F interconnects allow fabricated circuits to have smaller frequency shifts and minimal degradation on the transfer of power previously optimized in circuit schematic simulations when interconnects are not taken into considerations. Monitoring the parasitic resistance, inductance, and capacitive substrate loss (i.e., real and imaginary impedances) individually is cumbersome. Thus, exploiting the proposed I_F provides a quick benchmarking indicator for circuit designers.

As frequency increases, interconnects behave like transmission lines with distributed characteristics. Hence, a simple lumped element π model is insufficient to describe its behavior. It was found that when the single-π lumped element model is adopted for the same interconnect, the model deviation between the measured and simulated values for R and I_F worsen from about ±8% to more than ±25%.

A lumped circuit model is sufficient to predict the characteristics of the interconnect line if it is less than 1/20th of the wavelength λ of the signal. However, the velocity of the signal propagating in the interconnect will be reduced when it is embedded within a dielectric material g. Relative permittivity ε_r and permeability μ_r of material g have to be considered when calculating the wavelength λ_g of the signal. Therefore, critical lengths of interconnects in silicon process will be smaller because they are embedded within silicon dioxide resting on a lossy silicon substrate. The critical physical lengths for RF interconnects in silicon-based process with silicon and silicon dioxide as dielectric materials can be calculated as follows:

$$\text{Critical length} < \lambda_g / 20 \quad (2.101)$$

$$\text{Critical length} < \frac{c}{\sqrt{\varepsilon_r} \times \sqrt{\mu_r} \times \text{frequency} \times 20} \quad (2.102)$$

where λ_g is the wavelength in material g and c = speed of light (in free space ≅ 3E8 *m/sec*). Computations of the critical lengths for interconnects embedded within different dielectrics using Eqns. (2.101) and (2.102) are presented in Table 2.2 below.

Table 2.2 Interconnect critical lengths in free space, SiO_2 and silicon [5].

Frequency (GHz)	2.5	5.0	7.5	10	20	50
Critical Length=λ/20 in free space (μm)	6000	3000	2000	1500	750	300
Critical Length=λ/20 in SiO_2, ε_r=3.9 (μm)	3038	1519	1013	759	380	152
Critical Length=λ/20 in Si, ε_r=11.9 (μm)	1739	870	580	435	217	87

2.6 Varactors

The varactor is a semiconductor diode with the properties of a voltage-dependent capacitor. Specifically, it is a variable-capacitance, *pn*-junction diode that makes good use of the voltage dependency of the depletion-area capacitance of the diode. The variable capacitance property of the varactor allows it to be used in circuit applications, such as amplifiers, that produce much lower internal noise levels than circuits that depend upon resistance properties. Since noise is of primary concern in receivers, circuits using varactors are an important development in the field of low-noise amplification, mainly in the tunable LC tank circuits [100] for standards such as GSM, GPS, UMTS, Bluetooth, and WLAN. The most significant use of varactors to date has been as the basic component in parametric amplifiers.

2.6.1 Functions of Varactors

Varactors are also critical ingredients in the design of RF voltage-controlled oscillator. Because of the thin-film nature of SOI devices, traditional reverse-biased *pn* diode used in bulk design is not area efficient [13][101]. Recently there have been a number of discussions in using accumulation mode MOS capacitor as varactors in bulk technologies [102] as well as gated varactors [103] where a circuit model is drawn to calculate the resistances and capacitances.

According to [104], the measurement of various varactors at different frequencies provides a useful library to decide the best-fitting varactor for a particular design. There are several architectures for implementing integrated varactors, the most common being *pn* junction [105], MOS [106], and gated varactors [107]. The principal parameters of a varactor are the quality factor Q and the tuning range TR. Q factor is a fundamental parameter required for a good design of any circuit with passive elements and should be maximized. It measures the ability of the component to preserve the energy received during circuit operation. TR is the parameter that characterizes the relative capacitance variation with the biasing voltage and is defined as

$$TR\left(\%\right)=\pm\frac{C_{\max}-C_{\min}}{C_{\max}+C_{\min}}\times 100 \qquad (2.103)$$

where c_{max} is the varactor capacitance for zero bias and C_{min} is the varactor capacitance for maximum reverse voltage. Gutierrez's varactor [104] is based on the *pn* junction. This is because this structure presents some advantages over MOS and gated varactors. the quality factor is higher and it is easily scalable than others. In addition, the capacitance variation is smoother than that exhibited by MOS structures [105]. In his varactor shown in Figure 2.21, the operation of these varactors is based on the junction capacitance associated with the depletion region between the p+ diffusion (anode A) and the n-well (cathode C). *N*+ diffusions are placed on the n-well for decreasing the series resistance of the device. The p-sub is grounded. Under reverse biasing, a capacitance appears in this junction due to the depletion zone. When the bias voltage changes, the depletion zone changes and so does the capacitance.

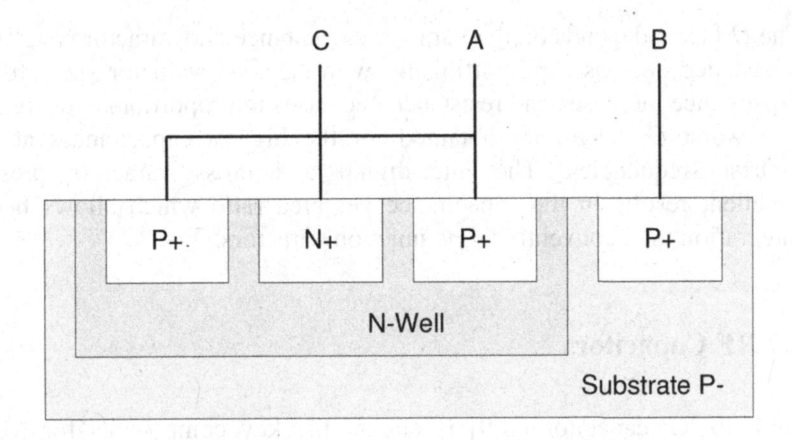

Figure 2.21 *PN* junction varacter [104].

2.6.2 Varactor Design

The square island configuration is normally used when a varactor is implemented with *pn* junction. A large periphery per unit area is desirable [108] as they will present longer capacitance values and can offer the typical tuning range of 10% to 20%. The square island configuration is normally used when a varactor is implemented with *pn* junction. The design of the varactors can be inter-digital or cross geometry [104]. The inter-digital varactors are designed with large *p+* islands in n-well, and in cross varactors, the *p+* contacts and *n+* contact crosses in an n-well zone.

In [104], capacitance values are obtained for the frequencies of GSM, GPS, UMTS, Bluetooth, and WLAN. Except for WLAN which gives larger capacitance values, the rest show very similar capacitances. For a given increase in working frequency under a bias voltage, parasitic impedance due to the inductance of the connection layers also increases; also, a resonance effect appears between the varactor capacitance and the parasitic inductance. As varactor size grows, the capacitance and the parasitic impedance of the connection layers also increase while the resonance frequency decreases.

The Q factor depends on the varactor capacitance and varactor resistance. Resistance decreases proportionally with size. As varactor size grows, capacitance increases and resistance decreases in proportion. Q decreases. The worst Q values are obtained for the highest capacitances at the highest frequencies. The inter-digital and cross varactors present excellent results in the capacitance per area ratio which allows better integration than conventional *pn* junction varactors.

2.7 RF Capacitors

On-chip RF capacitor [109] is one of the key components for RFIC designs such as filters, oscillators, mixers, data converters, sample and holds, and switched capacitor circuits. Geometry design variables include the number of fingers, length, width, gap, end gap, terminal width, strip thickness, substrate height, metal types, and dielectric constant. The optimum design is identified through the contour plot of the Q factor.

A capacitor is actually a lumped circuit element that stores energy through electrical fields. The amount of energy stored and for how long will depend on the design of the capacitor. For RF applications, the capacitor should have high Q factor. Capacitance is required to be linear and low in parasitic capacitance.

2.7.1 Capacitance

The capacitance of the well-known parallel plate capacitor is expressed as [11][109]

$$C = \frac{\varepsilon_0 A}{d} \tag{2.104}$$

where ε_0 is the permittivity of free space; A is the plate area; and d is the separation between the plates. If material medium exists between the plates, the charge storage ability per unit voltage is increased by a factor of ε_r. The capacitance is then given by

$$C = \frac{\varepsilon_r \varepsilon_0 A}{d} \tag{2.105}$$

where ε_r is the relative permittivity of the medium.

Among the wide variety of capacitors for RF circuits, a designer has to choose between them based on capacitor properties such as linearity, quality factor, breakdown voltage, area, matching, cost, performance, temperature coefficient, and sensitivity to process variation. The ideal capacitor is low cost, possesses small process variation, low temperature coefficient, low leakage, low noise, low parasitic ground capacitance, high capacitance density, good matching, high quality factor, high yield, high self-resonance frequency, high breakdown voltage, and constant capacitance over the usage frequency range [110].

2.7.2 Geometry

An inter-digital capacitor is a meandered gap in a microstrip or stripline center conductor. The dimensions are less than a quarter wavelength. The capacitance depends on finger width, finger length, finger spacing, and the number of fingers (Figure 2.22).

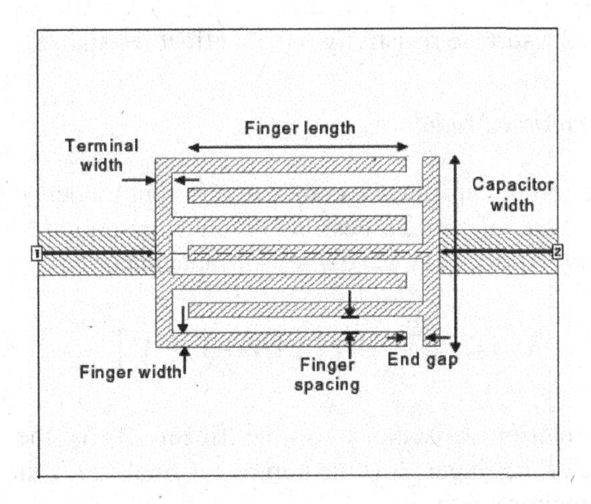

Figure 2.22 An inter-digital capacitor (top view) [11].

2.7.3 Quality Factor and Series Resistance

We evaluate a HF on-chip capacitor by its Q factor or series resistance. A perfect capacitor would exhibit an ESR of zero ohms and would be purely reactive with no real or resistive component. It would return all the stored energy to its network while real capacitors incur losses during the charge/discharge cycle. These losses are due to the polarization of diploes or charge leakage within the dielectric material. The Q factor of the inter-digital capacitor is given by [11]

$$Q = \frac{Q_c Q_d}{Q_c + Q_d} \tag{2.106}$$

Q_c is the quality factor due to the conductor losses and Q_d is the quality factor due to the dielectric losses.

The series resistance of an inter-digital capacitor increases with frequency due to skin effect resistance. The series resistance R can be expressed as

$$R = \frac{4}{3}\frac{R_s l}{w} \text{ ohm} \tag{2.107}$$

where R_s is the surface resistivity or skin effect resistance.

2.7.4 Capacitance Modeling

The calculation of capacitance comes in different models due to different researches. One of the earliest work on inter-digital capacitors was performed by Alley [111]. He expressed capacitance as

$$C = \frac{\varepsilon_r + 1}{w} l \left[A_1 (N-3) + A_2\right] \tag{2.108}$$

A_1 is the interior capacitance of the finger. A_2 is the two exterior capacitance of the finger. N is the number of fingers. l is the finger length; w is the capacitor width.

Esfandiari [112] formulated capacitance as

$$C = (N+1)\,C_g\,l \tag{2.109}$$

where C_g is the static capacitance.

Naghed [113] expressed the capacitance of the coplanar inter-digital capacitor in the form of an equivalent circuit. F. Aryanfar [114] developed a model for the coplanar over the entire W-band frequency range. Casares-Miranda [109] developed a wire-bonded inter-digital capacitor by short circuiting across alternate fingers. Improvement is obtained without any diminution in capacitance.

2.7.5 Impedances

At a specified angular frequency, the characteristic admittance and propagation constant for the terminal strip are expressed as [112]

$$Z_{Te} = \sqrt{\frac{R_T + j\omega L_{Te}}{j\omega C_{Te} + N(Y_{11} + Y_{21})/2l_T}} \tag{2.110}$$

$$Z_{T0} = \sqrt{\frac{R_T + j\omega L_{T0}}{j\omega C_{Te} + N(Y_{11} + Y_{21})/2l_T}} \tag{2.111}$$

$$\text{where } \gamma_{Te} = \sqrt{(R_T + j\omega L_{Te})[j\omega C_{Te} + N(Y_{11} + Y_{21})/2l_T]} \tag{2.112}$$

$$\text{and } \gamma_{T0} = \sqrt{(R_T + j\omega L_{Te})[j\omega C_{T0} + N(Y_{11} + Y_{21})/2l_T]} \tag{2.113}$$

where C_{Te}, C_{TO}, L_{Te}, and L_{TO} represent the even- and odd-mode capacitances and inductances for the terminal strip; R_T is the resistance of the inductors and N is the number of fingers. Y_{11} and Y_{21} are the elements of the admittance matrix for $N/2$ inter-digital sections in parallel. These admittances are averaged over the terminal strip length in the above formula. Thus

$$Z_{11} = Z_{22} = \frac{1}{2}[Z_{0e} \coth \gamma_e l + Z_{00} \coth \gamma_0 l] \tag{2.114}$$

$$Z_{21} = Z_{12} = \frac{1}{2}[Z_{0e} \csc \gamma_e l + Z_{00} \csc \gamma_0 l] \tag{2.115}$$

In matrix form,

$$[Y] = [Z]^{-1} \tag{2.116}$$

where Z_{0O}, Z_{0e}, γ_O, and γ_e are the odd and even-mode impedances and propagation constants of the fingers. l is the length of overlap of the fingers.

2.7.6 Design Considerations

During the design of the capacitor, a few aspects need to be noted by the designer. Shorter finger length reduces the series resistance which eventually leads to a higher quality factor Q. From observations by [11][109], Q_{max} decreases when the finger length increases from 0.06mm to 0.04mm. C increases when finger length increases from 0.06mm to 0.40mm. Increasing the width will also increase the capacitance. The number of fingers gives huge impact to the inter-digital capacitor performance. Capacitance increases when the number of fingers increases from 4 to 10. However, when the number of fingers increases, Q decreases. As finger spacing or finger width increases, the number of fingers will decrease. The number of fingers is proportional to the finger length. Silver has the highest conductivity (and highest Q_{max}) followed by copper and gold. Therefore price is the reason copper is used for RFIC capacitors.

2.8 Process Design Kits

A design kit is a collection of technology files that enables designers to create, simulate and verify their designs from front-end to back-end using foundry-specific processes. It contains device libraries, technology files, simulation models, parameterized device generators, DRC, LVS and PEX files, and documentations. These components provide a complete, integrated design environment to accelerate designs from front-end to back-end. It saves setup time, reduces design cycle and time-to-market, and reduces cost.

2.8.1 Benefits

Consolidating all the data from the foundries and putting them together in a design environment can be very time-consuming and error-prone. With a design kit, all the files are co-related as shown in Fig. 2.23 and 2.24, and the technology files and scripts matched to the software tools. It is necessary to accelerate design cycle as product life cycles are short. Device symbols with integrated models and net-list configuration files aid circuit designers to accelerate their circuit creation.

The Schematic Driven Layout (SDL) flow and parameterized device generators enable layout designers to increase productivity and focus on optimizing the design. There is increasing important for designers to achieve first-pass silicon and accuracy due to high mask cost. Rework cycles due to errors in the DRC and LVS decks and mistakes that occur from inaccurate models are reduced as these files are often foundry validated and built by specialized teams. Basically, design kits require the appropriate tool set from vendors and help to increase the adoption of the vendors' tool within the design flow. They provide easy access to required foundry information for design companies to start their design cycles and help to showcase the processes available in the foundry and promote the foundry as a first tier source. The whole process and application of an example of a PDK is nicely summarized in Figures 2.23 and 2.24 below.

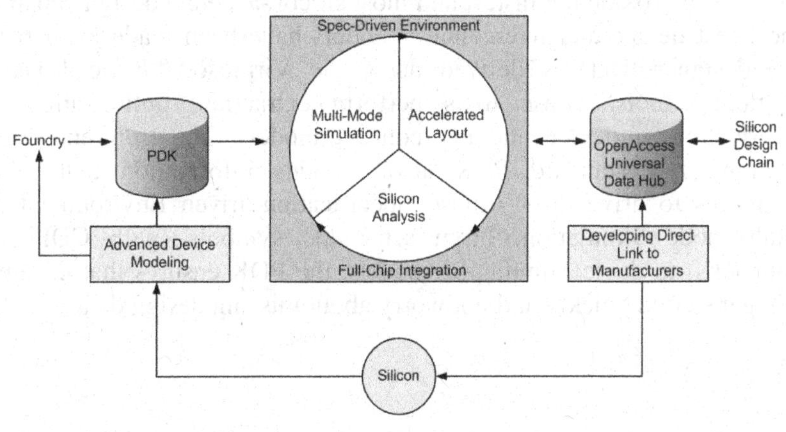

Figure 2.23 The Cadence Virtuoso custom design platform derives its silicon accuracy directly from a PDK. The PDK contains information specific to each design task and to specific design tools.

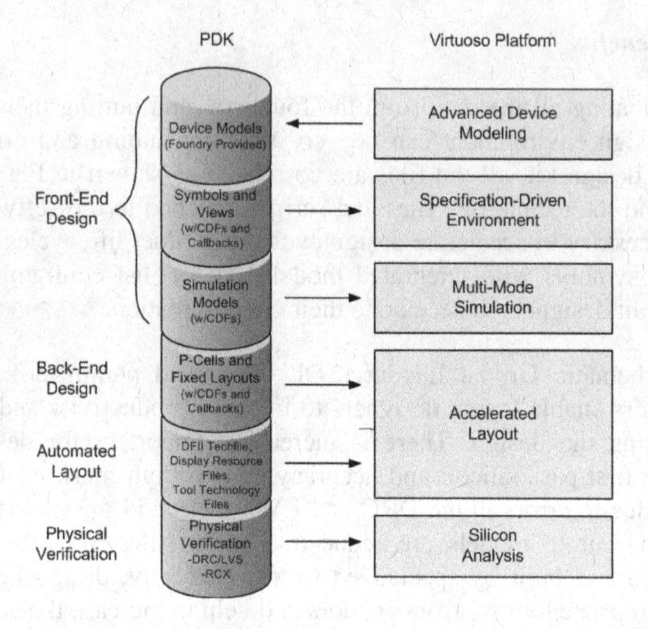

Figure 2.24 How a PDK data maps to the Cadence Virtuoso platform's design tasks.

2.8.2 Advanced Device Modeling and Front-end Design

In order to provide the fastest and most silicon-accurate design capability, the latest device and interconnect models have been made available to the designer. Data is derived using the Virtuoso Advanced Device Modeling tools. These tools perform extraction, optimization and verification of device and interconnect models. The front-end design section of a standard PDK also contains information that allows designers to drive the Virtuoso Specification-driven Environment and Multi-mode Simulation. From schematic symbols (with CDFs and callbacks) to device simulation models, the PDK ensures that designers can get started quickly and not worry about missing design data.

2.8.3 Back-end Design and Accelerated Layout

PDKs contain the data necessary to drive back-end design tasks using the Virtuoso Accelerated Layout capabilities. They also include P-Cells and layouts for a variety of components, and this data is matched to both the front-end design data as well as Virtuoso platform's physical verification design tasks.

2.8.4 Physical Verification and Silicon Analysis

The PDK physical verification section includes Assura DRC/LVS and leading-edge 3D device-level parasitic extraction from Assura RCX. Rule decks for DRC and LVS are developed collaboratively with Chartered and are fully qualified for analog and mixed-signal/RF design flow with the widely used Virtuoso platform, providing a seamless front-to-back design solution customers can rely on. The heart of the physical verification section is Assura RCX, which is a comprehensive parasitic RLCK extraction solution for analog and mixed-signal/RF design. Assura RCX provides the foundation for silicon-accurate simulation and analysis, allowing designers to optimize layouts interactively to increase chip performance and silicon yield.

2.8.5 Future of Process Design Kits

The advances in the semiconductor design domain raise the need to establish some fundamentals, like naming conventions and delivery mechanisms. This led to the formation of a technical subcommittee for the standardization of custom design kit data – Accellera OpenKit Initiative (OKI). This committee includes representatives from different sectors in the industry, such as the foundries, EDA and intellectual property (IP) companies.

Generally, a shorter and less complex kit development saves significant amounts of engineering time currently spent in integrating tools with design kits as well as simplifying the use of multiple tools, flows and different IC processes.

To improve qualification, test vectors can be created to simplify the complex processes in validating the kits and help in delivering high-quality kits.

With improved capabilities and disciplined methodologies, third party vendors may be able to build on the existing framework and develop additional features, like screen videos and scalable RF components

Foundries can understand the requirements of design kits and start to populate these kits with the necessary information earlier in the process development cycle, thereby giving their customers as early access to new processes as possible

Interoperability of tools and kits are greatly needed as design kits are specific to a particular vendor's tools. This results in difficulties in porting design across tool sets.

The OpenAccess Coalition (OAC) is formed which consists of leading semiconductor, systems, and EDA supplier companies. Today, the OpenAccess database and OpenAccess API (Application Programming Interface) are available for OAC partners. This is a C++ program interface to IC design data and provides a high performance, high capacity electronic design database with architecture designed for easy integration and fast application development.

With the advent of semiconductor process technology, circuit complexity, and continual scaling of devices, the need for a robust and reliable process design kit becomes ever more imperative.

2.9 Summary

Currently, most foundries provide discrete sizes for RF models in their PDKs and this will limit the device selection available for circuit designers and result in non-optimized RF circuits. By generating accurate and scalable RFCMOS models, the designed circuits can be optimized during the simulation process and achieve single tape-out success. This will certainly reduce the time to market for the developed RF chips.

In a device perspective, short-channel MOSFETs [4] have high-frequency figure of merit (FOM) such as cutoff frequency (f_T) and maximum oscillation frequency (f_{MAX}) which are much higher than 100GHz in the advanced deep-submicron technology. The consideration of the W_f effect on f_T and f_{MAX} is important as the wrong selection of the W_f for transistor layout will affect either f_T or f_{MAX}. Since the behavioral trend of f_T and f_{MAX} versus W_f is different, some trade-off is needed when selecting the best W_f value for transistor to be used in a specific application. Although both f_T and f_{MAX} are commonly used to compare the performance of the transistor, in circuit design such as VCO and transistor gain stages, the power amplification capability is more important. Hence the f_{MAX} optimization for the transistor W_f will be very crucial in these applications.

Evaluating the pros and cons of various approaches published in the literature, physical design parameter optimization of spiral inductors offers huge impact and benefits the industry since these passive integrated inductors are widely adopted in commercial products. From [5], it is found that the inductor's core diameter has immense influence on its performance and must be larger than 100μm to minimize the formation of conductor eddy current. On the other hand, small inductors with narrow conductor widths, such as 4μm, are less susceptible to the conductor's eddy current effect; hence, using small core diameters will reduce the circuit silicon real estate required. Conductor spacing should be small as it offers larger inductance value and consume smaller silicon area. Optimization of conductor width allows peak Q-factor of spiral inductors to be tuned at the circuit's operating frequency, and in doing so, permits inductors over a wide range of inductance values to have much larger Q-factors at that particular frequency of interest.

It is generally necessary to control terminating impedances of RF signal paths, especially in wide-band applications where path lengths are not negligible relative to wavelength. Wideband RF transformers are wound using twisted wires which behave as transmission lines, and the required coupling occurs along the length of these lines as well as magnetically via the core. Optimum performance is achieved when primary and secondary windings are connected to resistive terminating impedances for which the transformer is designed [7]. The performance of a transformer is affected by the metal and oxide thickness, transformer area, and the number of turns. Besides, there are generally three contributions to the core loss: (a) eddy-current loss, which increases with frequency; (b) hysteresis loss, which increases with flux density (applied signal level); (c) residual loss, due partially to gyromagnetic resonance. Insertion loss of a transformer is the fraction of input power lost when the transformer is inserted into an impedance-matched transmission system in place of an ideal (theoretically lossless) transformer having the same turns ratio. Actual insertion loss is affected by non-ideal characteristic impedance of transformer windings, as well as winding and core losses. Insertion loss at low frequency is affected by the parallel (magnetizing) inductance. At low temperature, low-frequency insertion loss tends to increase because of decreasing permeability of the magnetic core. High-frequency insertion loss is attributed to inter-winding capacitance and series (leakage) inductance. At high temperature it tends to become greater due to increase in the loss component of core permeability.

There are proliferations of parasitic inductance, resistance and substrate losses associated with interconnects [5] as operating frequency escalates. Long narrow-width interconnects should be avoided in RFIC since short wide-width interconnects are less lossy. The continuous scaling down of the physical dimensions of transistors, which has beneficial effects on their performance and enables a larger number of transistors per unit area, requires a growing number of interconnects. Unfortunately, since the performance of interconnects does not improve by scaling down their dimensions, they are becoming the performance limiter of high performance microprocessors. Presently they are still a big problem. The dielectric material, the signal wavelength, the permittivity, and the permeability must be considered when deciding the critical length.

For inter-digital varactors [103][104] at GSM frequency, when the varactor size grows, the capacitance increases and the resistance decreases proportionally to the size. Generally, varactors made by technologies with thicker metal layers are expected to provide better Q. The capacitance at WLAN frequency is larger because this frequency is near the frequency of resonance of the varactor. When the varactor size grows, the capacitance and the parasitic impedance of the connection layers also increase and the resonance frequency decreases. For the gated varactor, it has one additional terminal compared to either the reverse-biased pn junction varactor or the accumulation-mode varactor. Therefore it has a high Q factor and a high tuning range.

The material properties and geometrical parameters of the on-chip inter-digital capacitor are very important in order to optimize the quality factor and capacitance [11]. Q_{max} will increase with lesser number of fingers, smaller finger length, and higher metal conductivity. In order to get high Q_{max}, finger length plays an important role because its gives a big impact compared to other parameters. However, the design with small finger length produces low capacitance. The capacitance C increases with more number of fingers and larger finger length. The metal conductivity gives low impact to the capacitance. A larger finger length and more number of fingers will consume a larger cost to RFIC design.

References

[1] C. H. Diaz, "CMOS Technology for MS/RF SoC", IEEE Workshop on Microelectronics and Electron Devices 2004, pp. 24-27, 2004.
[2] W. K. Chen, "The Circuits and Filters Handbook", CRC Press, 2003.
[3] Y. H. Cheng and C. M. Hu, "MOSFET Modeling and BSIM3 User's Guide", Springer, 1999.
[4] A. F. Tong, "Design of a Scalable Model for Deep Submicron MOSFETs", PhD Thesis, Nanyang Technological University, 2009. [36]
[5] C. B. Sia, "Device Design, Characterization and Modeling of Inductors and Interconnects for RFIC Applications", PhD Thesis, Nanyang Technological Univ., Singapore, 2008.
[6] M. H. Cho and L. K. Wu, "A Novel Electrically Tunable RF Inductor with Ultra-Low Power Consumption", IEEE Microwave and Wireless Components Letters, vol. 18, no. 4, pp. 242-244, April 2008.
[7] H. T. Gan, "On-Chip Transformer Modeling, Characterization, and Applications in Power and Low Noise Amplifiers", PhD Thesis, Stanford Univ., 2006.
[8] M. F. Chang, V. P. Roychowdhury, L. Y. Zhang, H. C. Shin, and Y. X. Qian, "RF/Wireless Interconnect for Inter and Intra-Chip Communications", Proc. of the IEEE, vol. 89, no. 4, pp. 456-466, April 2001.
[9] J. Chen and L. He, "Transmission Line Modeling and Synthesis for Multi-Channel Communication", Proc. of the 2005 IEEE International Behavioral Modeling and Simulation Workshop (BMAS 2005), pp. 94-99, 22-23 September 2005.
[10] S. C. Kelly, J. A. Power, Oapos, and M. Neill, "Optimization of Integrated RF Varactors on a 0.35 mum BiCMOS Technology", International Conference on Microelectronic Test Structures 2003, pp. 113-117, 17-20 March 2003.
[11] M. N. Mariyatul, "Design and Modeling of On-Chip Planar Capacitor for RF Application", UTM, Malaysia, May 2006.
[12] C. T. Sah, "Fundamentals of Solid-State Electronics", World Scientific, 1991.

[13] S. M. Sze and K. K. Ng, "Physics of Semiconductor Devices (3rd Edition)", John Wiley and Sons, Inc., 2007.
[14] W. Liu, "MOSFET Models for SPICE Simulation including BSIM3v3 and BSIM4", John Wiley and Sons, Inc., 2001.
[15] W. Liu et. al., "BSIM3v3.3 MOSFET Model - Users' Manual", Dept. of Electrical Engineering and Computer Sciences, UC Berkeley. 2005.
[16] S. F. Tin, A. A. Osman, K. Mayaram, and C. Hu, "A Simple Sub-circuit Extension of the BSIM3v3 Model for CMOS RF Design," IEEE J. Solid-State Circuits, vol. 35, no. 4, 612-624, Apr. 2000.
[17] L. C. Png, K. W. Chew, and K. S. Yeo, "Impact of forward and reverse deep n-well biasing on the *1/f* noise of 0.13 μm n-channel MOSFETs in triple well technology", Solid-State Electronics, vol. 53, no. 6, pp. 599-606, June 2009.
[18] T. E. Kolding, "Calculation of MOSFET Gate Impedance", Technical Report R98-1009, ISSN 0908-1224.
[19] S. H. Lee, H. K. Yu, C. S. Kim, J. G. Koo, and S. N. Kee, "A Novel Approach to Extracting Small-Signal Model Parameters of Silicon MOSFET's", IEEE Microwave and Guided Wave Letter, vol. 7, no. 3, March 1997.
[20] M. J. Hung, C. C. Enz, D. R. Pehlke, M. Schroter, B. J. Sheu, "Accurate modeling and parameter extraction for MOS transistors valid up to 10GHz", IEEE Trans. Electron Devices, vol. 46, no. 11, Nov 1999.
[21] C. C. Enz, Y. H. Cheng, "MOS Transistor Modeling for RF IC Design", IEEE Trans. Solid-State Circuits, vol. 35, no. 2, pp. 186-201, Feb 2000.
[22] S. H. Lee, C. S. Kim, and H. K. Yu, "A Small RF Model and its Parameter Extraction for Substrate Effects in RF MOSFETs", IEEE Trans. On Electron devices, vol. 48, no. 7, July 2001.
[23] I. Kwon, M. Je, K. Lee, and H. C. Shin, "A Simple and Analytical Parameter-Extraction Method of a Microwave MOSFET", IEEE Trans. On Microwave Theory and Techniques, vol. 50, no. 6, June 2002.
[24] M. Je, I. Kwon, J. H. Han, H. C. Shin, and K. Lee, "On the Large-Signal CMOS Modeling and Parameter Extraction for RF Application", International Conference on Simulation of Semiconductor Processes and Devices, 2002. SISPAD 2002.
[25] Y. H. Cheng, C. H. Chen, C. Enz, M. Matloubian, and M. J. Deen, "MOSFET Modeling for RF Circuit Design", Proceedings of the Third IEEE International Caracas Conference on Devices, Circuits and Systems (ICCDCS 2000), Cancun, Mexico, pp. D23, pp.1-8, March 2000.
[26] H. S. Bennett, R. Brederlow, J. C. Costa, P. E. Cottrell, W. M. Huang, A. A. Immorlica, J. E. Mueller, M. Racanelli, H. Shichijo, C. E. Weitzel, and B. Zhao, "Device and Technology Evolution for Si-based RF Integrated Circuits", IEEE Trans. Electron Devices, vol. 52, no.7, Jul. 2005, pp. 1235-1257.
[27] J. Pekarik, D. Greenberg, B. Jagannathan, R. Groves, J. R. Jones, R. Singh, A. Chinthakindi, X. Wang, M. Breitwisch, D. Coolbaugh, P. Cottrell, J. Florkey, G. Freeman, and R. Krishnasamy, "RFCMOS Technology from 0.25μm to 65nm: The State of the Art," IEEE Proceedings, Custom Integrated Circuits Conference, pp. 217-224, 3-6 Oct. 2004.
[28] Y. H. Cheng, "An Overview of Device Behavior and Modeling of CMOS Technology for RF IC Design", IEEE Int. Symposium on Electron Devices for Microwave and Optoelectronic Applications, Nov. 2003, pp. 109-114, Nov. 2003.

[29] K. Lee, I. Nam, I. Kwon, J. Gil, K. Han, S. Park, and Bo-Ik Seo, "The Impact of Semiconductor Technology Scaling on CMOS RF and Digital Circuits for Wireless Application", IEEE Trans. Electron Devices, vol. 52, no. 7, pp. 1415-1422, July 2005.

[30] T. Manku, "Microwave CMOS—Device Physics and Design", IEEE J. of Solid-State Circuits, vol. 34, No. 3, March 1999.

[31] I. Bahl and P. Bhartia, "Microwave Solid-State Circuit Design", John Wiley and Sons, Inc., New York, 1988.

[32] A. F. Tong et. al., "RFCMOS Unit Width Optimization Technique", IEEE Transactions on Microwave Theory and Techniques, vol. 55, no. 9, pp. 1844 - 1853, Sept. 2007.

[33] H. Rothe and W. Dahlke, "Theory of Noisy Fourpoles", Proc. IRE, vol. 44, no. 6, pp. 811-818, June 1956.

[34] T. H. Lee, "The Design of CMOS Radio-Frequency Integrated Circuits (1st Edition)", Cambridge Univ. Press, ch. 11, p. 276, 2001.

[35] Y. Cheng et. al., "A Physical and Scalable BSIM3v3 I-V Model for Analog/Digital Circuit Simulation," IEEE Trans. Electron Devices, vol. 44, pp. 277-287, Feb 1997.

[36] K. Kuhn, R. Basco, D. Becher, M. Hattendorf, P. Packan, I. Post, P. Vandervoorn, I. Young, "A Comparison of State-of-the-Art NMOS and SiGe HBT Devices for Analog/Mixed-Signal/RF Circuit Applications", Symposium on VLSI Technology Digest of Technical Papers, pp.224-225, June 2004.

[37] D. R. Greenberg et al., "Noise Performance of a Low Base Resistance 200GHz SiGe Technology", 2002 IEDM Digest, pp. 787-790, 2002.

[38] A. A. Abidi, "RF CMOS Comes of Age", IEEE J. of Solid-State Circuits, vol. 39, No. 4, pp. 549-561, April 2004.

[39] F. Opt'Eynde, J. Schmit, V. Charlier, R. Alexandre, C. Sturman, K. Coffin, B. Mollekens, J. Craninckx, S. Terrij, A. Monterastelli, S. Beerens, P. Goetschalckx, M. Ingels, D. Joos, S. Guncer, and A. Pontioglu, "A Fully Integrated Single-Chip SOC for Bluetooth", IEEE Int. Solid-State Circuits Conf. Dig. Tech. Papers, pp. 196-197, Feb 2001.

[40] H. Darabi. S. Khorram, E. Chien, M. Pan, S. Wu, S. Moloudi, J. C. Leete, J. J. Rael, M. Syed, R. Lee, B. Ibrahim, M. Rofougaran, and A. Rofougaran, "A 2.4GHz CMOS Transceiver for Bluetooth", IEEE Int. Solid-State Circuits Conf. Dig. Tech. Papers, pp. 200-201, February 2001.

[41] P. T. M. van Zeijl, J. Eikenbroek, P. Vervoort, S. Setty, J. Tangenberg, G. Shipton, E. Kooistra, I. Keekstra, and D. Belot, "A Bluetooth Radio in 0.18-µm CMOS", IEEE Int. Solid-State Circuits Conf. Dig. Tech. Papers, pp. 86-87, Feb 2002.

[42] A. Leeuwenburgh, J. ter Laak, A. Mulders, A. Hoogstraate, P. van Laarhoven, M. Nijrolder, J. Prummel, and P. Kamp, "A 1.9GHz Fully Integrated CMOS DECT Transceiver", IEEE Int. Solid-State Circuits Conf. Dig. Tech. Papers, pp. 450-507, Feb 2003.

[43] S. Lee and H. K. Yu, "A New Extraction Method for BSIM3v3 Model Parameters of RF Silicon MOSFETs", Proc. IEEE 1999. Int. Conf. On Microelectronic Test Structures, vol. 12, pp.95-9, March 1999.

[44] H. M. Chen et. al., "BSIM4.2.1 MOSFET Model – User's Manual", UC Berkeley, 2001.

[45] M. Lee, R. B. Anna, J. C. Lee, S. M. Parker, and K. M. Newton, "A Scalable BSIM3v3 RF Model for Multi-Finger NMOSFETs with Ring Substrate Contact", IEEE Int. Symp. On Circuits and Systems, vol. 5, pp. 221-224, May 2002.

[46] S. P. Voinigescu, M. Tazlauanu, P. C. Ho, and M. T. Yang, "Direct Extraction Methodology for Geometry-Scalable RF-CMOS Models", Proc. IEEE 2004. Int. Conf. On Microelectronic Test Structures, vol. 17, pp.235-240, March 2004.
[47] S. Yoshitomi, A. Bazigos, and M. Bucher, "EKV3 Parameter Extraction and Characterization of 90nm RF-CMOS Technology", Int. Conf. On Mixed Design of Integrated Circuits and Systems, pp.74-79, June 2007.
[48] X. Jin, J. Ou, C. H. Chen, W. Liu, M. J. Deen, P. R. Gray, and C. Hu, "An Effective Gate Resistance Model for CMOS RF and Noise Modeling," Technical Digest of International Electron Devices meeting, pp. 961-964, Dec 1998.
[49] A. Nakamura, N. Yoshikawa, T. Miyazako, T. Oishi, H. Ammo, and K. Takeshita, "Layout Optimization of RF CMOS in the 90nm Generation", IEEE Radio Frequency Integrated Circuits Symp., June 2006.
[50] F. R. Gleason, "Thin-Film Microelectronic Inductors", Proceedings of the National Electronics Conference, vol. 20, pp. 197-198, 1964.
[51] H. G. Dill, "Designing Inductors for Thin-Film Applications", Electronic Design, pp. 52-59, Feb. 1964.
[52] A. Olivei, "Optimized Miniature Thin-Film Planar Inductor Compatible with Integrated Circuits", IEEE Transactions on Parts, Materials, and Packaging, vol. 5, no. 2, pp. 71-88, 1969.
[53] H. M. Greenhouse, "Design of Planar Rectangular Microelectronic Inductors", IEEE Trans. on Parts, Hybrids and Packaging, vol. php-10, no. 2, pp. 101-109, Jun. 1974.
[54] F. W. Grover, "Inductance Calculations", Dover Publications, N. Y., 1962.
[55] J. C. Maxwell, "A Treatise on Electricity and Magnetism, Parts III and IV, 1st Edn", Dover Publications, N. Y., 1954.
[56] S. C. Gupta, J. W. Bayless, and B. Peikari, "Circuit Analysis with Computer Applications to Problem Solving", Intext Educational Publishers, 1972.
[57] J. D. Kraus, "Electromagnetics (4th Edition)", McGraw-Hill, 1991.
[58] P. Min, S. Lee, S. K. Cheon, K. Y. Hyun, and S. N. Kee, "The Detailed Analysis of High Q CMOS Compatible Microwave Spiral Inductors in Silicon Technology", IEEE Trans. on Electron Devices, vol. 45, no. 9, pp. 1953-1959, Sep. 1998.
[59] B. Breen, "Multi-layer Inductor for High Frequency Applications", Proc. IEEE Electronic Components & Technology Conf., pp. 551-554, 1991.
[60] K. Kamogawa, K. Nishika, I. Toyoda, T. Tokumitsu, and M. Tanaka, "A Novel High-Q and Wide-Frequency-Range Inductor Using Si 3D MMIC Technology", IEEE Microwave and Guided Wave Letter, vol. 9, no. 1, pp. 16-18, Jan. 1999.
[61] J. N. Burghartz, M. Soyuer and K. A. Jenkins, "Microwave Inductors and Capacitors in Standard Multilevel Interconnect Silicon Technology", IEEE Trans. Microwave Theory Tech., vol. 44, no. 1, pp. 100-104, Jan. 1996.
[62] D. C. Edelstein and J. N. Burghartz, "Spiral and Solenoidal Inductor Structures on Silicon using Cu-Damascene Interconnects", Proc. IEEE International Interconnect Technology Conf., pp. 18-20, 1998.
[63] R. Yang, H. Qian, J. Li, Q. Xu, C. Hai and Z. Han, "SOI Technology for Radio-Frequency Integrated-Circuit Applications", IEEE Trans. Electron Devices, vol. 53, no. 6, pp. 1310-1316, Jun. 2006.
[64] A. Tekin, E. Zencir, D. Huang, W. Liu, and N. S. Dogan, "A 700-MHz VCO using high-Q silicon on insulator (SOI) inductors", Proc. IEEE Radio and Wireless Symposium, pp. 427-429, Jan. 2006.

[65] V. Malba, D. Young, J. J. Ou, A. F. Bernhardt and B. E. Boser, "High-Performance RF Coil Inductors on Silicon", Proc. Electronic Components and Technology Conf., pp. 252-255, 1998.
[66] A. Nakayama and K. Shimizu, "An Improved Reflection Wave Method for Measurement of Complex Permittivity at 100MHz – 1GHz", IECE Transactions Electron, vol. E77-C, no.4, pp 633-638, Apr. 1994.
[67] V. Ermolov, T. Lindström, H. Nieminen, M. Olsson, M. Read, T. Ryhänen, S. Silanto, and S. Uhrberg, "Microreplicated RF Toroidal Inductor", IEEE Trans. Microwave Theory Tech., vol. 52, no. 1, pp. 29-37, Jan. 2004.
[68] F. Mernyei, F. Darrer, M. Pardoen, and A. Sibrai, "Reducing the Substrate Losses of RF Integrated Inductors", IEEE Microwave and Guided Wave Letter, vol. 8, no. 9, pp 300-301, Sep. 1998.
[69] C. P. Yue and S. S. Wong, "On-Chip Spiral Inductor with Patterned Ground Shield for Si-Based RF IC's", IEEE J. Solid-State Circuits, vol. 33, pp. 743-752, May 1998.
[70] J. N. Yang, Y. C. Cheng, C. Y. Lee, "A design of CMOS broadband amplifier with high-Q active inductor", The 3rd IEEE International Workshop on System-on-Chip for Real-Time Applications, 2003.
[71] D. R. Pehlke, A. Burstein, M. F. Chang, "Extremely High-Q Tunable Inductor for Si-based RF Integrated Circuit Applications", Int. Electron Devices Meeting Tech. Dig., pp. 63-66, 1997.
[72] J. M. Lopez-Villegas, J. Samitier, C. Cane, P. Losantos, J. Bausells, "Improvement of Quality Factor of RF Integrated Inductors by Layout Optimization", IEEE Trans. Microwave Theory Tech., vol. 48, no.1, pp. 76-83, Jan. 2000.
[73] J. E. Post, "Optimizing the design of spiral inductors on silicon", IEEE Trans. Circuits Syst. II., vol. 47, no. 1, pp. 15-17, Jan. 2000.
[74] C. P. Yue, C. Ryu, J. Lau, T. H. Lee, and S. S. Wong, "A Physical Model for Planar Spiral Inductors on Silicon", Int. Electron Devices Meeting Tech. Dig., pp. 155-158, Dec. 1996.
[75] C. B. Sia, B. H. Ong, K. W. Chan, K. S. Yeo, J. G. Ma and M. A. Do, "Physical Design Optimization of Integrated Spiral Inductors for Silicon-Based RFIC Applications", IEEE Transactions on Electron Devices, vol. 52, no. 12, pp. 2559-2567, Dec. 2005.
[76] C. B. Sia, B. H. Ong, W. M. Lim, K. S. Yeo and T. Alam, "Modeling and Layout Optimization of Differential Inductors for Silicon-based RFIC Applications", IEEE Trans. Electron Devices, vol. 55, no. 4, pp. 1058-1066, Apr 2008.
[77] N. Camilleri, D. Lovelace, J. Costa and N. David, "New Development Trends for Silicon RF Device Technologies", IEEE Microwave and Millimeter-Wave Monolithic Circuits Symposium, pp. 5-8, 1994.
[78] L. E. Larson, "Integrated Circuit Technology Options for RFIC's - Present Status and Future Directions", IEEE Journal of Solid-State Circuits, vol. 32, no. 3, pp. 387-397, Mar. 1998.
[79] B. Razavi, "CMOS Technology Characterization for Analog and RF Design", IEEE Journal of Solid-State Circuits, vol. 34, no. 3, pp. 268-276, Mar. 1999.
[80] M. T. Reiha, T. Y. Choi, J. H. Jeon, S. Mohammadi, L. P. B. Katehi, "High-Q differential inductors for RFIC design", Proc. 33rd European Microwave Conf., vol. 1, pp. 127-130, Oct. 2003.

[81] M. Politi, V. Minerva, S. C. d'Oro, "Multi-layer realization of symmetrical differential inductors for RF silicon IC's", Proc. 33rd European Microwave Conf., vol. 1, pp. 159-162, Oct. 2003.

[82] M. Danesh and J. R. Long, "Differentially Driven Symmetric Microstrip Inductors", IEEE Trans. Microwave Theory Tech., vol. 50, no. 1, pp. 332-341, Jan. 2002.

[83] M. Drakaki, A. Hatzopoulos, A. Siskos, "CMOS Inductor Performance Estimation using Z and S-parameters", Proc. IEEE International Symposium on Circuits and Systems, pp. 2256-2259, 2007.

[84] J. Rogers and C. Plett, "A 5GHz Radio Front-End with Automatically Q Tuned Notch Filter", IEEE BCTM, pp. 69-72, Sept. 2002.

[85] M. S. J. Cabanillas and G. Rebeiz, "A Low-Noise Transformer-Based 1.7GHz CMOS VCO", ISSCC Digest of Tech. Papers, pp. 286-287, Feb 2002.

[86] S. S. X. Li and D. J. Allstot, "Gm-Boosted Common-Gate LNA and Differential Colpitts VCO/QVCO in 0.18 μm CMOS", IEEE J. Solid-State Circuits, vol. 40, pp. 2609-2619, Dec 2005.

[87] A. Zolfaghari et. al., "Stacked Inductors and Transformers in CMOS", IEEE J. Solid-State Circuits, vol. 36, pp. 620-628, Apr. 2001.

[88] S. S. Mohan, "The Design, Modeling, and Optimization of On-Chip Inductor and Transformer Circuits", PhD Thesis, Stanford Univ., 1999.

[89] O. El-Gharniti, et. al., "Characterization of Si-Based Monolithic Transformers with Patterned Ground Shield", Proc. RFIC Symp. Dig., pp. 261-264, June 2006.

[90] S. M. Yim, et. al., "The Effects of a Ground Shield on the Characteristics and Performance of Spiral Inductors", IEEE J. Solid-State Circuits, vol. 37, no. 2, pp. 237-244, Feb 2002.

[91] K. T. Ng, B. Rejaei, and J. Burghartz, "Substrate Effects in Monolithic RF Transformers on Silicon", IEEE Trans. Microwave Theory Tech., vol. 50, no. 1, pp. 377-383, Jan 2002.

[92] I. Aoki, S. Kee, D. B. Rutledge, and A. Hajimiri, "Distributed Active Transformer — A New Power-Combining and Impedance-Transformation Technique", IEEE Trans. Microwave Theory and Techniques, vol. 50, pp. 316-331, Jan 2002.

[93] C. B. Sia, B. H. Ong, K. S. Yeo, J. G. Ma and M. A. Do, "Accurate and Scalable RF Interconnect Model for Silicon-based RFIC Applications", IEEE Trans. Microwave Theory and Tech., vol. 53, no. 9, pp. 3035-3044, Sep. 2005.

[94] Ansoft's High Frequency Structure Simulator, http://www.ansoft.com/hfss/

[95] D. M. Pozar, "Microwave Engineering, 3rd Edition", John Wiley & Sons Ltd, 2005.

[96] T. C. Edwards and M. B. Steer, "Foundations of Interconnect and Microstrip Design, 3rd Edition", John Wiley & Sons Ltd, 2000.

[97] E. Chiprout, "Interconnect and Substrate Modeling and Analysis: An Overview", IEEE J. of Solid-State Circuits, vol. 33, no. 9, pp. 1445-1452, Sep. 1998.

[98] C. P. Yue, C. Ryu, J. Lau, T. H. Lee, and S. S. Wong, "A Physical Model for Planar Spiral Inductors on Silicon", Int. Electron Devices Meeting Tech. Dig., pp. 155-158, Dec. 1996.

[99] M. Park, C. H. Kim, C. S. Kim, M. Y. Park, S. D. Kim, Y. S. Youn, and H. K. Yu, "Frequency-Dependent Series Resistance of Monolithic Spiral Inductors", IEEE Microwave and Guided Wave Letter, vol. 9, no. 12, pp. 514-516, Dec. 1999.

[100] P. Adreani, "A Comparison between Two 1.8GHz CMOS VCOs Tuned by Different Varactors", Proc. ESSCIRC'98, pp. 380-383, Sept. 1998.

[101] J. N. Burghartz, M. Soyeur, K. A. Jenkins, "Integrated RF and Microwave Components in BiCMOS Technology", IEEE Trans. Electron Devices, vol. 43, pp. 1559-1570, Sept. 1996.
[102] R. Castello et. al., "A +/- 30% Tuning Range Varactor Compatible with Future Scaled Technology", Dig. Tech. Papers, Symp. VLSI Circuits, pp. 34-35, 1998.
[103] K. Q. Shen et. al., "A Three-Terminal SOI Gated Varactor for RF Applications", IEEE Trans. on Electron Devices, vol. 48, no. 2, pp. 289-293, Feb 2001.
[104] I. Gutierrez et. al., "PN Junction Integrated Varactors for RF Applications at Different Standard Frequencies", Silicon Monolithic Integrated Circuits in RF Systems, 2003. Digest of Papers. 2003 Topical Meeting on, vol. 9, no. 11, pp. 118-121, Apr. 2003.
[105] E. Pedersen, "RF CMOS Varactors for Wireless Applications", PhD Thesis, RISC Group, Denmark, Aalborg Univ., 2000.
[106] F. Svelto et. al., "A Metal-Oxide-Semiconductor Varactor", IEEE Electron Device Letters, vol. 20, no. 4, pp. 164-166, Apr. 1999.
[107] W. M. Y. Wong, "A Wide-Tuning Range Varactor", IEEE J. of Solid-State Circuits, vol. 35, pp. 773-778, May 2000.
[108] H. Samavati et. al., "Fractal Capacitors", IEEE J. of Solid-State Circuits, vol. 33, pp. 2035-2041, Dec. 1998.
[109] T. Ytterdal et. al., "Device Modeling for Analog and RF CMOS Circuit Design", John Wiley and Sons, Inc., 2003.
[110] F. P. Casares-Miranda et. al., "Wire-Bonded Inter-digital Capacitor", IEEE Trans. Microwave and Wireless Components, vol. 15, no. 10, 2005.
[111] G. D. Alley, "Integrated Capacitor and Their Application to Lumped-Element Microwave Integrated Circuits", IEEE Trans. Microwave Theory, vol. MTT-18, 1970.
[112] R. Esfandiari et. al., "Design of Inter-digital Capacitor and Their Application to GaAs Filters", IEEE Trans. Microwave Theory, vol. MTT-31, 1983.
[113] M. Naghed and I. Wolff, "Equivalent Capacitances of a Coplanar Waveguide Discontinuities and Inter-digital Capacitors using a Three Dimensional Finite Difference Method", IEEE Trans. Microwave Theory, vol. 38, no 12, 1990.
[114] F. Aryanfar and K. Sarabandi, "Characterization of Semi-lumped CPW Elements for Millimeter-Wave Filter Design", IEEE Trans. Microwave Theory, vol. 53, no. 4, 2005.

When you read, don't think. Try to understand.

When you think, don't read. Try to imagine.

Kiat Seng YEO

CHAPTER 3

RF CMOS Low Noise Amplifiers

The low noise amplifier (LNA) is one of the most important building blocks in a receiver. In this chapter, the basic concepts and terminologies used in LNA will be described and defined in Section 3.1. In Section 3.2, the various input matching architectures of LNAs will be shown. This will be followed by the input matching analysis of single-band LNA in Section 3.3. Subsequently, the design of a single-band LNA will be presented as a design example in Section 3.4. In this LNA, a parallel *LC* network will be used to replace the large gate inductor, which is normally required for the common LNA input architecture.

3.1 Basic Concepts of LNAs

Concepts and terminologies used in LNA, for example, operating frequency, sensitivity, noise figure (*NF*), 1-*dB* compression point, intermodulation distortion and so on will be described and defined in this section. It is important to understand that the specifications of the LNA will affect significantly the performance of the entire system. Analysis of these parameters helps to optimize the performance of LNAs for different applications. Table 3.1 shows the specifications of a typical LNA for Global System for Mobile communications (GSM)-1800 application.

Table 3.1 Specifications of a typical GSM-1800 LNA.

Parameter	Typical Value
Operating Frequency (GHz)	1.8–1.9
NF (*dB*)	2–3
Voltage Gain (*dB*)	18–22
S_{11} (*dB*)	-12– -15
S_{12} (*dB*)	-30 – -40
IIP3 (dBm)	-4–0

3.1.1 Operating Frequency

As the LNA is located at the front end of the receiver, the operating frequency of the LNA must at least cover the receiving band, e.g. 1805-1880MHz for the GSM-1800 receiver. The LNA has to satisfy all the requirements as shown in Table 3.1 over the whole operating frequency range. For applications where the ratio of (operating frequency range) / (centre of operating frequency) is large, this could pose a great design challenge. For example, both the input voltage reflection coefficient (S_{11}) and voltage gain requirements of an ultra-wideband (UWB) LNA operating from 3.1GHz to 6GHz will be hard to meet.

3.1.2 Sensitivity

A receiver's sensitivity is defined as the minimum input level that the system can detect with acceptable signal-to-noise ratio, which is determined by the receiver's demodulation scheme. Sensitivity is often expressed in terms of the minimum detectable signal (MDS) level at the receiver's input, which produces acceptable level of signal-to-noise ratio (*SNR*) at the receiver's output. MDS is given by

$$P_{in,\min}(dBm) = -174dBm/Hz + NF + 10\log_{10} B + SNR_{\min} \tag{3.1}$$

where SNR_{min} is the acceptable signal-to-noise ratio at the receiver's output; $P_{in,min}$ is the minimum input level that achieves SNR_{min}, and B is the effective noise bandwidth which is often approximated by the channel bandwidth. Note that the sum of the first three terms is the total integrated noise of the system and is sometimes called the *noise floor*.

Example

For a class 3 GSM-1800 receiver, find the maximum allowable noise figure of the receiver.

Solution

The MDS of a class 3 GSM-1800 receiver is specified at -102dBm. Assuming that the SNR_{min} is 7dB and the channel bandwidth is 200kHz. Using equation (3.1)

$$P_{in,\min}(dBm) = -174 dBm/Hz + NF + 10\log_{10} B + SNR_{\min}$$
$$-102\ dBm\ = -174 dBm/Hz + NF(dB) + 10\log_{10} 200k(dB) + 7dB$$

Maximum allowable noise figure of the whole receiver, $NF = 12dB$

3.1.3 Noise Figure and Voltage Gain

The noise factor (F) is a measurement of the noise performance of a circuit. It is frequently expressed in decibels and commonly referred to as the noise figure (NF).

$$NF = 10\log_{10} F \tag{3.2}$$

The noise figure is defined in many ways. The original definition is

$$NF = SNR_{in}(dB) - SNR_{out}(dB) \tag{3.3}$$

where SNR_{in} and SNR_{out} are the signal-to-noise ratios measured in dB at the input and output respectively. In other words, NF is a measure of the SNR degradation due to noise generated by the circuit itself.

For a cascade of N stages, the overall noise factor can be obtained in terms of F and gain at each stage. The total noise factor can be expressed by the Friis equation

$$F_{tot} = F_1 + \frac{F_2 - 1}{A_{p1}} + \frac{F_3 - 1}{A_{p1} A_{p2}} ... + \frac{F_N - 1}{A_{p1} A_{p2} ... A_{p(N-1)}} \tag{3.4}$$

where F_m is the noise factor of stage m and A_{pm} is the available power gain of the mth stage. This equation indicates that the noise contributed by each stage decreases as the gain of the preceding stage increases. Thus, the first few stages in a cascade are the most critical stages. In practice, the LNA is the first active block in the receiving chain. Therefore, its noise figure sets the minimum NF of the system. An LNA should provide enough gain to overcome the noise contribution of the subsequent stages and add as little noise as possible.

Despite the popularity of Friis equation, in the context of modern integrated RF circuit, it should be applied with a pinch of salt. The reason is that Friis equation is defined for the available signal and noise powers, while power matching is often not necessary in integrated circuit design. As a rule of thumb, power matching is only necessary if the interconnect distance between two components is longer than 1/10 of the wavelength. For example, the interface to an external antenna-filter that requires 50Ω input/output matching must be matched properly due to the long interconnect distance. Power matching is thus necessary in order to maintain the shape of the transfer function and to minimize signal loss. On the other hand, no power matching is generally necessary for on-chip blocks as the interconnect distance between the on-chip blocks are usually in the range of tens of micrometers while the wavelength at 10GHz is in the range of few millimeters.

In a CMOS integrated circuit receiver, the gain and noise level should be defined in terms of voltage and impedance instead of power as the signal is transferred in the voltage mode. With this understanding, the Friis equation can be modified with reference to Figure 3.1[1].

$$F_{tot} = 1 + \frac{v_{ST1NIN}^2 + \dfrac{v_{ST2NIN}^2}{A_{v1}^2} + \dfrac{v_{ST3NIN}^2}{A_{v1}^2 \cdot A_{v2}^2}}{v_{SOURCEIN}^2} = F_1 + \frac{F_2 - 1}{A_{v1}^2} + \frac{F_3 - 1}{A_{v1}^2 \cdot A_{v2}^2} \quad (3.5)$$

where $v_{SOURCEIN}^2$ refers to the input noise at reference plane ® in V^2/Hz. V_{STXNIN} and A_{VX} refer to the input referred voltage noise and voltage gain for each stage respectively, where X is the number of the stage. For the case where the source resistance R_s is properly power matched to Stage 1, or in other words, the input impedance of Stage 1 is equal to the conjugate of Z_s, the equation for $v_{SOURCEIN}^2$ is

$$v_{SOURCEIN}^2 = kTR_s \quad (3.6)$$

where k is the Boltzmann's constant, T is the temperature defined at 290K, R_s is the source resistance usually at 50Ω. Equation (3.5) is based on an approximation that the input impedance of each stage is large compared to the output impedance of the previous stage.

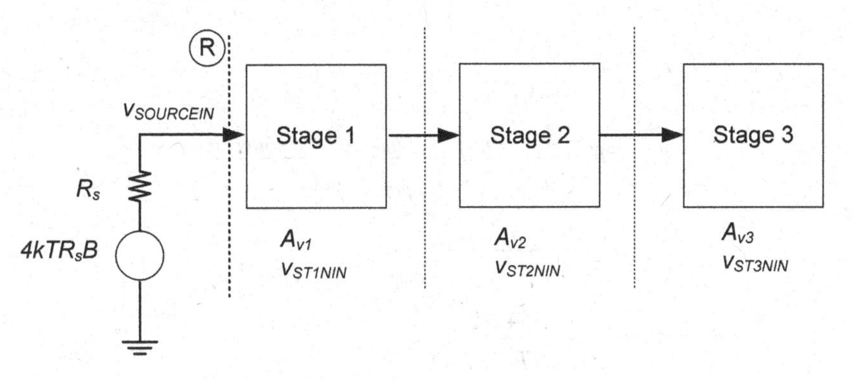

Figure 3.1 Integrated blocks noise figure calculation.

Example

In order to gain insight into the relationship between noise figure and voltage gain, consider an example where the total noise figure of the receiver NF_{tot} is given to be 20dB, the LNA has a voltage gain of 18dB (or 7.94) and the rest of the receiver has an input-referred noise voltage of 34nV/√Hz as shown in Figure 3.2.

Find the maximum allowable noise figure of the LNA.

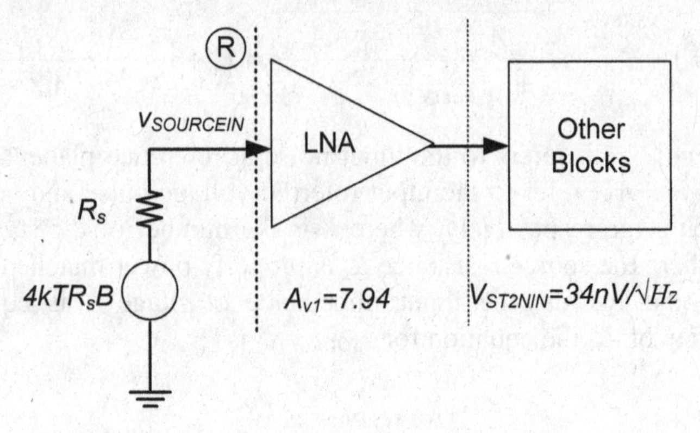

Figure 3.2 Calculation of LNA Noise Figure.

Solution

The maximum allowable noise figure for the LNA can be obtained using equation (3.5)

$$F_{tot} = F_1 + \frac{F_2 - 1}{A_{v1}^2}$$

$$F_{tot} = F_1 + \frac{v_{ST2NIN}^2 + v_{SOURCEIN}^2 - v_{SOURCEIN}^2}{v_{SOURCEIN}^2 \cdot A_{v1}^2}$$

$$F_{tot} = F_1 + \frac{v_{ST2NIN}^2}{v_{SOURCEIN}^2 \cdot A_{v1}^2}$$

$$100 = F_1 + \frac{(34 \cdot 10^{-9})^2}{kTR_s \cdot (7.94)^2}$$

Noise factor of the LNA, $F_1 = 8.4$ $(9.24dB)$

In this example, the total noise figure (F_{tot}) is rather relaxed. It can be deduced that the gain of the LNA (A_{v1}) has a larger impact on F_{tot} than its noise factor (F_1), as $F_{tot} \gg F_1$, F_{tot} is almost inversely proportional to A_{v1}^2. It is tempting to design for as large LNA gain as possible. However, higher gain at the LNA stage leads to tougher linearity requirements of the following stages, e.g. mixer, channel select filter, etc.

3.1.4 1-dB Compression Point

The linearity of a system determines the maximum allowable signal level at its input. All real-life systems exhibit some degree of nonlinearity. Signal distortion is a direct consequence of the nonlinear behavior of solid-state devices in circuits. The most common measures of non-linearity are the 1-dB compression point ($P_{1\text{-dB}}$) and the third-order intercept point (IP_3).

If a sinusoid is applied to a nonlinear system, the output generally exhibits frequency components that are integer multiples of the input frequency. When the input signal is $x(t) = A\cos\omega \cdot t$, the output through the system will be

$$y(t) = \alpha_1 x(t) + \alpha_2 x(t)^2 + \alpha_3 x(t)^3 \qquad (3.7)$$

$$y(t) = \alpha_1 A\cos\omega \cdot t + \alpha_2 A^2 \cos^2 \omega \cdot t + \alpha_3 A^3 \cos^3 \omega \cdot t$$

$$\approx \frac{\alpha_2 A^2}{2} + (\alpha_1 A + \frac{\alpha_3 A^3}{4})\cos\omega \cdot t + \frac{\alpha_2 A^2}{2}\cos 2\omega \cdot t + \frac{\alpha_3 A^3}{4}\cos 3\omega \cdot t \qquad (3.8)$$

where α_1, α_2, α_3 and so on are the gain coefficients of different nonlinear orders; A is the amplitude of the input signal $x(t)$.

In equation (3.8), the term at the input frequency is the fundamental and the higher-order terms are the harmonics. For most circuits of interest, $\alpha_3 < 0$. Therefore, the gain ($\alpha_1 A + \frac{\alpha_3 A^3}{4}$) is a decreasing function of A (amplitude). As the input power increases, the circuit components become saturated and the fundamental output fails to respond linearly to the input.

Figure 3.3 shows that the gain compression due to nonlinearities in the system causes the power gain to deviate from its ideal curve.

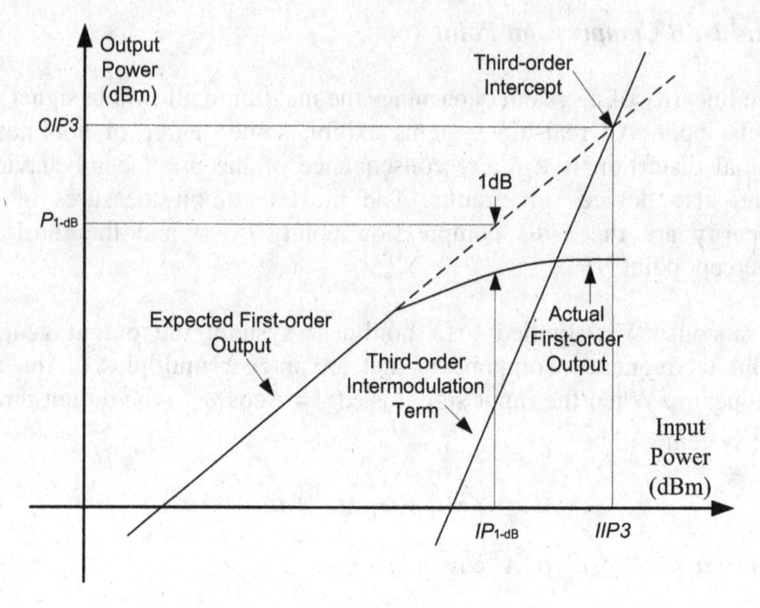

Figure 3.3 Illustrations of $IP_{1\text{-dB}}$ and $IIP3$.

The point at which the power gain is $1dB$ less than the ideal curve is referred to as the 1-dB compression point. The input power where $P_{1\text{-dB}}$ occurs is known as $IP_{1\text{-dB}}$. A system usually operates several decibels below this level to avoid the nonlinear region. The 1-dB compression point can be approximated as

$$IP_{1-dB} = 20\log_{10}\sqrt{0.145\left|\frac{\alpha_1}{\alpha_3}\right|} \tag{3.9}$$

3.1.5 The 3rd Order Intercept Point

The 3rd order intercept point (*IP3*) is another common measure of a circuit's nonlinearity in the presence of the intermodulation (IMD) of two interferers. When the frequencies of the two interferers are close enough, they can fall within the passband of the system and appear at the output as signal distortion.

If the two-tone interference inputs have the same amplitude, the 3rd order IMD power will grow at the rate that is equivalent to the cube of the input power as shown in Figure 3.3. In other words, the power of the third-order IMD increases three times faster (in *dB*) than that of the fundamental. The input signal level, where the power of the third-order IMD equals to that of the fundamental is defined as two-tone input-referred third-order intercept point *(IIP_3)*. The corresponding output level is called the output third-order intercept point (OIP_3). *IIP3* can be obtained by inserting a two-tone input, $x(t) = A\cos\omega_1 t + A\cos\omega_2 t$, into equation (3.7)

$$
\begin{aligned}
y(t) = (\alpha_1 + \frac{9}{4}\alpha_3 A^2)A\cos\omega_1 t + (\alpha_1 + \frac{9}{4}\alpha_3 A^2)A\cos\omega_2 t \\
+\frac{3}{4}\alpha_3 A^3\cos(2\omega_1 - \omega_2)t + \frac{3}{4}\alpha_3 A^3\cos(2\omega_2 - \omega_1)t + \cdots
\end{aligned}
\quad (3.10)
$$

In the case where $\alpha_1 \rangle\rangle \frac{9}{4}\alpha_3 A^2$, by the definition of *IIP3*, the input level (where the output components of the fundamental at ω_1 and ω_2 having the same amplitudes as those of the third-order IMD at $(2\omega_1 - \omega_2)$ and $(2\omega_2 - \omega_1)$) is

$$
|\alpha_1| A_{IIP3} = \frac{3}{4}|\alpha_3| A_{IIP3}^{\;3} \quad (3.11)
$$

IIP_3 can then be calculated as

$$
IIP_3 = 20\log_{10} A_{IIP3} = 20\log_{10}\sqrt{\frac{4}{3}\left|\frac{\alpha_1}{\alpha_3}\right|} \quad (3.12)
$$

Figure 3.3 shows that as input level increases, the actual first-order output starts to deviate from the expected output as the assumption $\alpha_1 \rangle\rangle \frac{9}{4}\alpha_3 A^2$ no longer holds. In order to overcome this, extrapolation on logarithmic scale can be use to find the intercept point, as shown in Figure 3.3.

From equation (3.10), equation (3.13) can be obtained by denoting the input level at each frequency as $A_{interference}$, the output components at ω_1 and ω_2 as A_ω, and the amplitude of the third-order IMD as A_{3IM}

$$\frac{A_\omega}{A_{3IM}} \approx \frac{|\alpha_1| A_{interference}}{\frac{3}{4}|\alpha_3| A_{interference}^3} = \frac{4|\alpha_1|}{3|\alpha_3|} \frac{1}{A_{interference}^2} \tag{3.13}$$

By substituting equation (3.11) into equation (3.13), the following equation can be obtained

$$\frac{A_\omega}{A_{3IM}} = \frac{A_{IIP3}^2}{A_{interference}^2} \tag{3.14}$$

From equation (3.14)

$$\begin{aligned} 20\log_{10} A_{IIP3}(V_{rms}) &= \frac{1}{2}[20\log_{10} A_\omega(V_{rms}) \\ &- 20\log_{10} A_{3IM}(V_{rms})] + 20\log_{10} A_{interference}(V_{rms}) \end{aligned} \tag{3.15}$$

If all signal levels are expressed in *dBm*

$$A_{IIP3}(dBm) = \frac{1}{2}[A_\omega(dBm) - A_{3IM}(dBm)] + A_{interference}(dBm) \tag{3.16}$$

Example

Find the required *IIP3* for a class 3 GSM-1800 receiver.

Solution

A class 3 GSM-1800 receiver requires that a wanted signal 3dB above the reference sensitivity (-102dBm+3dB = -99dBm) must be detectable with a bit error rate (BER) of 2% in the presence of two interferers with amplitudes of -45dBm. Assuming that both the interferers will reach the receiver input at -45dBm and an output SNR_{min} of 8dB is required, the 3rd order intermodulation must be smaller than an equivalent input signal power of (-99dBm – SNR_{min} = -107dBm). Using equation (3.16), the required *IIP3* for the receiver could be obtained

$$A_{IIP3}(dBm) = \frac{1}{2}[-45dBm - (-107dBm)] + (-45dBm) = -14dBm$$

For a cascaded N-stage network, the *IIP3* of the system ($IIP3_{tot}$) can be expressed as

$$\frac{1}{IIP3_{tot}} = \frac{1}{IIP3_1} + \frac{G_1}{IIP3_2} + \frac{G_1 G_2}{IIP3_3} + ... + \frac{G_1 G_2 ... G_{N-1}}{IIP3_N} \tag{3.17}$$

where $IIP3_i$ and G_i (i = 1, 2, ... N) are the *IIP3* and the available power gain of the ith stage respectively. It is observed from equation (3.17) that, for the *IIP3* calculation, the last stage contributes the most to the distortion of the system. It is different from the noise figure calculation, where the first stage is the most critical. Thus it is important to end the system with a high linearity block.

Similar to the discussion of noise figure, for integrated circuit calculations, the power gain should be replaced with square of the voltage gain when required.

3.1.6 S-Parameters

A two-port network is commonly used to represent a circuit with one input and one output. To characterize its behavior, measured data of both transfer function and impedance function of the two-port network must be obtained. At low frequencies, Z, Y, H and ABCD parameters are used to describe the two-port network. At high frequencies, these parameters cannot be measured accurately because the required short-circuit and open-circuit tests are difficult to achieve. Scattering parameters (S-Parameters), based on incident and reflected waves, are very useful in radio frequency range.

Figure 3.4 illustrates a two-port network, where a_1, a_2 are incident waves, b_1, b_2 are reflected waves.

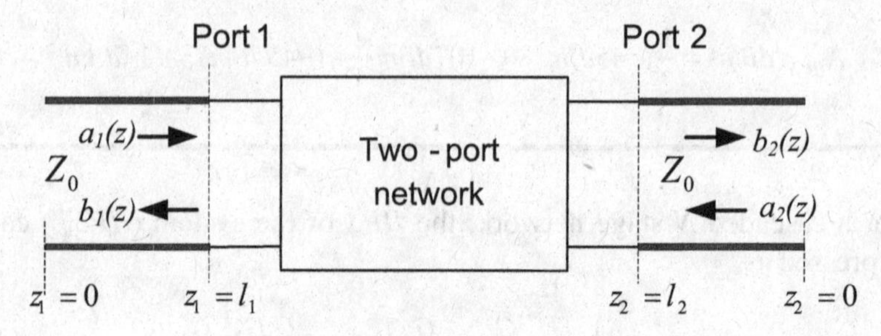

Figure 3.4 A two-port network.

Their relation can be expressed as

$$\begin{bmatrix} b_1(l_1) \\ b_2(l_1) \end{bmatrix} = \begin{bmatrix} S_{11} & S_{12} \\ S_{21} & S_{22} \end{bmatrix} \begin{bmatrix} a_1(l_1) \\ a_2(l_1) \end{bmatrix} = [S] \begin{bmatrix} a_1(l_1) \\ a_2(l_1) \end{bmatrix} \tag{3.18}$$

The matrix $[S] = \begin{bmatrix} S_{11} & S_{12} \\ S_{21} & S_{22} \end{bmatrix}$ is called the scattering matrix, where S_{11} is the input reflection coefficient, S_{12} is the reverse transmission coefficient, S_{21} is the forward transmission coefficient, and S_{22} is the output reflection coefficient. They can be calculated according to Figure 3.5 and equations (3.19a) – (3.19d), where both ports must be terminated with the characteristic impedance.

$$S_{11} = \frac{b_1(l_1)}{a_1(l_1)} \Big|_{a_2(l_2)=0} = \text{input reflection coefficient} \tag{3.19a}$$

$$S_{12} = \frac{b_1(l_1)}{a_2(l_2)} \Big|_{a_1(l_1)=0} = \text{reverse transmission coefficient} \tag{3.19b}$$

$$S_{21} = \frac{b_2(l_2)}{a_1(l_1)} \Big|_{a_2(l_2)=0} = \text{forward transmission coefficient} \tag{3.19c}$$

$$S_{22} = \frac{b_2(l_2)}{a_2(l_2)} \Big|_{a_1(l_1)=0} = \text{output reflection coefficient} \tag{3.19d}$$

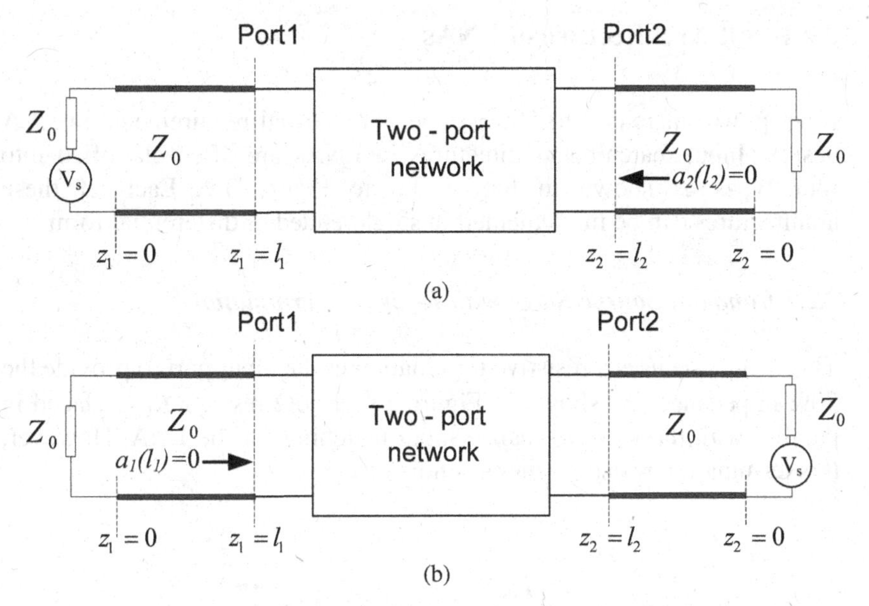

Figure 3.5 Definitions of S-parameters (a) S_{11} and S_{21} (b) S_{12} and S_{22}.

From the viewpoint of amplifier design, S_{11} and S_{22} denote how well the input and output impedances are matched to the reference impedance respectively. An integrated LNA's input is usually connected to either an external filter or an RF switch with 50Ω characteristic impedance. On the other hand, an LNA's output is usually connected to an on-chip mixer. Thus, usually S_{22} is only important for stand-alone LNA or for on-wafer measurement. Various methods of achieving input matching in order to get a good S_{11} and noise figure performance will be discussed in Section 3.2.

S_{21} measures the insertion effect (amplification gain) of the amplifier. It is only equal to the actual voltage gain if the LNA output is matched. S_{12} represents the isolation between output and input ports. High isolation is needed to prevent signal at the output to flow back to the LNA input terminal.

3.2 Input Architecture of LNAs

Input power matching to 50Ω is one of the usual requirements in LNA design. Input matching architectures in LNAs are often classified into four types as shown in Figure 3.6 to Figure 3.9. Each of these architectures can be implemented in single-ended or differential form.

3.2.1 Common Source Stage with Resistive Termination

This technique uses a resistive termination at the input port to provide the 50Ω impedance. As shown in Figure 3.6, a 50Ω resistor R_1 is placed in parallel with the input to realize input matching for the LNA. However, this termination resistor generates noise.

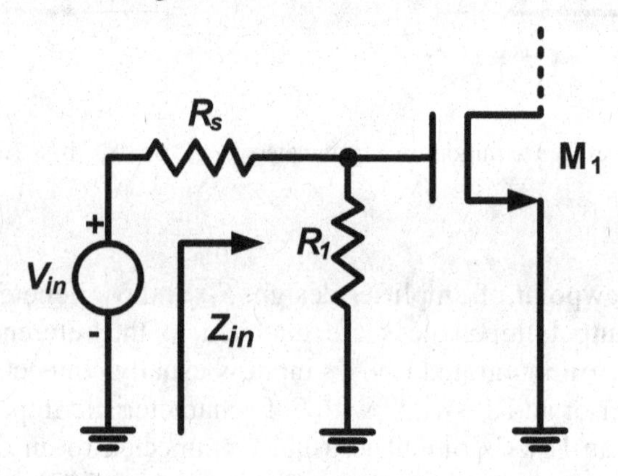

Figure 3.6 Common source with resistive termination.

The noise factor of the circuit can be calculated

$$F = \frac{\overline{V_{n,out}^2}}{\overline{V_{n,o,Rs}^2}} = \frac{\overline{V_{n,o,Rs}^2} + \overline{V_{n,o,R1}^2} + \overline{V_{n,o,M1}^2}}{\overline{V_{n,o,Rs}^2}} \approx \frac{4kT(R_s // R_1){g_m}^2 {R_L}^2 \Delta f + \overline{i_{n,d}^2} R_L^2}{\frac{1}{4} g_m^2 R_L^2} \frac{1}{4kTR_s \Delta f} \tag{3.20}$$

where $\overline{V^2_{n,out}}$ represents the total output noise; $\overline{V^2_{n,o,Rs}}$, $\overline{V^2_{n,o,R1}}$ and $\overline{V^2_{n,o,M1}}$ are the output noise due to R_s, R_1 and M_1, respectively; k is the Boltzmann's constant and T is the absolute temperature. MOSFET M_1 has various noise sources (refer to Chapter 5). Here, for simplicity in calculations, only the channel noise $\overline{i^2_{n,d}} = 4kT\gamma g_{d0}\Delta f$ is taken into account as it is the dominant noise source in most conditions. Equation (3.20) can be further simplified to

$$F = \frac{4(R_s // R_1)}{R_s} + \frac{4\gamma}{\alpha g_m R_s} = 2 + \frac{4\gamma}{\alpha g_m R_s} > 3dB \tag{3.21}$$

where $\alpha = \frac{g_m}{g_{d0}}$ and is typically less than one. For long-channel devices, γ is unity when the MOSFET operates in the triode region and 2/3 in the saturation region. For short-channel devices, γ can be significantly larger than 2/3. Taking $R_1 = R_s = 50\Omega$, $\gamma = 1$, $\alpha = 0.8$, $g_m = 0.06\Omega^{-1}$ as an example, F is calculated to be 3.67, which is 5.6dB. Thus, noise figure of this architecture can be much larger than 3dB. The poor noise figure makes this architecture unattractive for applications where a low *NF* as well as a good input matching are desired.

3.2.2 Common Gate Stage

Figure 3.7 shows a simplified $1/g_m$ termination architecture. This configuration requires that $1/(g_m+g_{mb}) \approx 1/g_m$ to be equal to R_s, where R_s is the 50Ω source resistance. This topology is suitable for wideband input matching designs, since g_m is basically constant over a quite wide frequency range. The main drawback of this method is that the transconductance of the input transistor cannot be arbitrarily high, thus imposing a lower bound on the noise figure. Here, for the simplicity of calculations, only the channel thermal noise $\overline{i^2_{n,d}} = 4kT\gamma g_{d0}\Delta f$ of the MOSFET is taken into account. Through derivation, F can be expressed as

$$F = \frac{4kTR_s g_m^2 R_L^2 \Delta f + \overline{i_{n,d}^2} R_L^2}{g_m^2 R_L^2} \cdot \frac{1}{4kTR_s \Delta f} \approx 1 + \frac{\gamma}{\alpha} \geq 1 + \frac{2/3}{1} = \frac{5}{3} = 2.2dB \tag{3.22}$$

where $\gamma = 2/3$ and $\alpha = 0.8$ are used in the example. The above example calculations suggest a reasonable noise figure. However, other noise sources such as gate induced noise and substrate noise can degrade the performance substantially. In short-channel MOSFETs, γ can be higher than 2/3, and α can be much less than 1, this leads to a large noise factor.

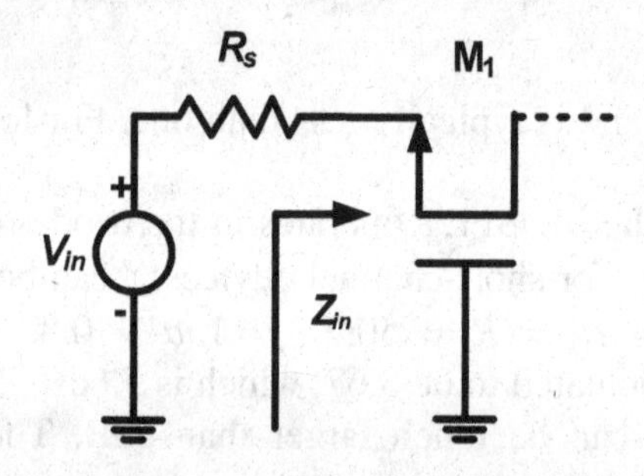

Figure 3.7 Common gate input architecture.

3.2.3 Common Source Stage with Shunt Feedback

Figure 3.8 illustrates another topology, which uses the resistive shunt and series feedback to set the input impedance to 50Ω. The input impedance can be expressed as

$$Z_{in} \approx R_{fb} / (1 + |A_v|) \tag{3.23}$$

where R_{fb} is the feedback resistor and A_v is the corresponding voltage gain. The topology has several disadvantages. Firstly, it is sensitive to process variations as the input impedance Z_{in} is dependent on R_{fb} and A_v as shown in equation (3.23) [12].

Secondly, this architecture often has high power dissipation. The high dissipation is partially due to the fact that shunt-series amplifiers of this type are naturally wideband; hence techniques that reduce the power consumption through LC tuning are not applicable. At last, the total phase shift around the loop may create instability for certain source and load impedances.

The noise factor for this configuration can be expressed as follows [12]

$$F = 1 + R_s \delta g_g + \left(\frac{G_s + G_{fb}}{g_m - G_{fb}} \right)^2 R_s (R_l + \gamma g_{d0}) + \left(\frac{G_s + g_m}{g_m - G_{fb}} \right)^2 R_s G_{fb} \quad (3.24)$$

where G_s, G_{fb} and G_1 are the conductance of the resistors R_s, R_{fb} and R_1, respectively.

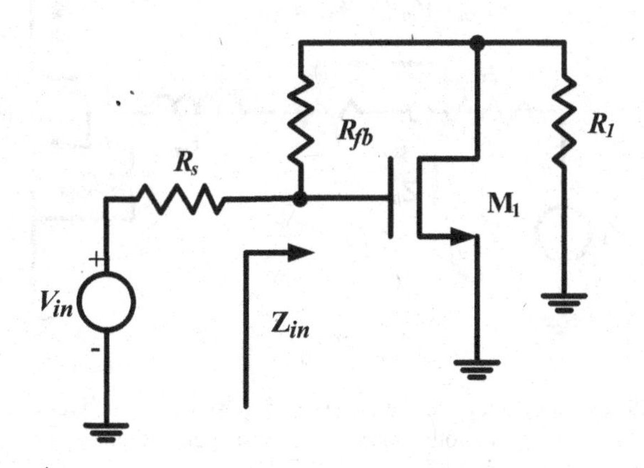

Figure 3.8 Common source input stage with shunt feedback.

3.2.4 Common Source Stage with Source Inductive Degeneration

The fourth architecture employs source inductive degeneration to generate a real term in the input impedance as shown in Figure 3.9(a).

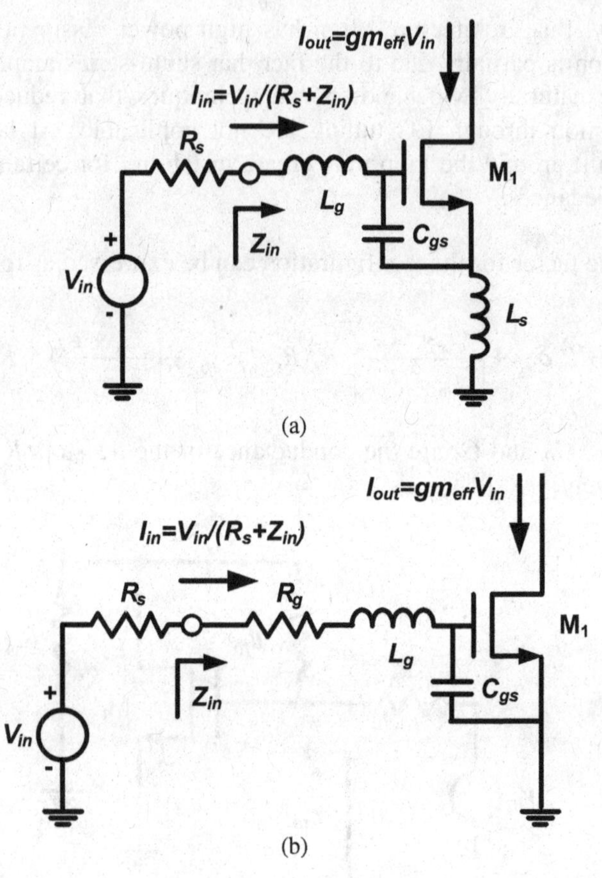

Figure 3.9 Common source input stage (a) With source inductive degeneration (b) Without source inductive degeneration.

The input impedance is

$$Z_{in} \approx j\left\{\omega L_g + \omega L_s - \frac{1}{\omega C_{gs}}\right\} + \left(g_{m1} / C_{gs}\right) L_s \qquad (3.25)$$

The impedance shown in equation (3.25) has a resistive term $(g_{m1}/C_{gs})L_s$, which is directly proportional to the inductance value. Whatever value this resistive term is, it does not generate thermal noise like an ordinary resistor does, because a pure reactance is noiseless. Therefore, this structure can be used to give specified input impedance without degrading the noise performance of the amplifier. For example, to get 50Ω input impedance, let the real part $(g_{m1}/C_{gs})L_s$ of equation (3.25) equal to R_s= 50Ω and the imaginary part $\omega L_g + \omega L_s - \frac{1}{\omega C_{gs}}$ be zero at the frequency of interest ω_0, which can be obtained through the following expression $\omega_0 = 1/\sqrt{(L_g + L_s)C_{gs}}$. In other words, when L_g and L_s are in resonant with C_{gs}, the input impedance Z_{in} can be made to be equal to R_s, thus input matching is achieved.

The transconductance gain is $gm_{eff} = i_{out}/V_{in}$, where $i_{out} = g_{m1}V_{gs1}$. It can be shown that $V_{gs1} = i_{in}(1/j\omega_0 C_{gs})$ where $i_{in} = V_{in}/(2R_s)$ can be obtained at matching condition, hence

$$gm_{eff} = g_{m1}\frac{1/j\omega_0 C_{gs}}{2R_s} = g_{m1}\frac{j\omega_0(L_s + L_g)}{2R_s} \tag{3.26a}$$

The transconductance gain changes by a factor (usually increases) of $\frac{j\omega_0(L_s + L_g)}{2R_s}$ from the transconductance gain of the transistor g_{m1}.

Example

If the source degeneration inductor is removed as shown in Figure 3.9(b), state the equations in order to achieve input matching.

Solution

The input impedance of the circuit shown in Figure 3.9(b) is

$$Z_{in} \approx j\left\{\omega L_g - \frac{1}{\omega C_{gs}}\right\} + R_g$$

where R_g is a resistor in series with L_g. In practice, part of R_g could be contributed from the equivalent series resistance due to the loss in L_g. The gate resistance of the transistor also contributes to R_g.

To achieve input matching, the real part of Z_{in}, which is R_g, must be equal to R_s, whereas the imaginary part $\omega L_g - \frac{1}{\omega C_{gs}}$ must be equal to zero.

It can be shown that the transconductance gain for Figure 3.9(b) at matching condition is

$$gm_{eff} = g_{m1}\frac{1/\,j\omega_0 C_{gs}}{2R_s} = g_{m1}\frac{j\omega_0 L_g}{2R_s} \tag{3.26b}$$

At matching condition, the transconductance gain in equation (3.26a) and (3.26b) are the same with same C_{gs}. In order words, removing L_s will not affect the gain of the LNA under these conditions.

For practical considerations, the input impedance in Figure 3.9(a) should also include R_g and the loss of L_s. Hence, in order to achieve input matching, the real part of the input impedance $R_g + (g_{m1}/C_{gs})L_s + R_L$ should be made equal to to R_s. R_g represents the gate resistance and the loss of L_g, whereas R_L represents the loss of L_s.

For Figure 3.9(a), the noise factor can be expressed as [3]

$$F = 1 + (\frac{\omega_0}{\omega_T})R_s\gamma g_{d0} + \left[(\frac{\omega_T}{\omega_0 g_m R_s})^2 + 1\right]R_s\delta g_g + 0.79R_s(\frac{\omega_0}{\omega_T})\sqrt{\gamma g_{d0}\delta g_g} \tag{3.27}$$

where $\omega_T \approx g_m / C_{gs}$ is the unity current gain frequency and ω_0 is the operating frequency. In practice, the gate inductor and source inductor are not ideal. Both have parasitic resistance, which will contribute thermal noise. Thus, their effects must be taken into account while doing noise optimization and calculation of input impedance, especially in the case where low-Q on-chip integrated inductors are used.

The common source input topology with source inductive degeneration provides a low noise figure, reasonable gain and low power consumption.

3.3 Input Matching Analysis

In this section, the input matching analysis of the single-band LNA modified from the traditional source inductive degeneration architecture will be first discussed. The 50Ω input matching is one of the considerations in LNA design. There are currently four input architectures in LNA designs: common source with resistive termination, common gate ($1/g_m$) termination, common source with shunt feedback topology and common source with source inductive degeneration topology. Of the mentioned four architectures, the fourth (as shown in Figure 3.10) is the most widely adopted. Its input impedance is given in Equation (3.25).

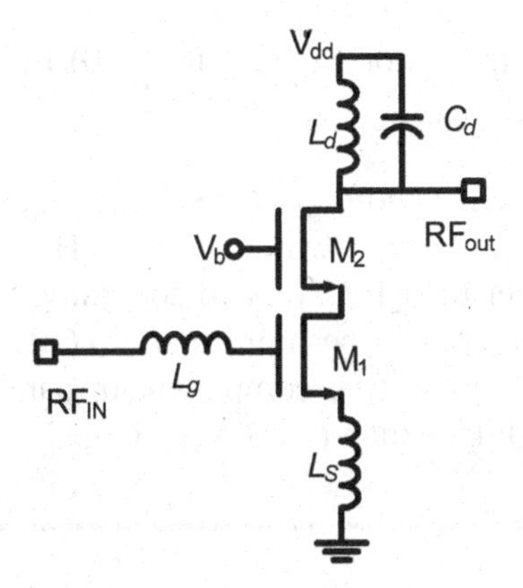

Figure 3.10 Typical single-band LNA using source inductive degeneration input topology.

To obtain input matching, the real part of equation (3.25) should be 50Ω, whereas the imaginary part should be zero at the frequency of interest

$$(g_{m1}/C_{gs})L_s = \omega_T L_s = 50 \tag{3.28a}$$

$$j\{\omega L_g + \omega L_s - \frac{1}{\omega C_{gs}}\} = 0 \tag{3.28b}$$

The LNA's operating frequency can be derived from Equation (3.28b). Vice versa, when the operating frequency ω_0 and the transistor size are known, inductor values L_s and L_g can be calculated as well. Hence the values of the components needed in the source inductive degeneration input topology can be obtained.

Example

Calculate the value of L_g in an LNA used for Bluetooth applications.

Solution

From equation (3.28b)

$$L_g + L_s = \frac{1}{(2\pi f_0)^2 C_{gs}} \tag{3.28c}$$

Let the size of the transistor M_1 (Figure 3.10) be 240μm/0.18μm; then the gate to source capacitance $C_{gs} \approx \frac{2}{3}WLC_{ox}$ [4] is about 0.23pF [5]. Substituting f_0 = 2.5GHz and C_{gs} = 0.23pF into equation (3.28c), the sum of $L_g + L_s$ can be obtained, which is 17.6nH. Generally, the cut-off frequency f_T is higher than 10GHz, and according to equation (3.28a), $L_s \leq$ 1nH. Therefore, L_g has to be more than 16.6nH in this example. The Miller effect adds a capacitive component in parallel with C_{gs}, and the final L_g is around 10nH, which is still very large.

However, there is a drawback for this topology. In order to realize the series resonance with the transistor M_1's small gate to source capacitance C_{gs} (about 0.23pF for a 240μm/0.18μm NMOS transistor), the gate inductor L_g is usually too large to be integrated on-chip (20nH in Gatta's work [2] and tens of nH in Shaeffer's work [3]). A large inductor implemented on-chip with the current standard CMOS technology usually has a low Q factor and occupies a considerable chip area.

A diameter of more than 250μm using CSM 0.18μm CMOS is required for a 10nH inductor. Moreover, to ensure the quality of the on-chip inductor, the distance between the inductor and any other components should generally be at least 50μm. Otherwise, the coupling effect and other electro-magnetic effects can degrade the inductor's performance, especially at gigahertz frequency. In other words, it means that a 10nH inductor will occupy an area with a diameter of at least 350μm. It is thus very costly.

Furthermore, in LNA design, the quality factor Q of L_g must be high, because its parasitic resistance contributes thermal noise. For instance, for a 10nH on-chip inductor with $Q = 7$ at 2.45GHz, the series parasitic resistance is 22Ω, which is 44 percent of the 50Ω source resistance. It can lead to quite a large noise figure (*NF*) increment. Therefore, L_g has been implemented off-chip [2, 3, 7] though this goes against the trend of system integration demanded by the competitive market. In short, the use of large-value inductors should be avoided, from either the perspective of system integration or *NF* improvement or chip area reduction.

There are several ways to reduce L_g. According to equation (3.28b), one way is to increase the input transistor's width W in order to increase C_{gs} ($C_{gs} \approx 2WLC_{ox}/3$ [4]). However, this could result in a large drain current I_d. Low I_d can be obtained by decreasing the over-drive voltage V_{od}. In practice V_{od} is only a fraction of a volt higher than the threshold voltage V_T. If V_{od} is reduced further, the circuit might not operate normally due to variations in the fabrication process.

Another method is to connect an additional capacitor C_d in parallel with C_{gs} [8], but the quality factor of the input stage ($Q_L = 1/\{2R_S\omega_0(C_{gs}+C_d)\}$ [8]) will be degraded. As a result, the LNA's performance will be degraded as well. A modified architecture that will neither increase power consumption nor degrade Q_L is preferable in most cases.

Considering the L_1C_1 parallel network illustrated in Figure 3.11, its impedance can be derived and simplified as

$$Z = \frac{j\omega L_1}{1-\omega^2 L_1 C_1} = j\omega\left(\frac{L_1}{1-\omega^2 L_1 C_1}\right) = j\omega\frac{L_1}{1-(\omega/\omega_0)^2} \tag{3.29}$$

where $\omega_0 = 1/\sqrt{L_1C_1}$ is the resonant frequency of the L_1C_1 parallel network. To guarantee normal operation of the circuit, ω_0 should be located outside the operating frequency band. It is easy to find that the L_1C_1 parallel network is equivalent to an inductor L_2 where $L_2 = \dfrac{L_1}{1-\omega^2 L_1 C_1} = \dfrac{L_1}{1-(\omega/\omega_0)^2}$. Under the condition that $0 < 1-\omega^2 L_1 C_1 < 1$ (to ensure L_2 to be positive), $L_2 > L_l$, and as ω approaches ω_{01}, L_2 will be significantly larger than L_1. Hence, a small L_1C_1 parallel network can be expected to generate a large inductance and to replace the large L_g [6].

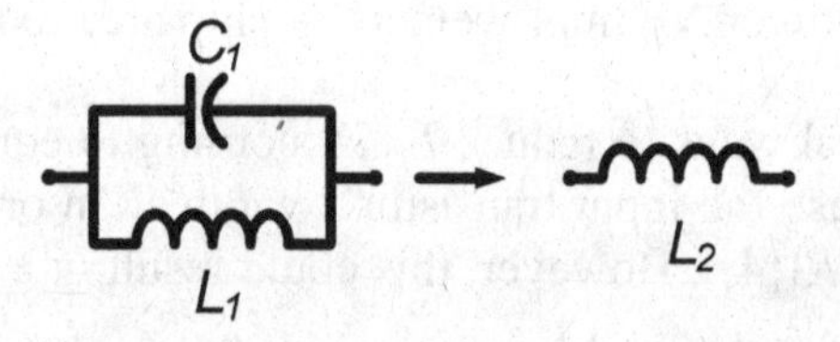

Figure 3.11 An L_1C_1 parallel network and its equivalent circuit.

Figure 3.12 shows a graph of L_2 versus frequency when L_1 = 3.2nH and C_1 = 0.8/0.9/1.0pF. The figure indicates that at 2.45GHz, the L_1C_1 parallel network can generate an inductance of 10nH when L_1 = 3.2nH and C_1 = 0.9pF.

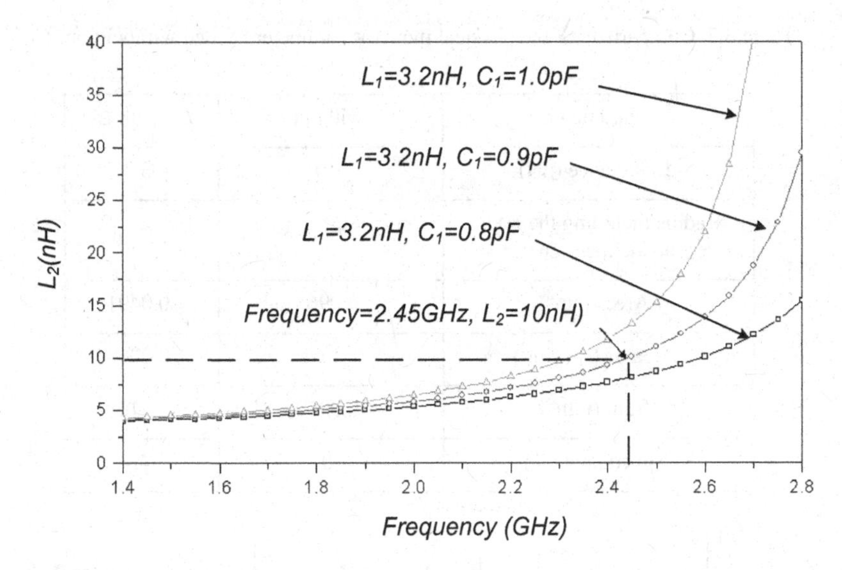

Figure 3.12 Relation of the generated L_2 vs. frequency (L_1=3.2nH, C_1=0.8/0.9/1.0pF).

Compared to the 10nH inductor, a 3.2nH inductor is easier to be implemented on-chip based on the current CMOS technology, and it consumes a chip area with a diameter of only around 150μm. The 0.9pF capacitor has an area of 30μm × 30μm, which is relatively small compared to that of the on-chip inductor.

Table 3.2 summarizes the area consumption before and after C_1 is introduced. From the table, it can be seen that the chip area of L_1C_1 decreases significantly compared to the large L_g. The simulation results also indicate that good input-reflection coefficient S_{11} can be achieved using the above technique as shown in Figure 3.13.

Table 3.2 Comparison of the physical inductor parameter needed with/without C_1.

Parameter	L_g (without C_1)	L_1 (with C_1)
Inductance (nH)	10	3.2
Radius including the 50μm outside the inductor	175μm	125μm
Area (mm^2)	0.0962	0.0491
Area of C_1 (mm^2)	0	0.0009
Sum (mm^2)	0.0962	0.0500
Saved area (%)	0	48

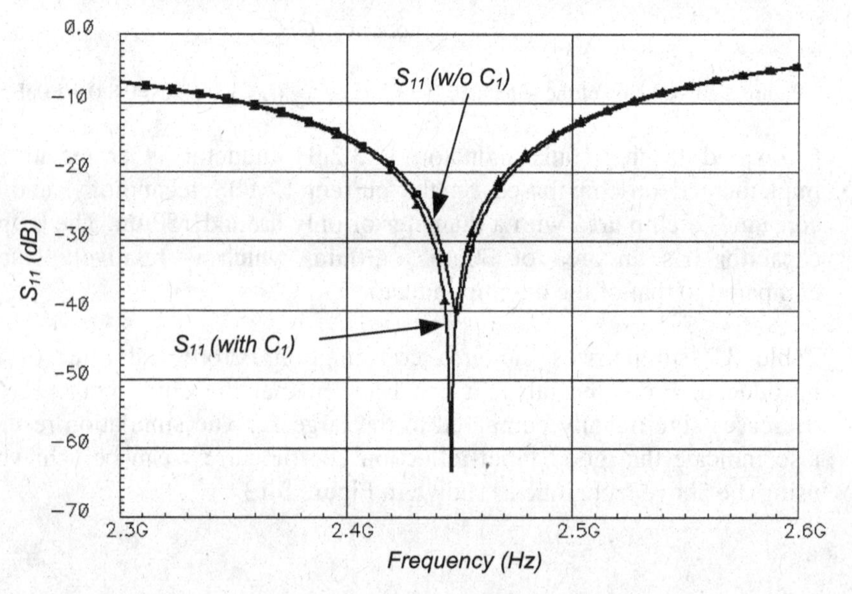

Figure 3.13 Simulated S_{11} with L_g=10nH (without C_1) and L_1=3.2nH, C_1=0.9pF.

In the previous analyses, the parasitic resistance of the inductor L_1 is ignored. In practice, the on-chip spiral inductor suffers from metal ohmic loss, parasitic capacitive loss and eddy current loss. In this case, for analytical simplicity, the inductor is simply modeled as an ideal inductor in series with its parasitic resistance. This is a good approximation when the spiral inductor is implemented with the top metal and has a relative small size. Here, for simplicity, the on-chip capacitor is considered as ideal. Practically, the quality factor of the on-chip capacitor is more than 100. Therefore, the influence of its parasitic is relatively insignificant compared to the parasitic of the on-chip inductor. Figure 3.11 can be modified into Figure 3.14.

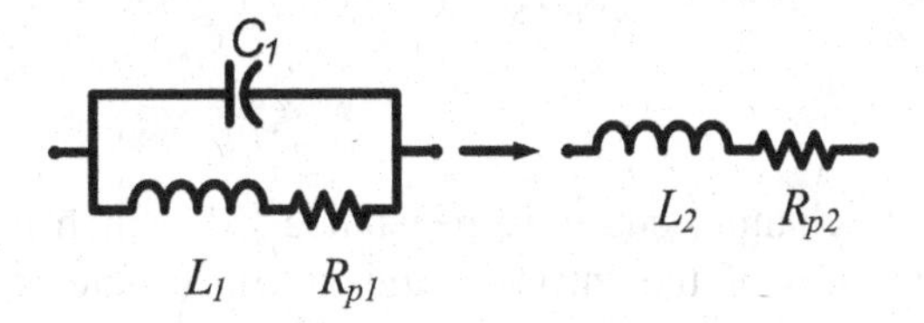

Figure 3.14 Modified L_1C_1 parallel network and its equivalent circuit.

Consequently, equation (3.29) can be rewritten as

$$Z = R_{p2} + j\omega L_2 \tag{3.30}$$

where

$$R_{p2} = \frac{R_{p1}}{(1-\omega^2 L_1 C_1)^2 + (\omega C_1 R_{p1})^2} \approx \frac{R_{p1}}{(1-\omega^2 L_1 C_1)^2} = \frac{R_{p1}}{1-(\omega/\omega_{01})^2} \tag{3.30a}$$

$$L_2 = \frac{L_1 - \omega^2 L_1^2 C_1 - C_1 R_{p1}^2}{(1-\omega^2 L_1 C_1)^2 + (\omega C_1 R_{p1})^2} \approx \frac{L_1 - \omega^2 L_1^2 C_1 - C_1 R_{p1}^2}{(1-\omega^2 L_1 C_1)^2} \approx \frac{L_1}{1-(\omega/\omega_{01})^2} \tag{3.30b}$$

$$\omega_0 = 1/\sqrt{L_1 C_1} \tag{3.30c}$$

Due to the influence of the parasitic resistance R_{p1}, the value of L_2 is decreased a little. The simplifications of equation (3.30a) and equation (3.30b) are reasonable for estimation. Figure 3.15 illustrates the characteristics of the network formed by $L_1 = 3.2\text{nH}$ ($Q = 10$) and $C_1 = 0.9\text{pF}$. The figure shows that $L_{2\_simplify}$ is around 10nH and L_2 is around 9.4nH at 2.45GHz, which indicates a 6% difference. This confirms that a high inductance can be achieved with a small on-chip L_1C_1 parallel network, even if L_1 is not an ideal inductor. Using the lossy L_1C_1 parallel network (Figure 3.14) to replace the large gate inductor L_g (Figure 3.10), the input impedance can be re-written as

$$Z_{in} = \{ j\omega L_2 + j\omega(1+g_m R_i)L_s - j\frac{1}{\omega C_{gs}}\} + (\frac{g_{m1}}{C_{gs}}L_s + R_{p2} + R_g + R_i) \tag{3.31}$$

where R_i is the channel charging resistance [9], which is about $1/(5g_{d0})$ [4]; R_g is the sum of the intrinsic and extrinsic gate resistance. For a multi-finger layout, the typical value of R_g is 3Ω to 5Ω (for a transistor width around 200μm). The value of R_i depends on g_m and it is about 2Ω to 4Ω for general operations and around 10Ω for a small-sized transistor. R_{p2} is expressed in equation (3.30a) and is larger than R_{p1}. Figure 3.15 shows that R_{p2} is around 48Ω at 2.45GHz.

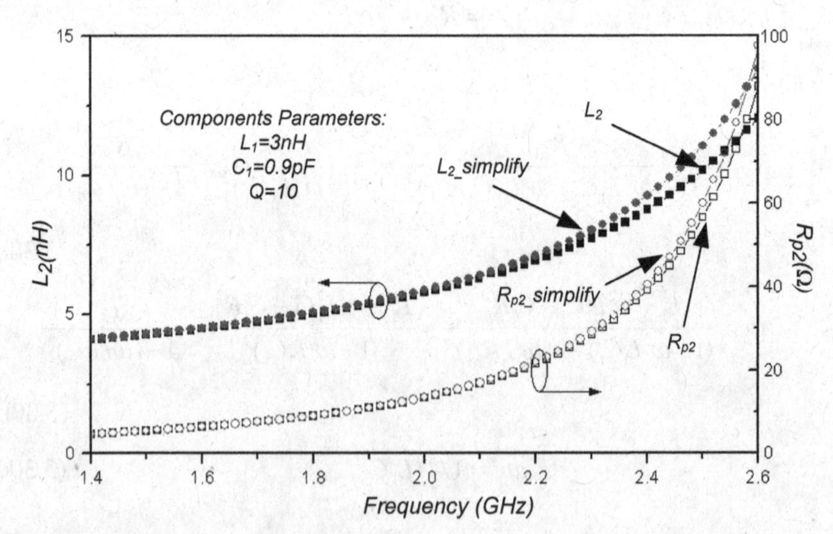

Figure 3.15 Relation of the generated L_2, $L_{2\_simplify}$ and R_{p2}, $R_{p2\_simplify}$ vs. frequency.

Consequently, it can be seen that $(R_{p2} + R_i + R_g)$ leads to an R_{in} of 53Ω to 57Ω, which is very close to the required 50Ω input impedance. It means that through the introduction of L_1C_1, the unavoidable harmful parasitic resistance R_{p1} of L_1 can be utilized to generate a real term to satisfy the 50Ω input matching.

The original purpose of adding L_s (lossless) is to optimize input matching [3] and noise matching [10]. However, the on-chip L_s is lossy. It generates thermal noise and affects noise or gain matching conditions. Hence, it makes sense to remove L_s which also saves chip area. After the removal of L_s, equation (3.31) is modified to be

$$Z_{in} = \{j\omega L_2 - j\frac{1}{\omega C_{gs}}\} + (R_{p2} + R_g + R_i) \tag{3.32}$$

The resonant frequency is changed to $\omega_0 = 1/\sqrt{L_2C_{gs}}$. Thus at frequency ω_0 = 2.45GHz, S_{11} is calculated to be between -31dB and -23.7dB without L_s. In practice, the required S_{11} is typically below -10dB [11,13]. Hence, S_{11} being in the range from -31dB to -23.7dB is more than enough for practical applications. The final modified input architecture is illustrated in Figure 3.16.

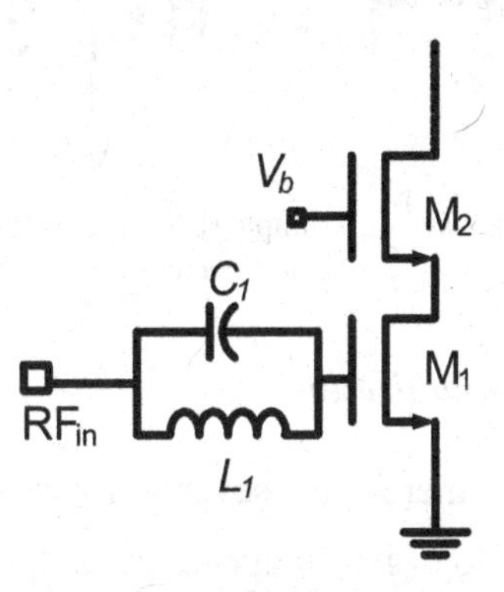

Figure 3.16 The final modified input architecture.

3.4 Design of a Single-band LNA (LNA1)

A basic LNA consists of a transconductance stage and an RF load. The transconductance stage translates an input voltage into an output current, while the RF load translates the current into an output voltage. The noise performance of the transconductance stage affects the LNA's noise performance.

Figure 3.17 shows the schematic of the 2.4GHz design example (LNA1), which will be explained in detail in the following sections.

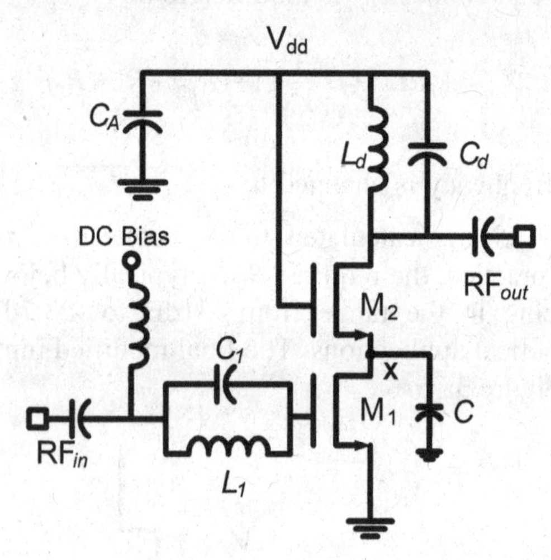

Figure 3.17 A 2.4GHz design example (LNA1) based on the parallel L_1C_1 input architecture.

3.4.1 Noise Figure Optimization

In a CMOS LNA, there are several main noise sources: channel noise $\overline{i_{n,d}^2} = 4kT\gamma g_{d0}\Delta f$, gate induced noise $\overline{i_{n,g}^2} = 4kT\delta g_g \Delta f$ [3], thermal noise of various parasitic resistors and so on. Figure 3.18 illustrates a simplified small-signal model for the LNA *NF* calculation. For the calculations, L_1C_1 network is to be replaced by the equivalent circuit in Figure 3.14.

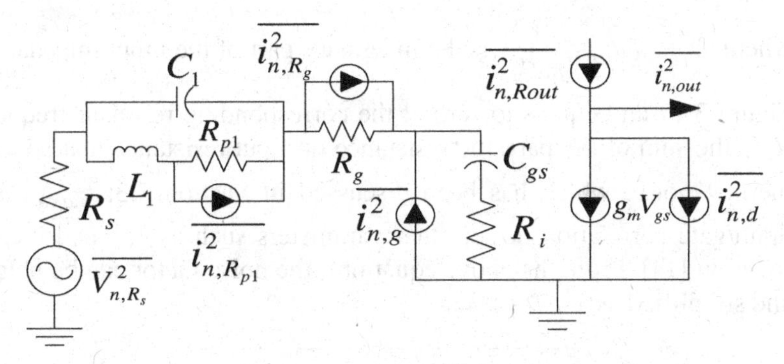

Figure 3.18 Simplified small-signal model for LNA *NF* calculation [6].

From Figure 3.18, the output noise can be derived and expressed in equations (3.33a) to (3.33f)

$$\overline{i^2_{n,o,R_S}} = \left| \frac{g_m R_S}{j\omega C_{gs}(R_1 + R_S) + j\omega C_{gs} D_0} \right|^2 \overline{i^2_{n,R_S}} \tag{3.33a}$$

$$\overline{i^2_{n,o,d}} = \overline{i^2_{n,d}} \tag{3.33b}$$

$$\overline{i^2_{n,o,R_1}} = \left| \frac{g_m R_1}{j\omega C_{gs}(R_1 + R_S) + j\omega C_{gs} D_0} \right|^2 \overline{i^2_{n,R_1}} \tag{3.33c}$$

$$\overline{i^2_{n,o,g}} = \left| \frac{g_m}{j\omega C_{gs}} (1 - \frac{1}{j\omega C_{gs}(R_S + R_1) + j\omega C_{gs} D_0}) \right|^2 \overline{i^2_{n,g}} \tag{3.33d}$$

$$\overline{i^2_{n,o,R_{out}}} = \overline{i^2_{n,R_{out}}} \tag{3.33e}$$

$$\overline{i^2_{n,o,corr}} = \frac{2 g_m c}{\omega C_{gs}} (1 - \frac{jD_0}{\omega C_{gs}((R_S + R_1)^2 + |D_0|^2)}) \sqrt{\overline{i^2_{n,g}} \cdot \overline{i^2_{n,d}}} \tag{3.33f}$$

where $D_0 = j\omega L_2 + \frac{1}{j\omega C_{gs}}$ is the imaginary part of the input impedance of Figure 3.17 and equals to zero at the corresponding resonant frequency; R_1 is the sum of the parasitic resistance or input resistance that generates thermal noise, which has been discussed in Section 3.3; $\overline{i_{n,o,corr}}$ is the drain/gate correlation noise; other parameters such as δ, γ and c can be found in [14]. From the above equations, the noise factor can be obtained and simplified as

$$F = \frac{\overline{i^2_{n,o,R_S}} + \overline{i^2_{n,o,g}} + \overline{i^2_{n,o,d}} + \overline{i^2_{n,o,corr}} + \overline{i^2_{n,o,R_1}} + \overline{i^2_{n,o,R_{out}}}}{\overline{i^2_{n,o,R_S}}}$$

$$\approx 1 + \frac{R_1}{R_S} + \frac{\delta\alpha}{5g_m R_S}\{(j\omega C_{gs} D_0 - 1)^2 + \omega^2 C_{gs}^2 (R_S + R_1)^2\} \tag{3.34}$$

$$+ \frac{\gamma}{\alpha g_m R_S}\{(j\omega C_{gs} D_0)^2 + \omega^2 C_{gs}^2 (R_S + R_1)^2\}$$

$$+ \frac{2c}{g_m R_s}\sqrt{\frac{\gamma\delta}{5}}\{(j\omega C_{gs} D_0)^2 - j\omega C_{gs} D_0 + \omega^2 C_{gs}^2 (R_S + R_1)^2\}$$

The noise factor at resonance frequency ω_0 ($D_0 = 0$ under this condition) can be achieved as well

$$F_0 = 1 + \frac{R_1}{R_S} + \left(\frac{\omega_0}{\omega_T}\right)^2 \frac{\gamma g_{d0}}{R_S}\left\{\left[(R_S + R_1)^2 + \frac{1}{\omega_0^2 C_{gs}^2}\right]\frac{\delta\alpha^2}{5\gamma} + (R_S + R_1)^2 + 2(R_S + R_1)^2 c\sqrt{\frac{\delta\alpha^2}{5\gamma}}\right\}. \tag{3.35}$$

In CMOS LNAs, the MOS transistor is usually designed to operate in the saturation region [14]. When the transistor operates in the saturation region [4], the equation for the drain current is

$$I_{ds} = \frac{W}{2\xi L_{eff}}\mu_{eff} C_{ox}(V_{GS} - V_T)^2 \tag{3.36}$$

where μ_{eff} is the carrier's effective mobility in the transistor channel and ξ is a factor which takes the short channel effect and other effects into account. ξ is often set to one in the simplified MOS transistor model, but can be significantly larger in short-channel MOS transistors.

Taking the derivative of equation (3.36) with respect to V_{GS} yields

$$g_m = \frac{W}{\xi L_{eff}} \mu_{eff} C_{ox} (V_{GS} - V_T) = \sqrt{\frac{2W}{\xi L_{eff}} \mu_{eff} C_{ox} I_{ds}} \quad (3.37)$$

Combining equation (3.35) and equation (3.37), F can be rewritten as a function of the transistor width

$$F = 1 + R_\Delta / R_S + k_1 W^{-1/2} + k_2 \omega_0^2 L_{eff}^2 W^{3/2} \quad (3.38)$$

where k_1 and k_2 are equation coefficients. Both are functions of power consumption $P_d = V_{dd} I_{ds}$.

Figure 3.19 illustrates the relation between NF, I_{ds} and W in equation (3.38). With the following parameter values: $\gamma = 2$, $\delta = 4$, $\xi = 1.25$, $\alpha = 0.8$, $\mu_{eff} = 0.03\text{m}^2/\text{V}$, $C_{ox} = 0.847\text{mF/m}^2$, $L_1 = 1.9\text{nH}$ ($Q = 7.5$) and $C_1 = 1.4\text{pF}$, the plot shows that the LNA has an optimum W_{opt} of around $240\mu m$, which corresponds to $NF_{\min}$ when I_{ds} is chosen to be 8mA.

W_{opt} can also be obtained through direct mathematic derivation. By differentiating equation (3.38) with respect to W and equalizing it to zero, the resulting equation is

$$W_{opt} = \sqrt{k_1 / (3 k_2 \omega_0^2 L_{eff}^2)} \quad (3.39)$$

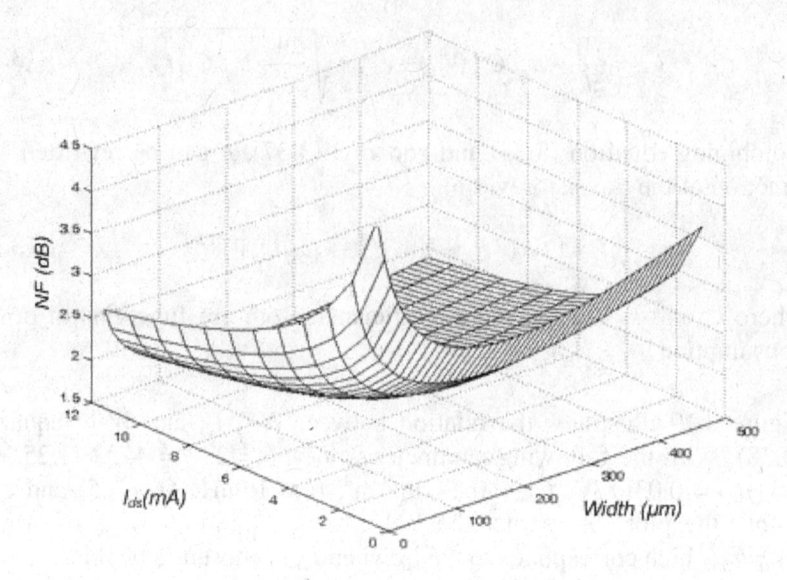

Figure 3.19 *NF* versus I_{ds} and W for γ=2, δ=4, ξ=1.25, α=0.8, μ_{eff}=0.03m^2/V, C_{ox}=0.847mF/ m^2, L_1=1.9nH (Q=7.5) and C_1=1.4pF.

3.4.2 Design Methodology

Generally, noise optimization is a process to find the optimum device size, so that a relative low *NF* can be obtained while giving consideration to the input reflection coefficient, power consumption and so on. In the design example shown in Figure 3.17, L_1 = 1.9nH (Q = 7.5 at 2.45GHz), C_1 = 1.4pF, I_{ds} = 8mA and W = 240μm are the chosen parameters, the simulated *NF* of about 2dB and S_{11} between -22dB and -18dB are obtained. The parasitic resistance and the terminal resistance in total contribute about 1dB to *NF*. Although *NF* is high compared to other designs using high-Q off-chip inductors, it has the advantages of being compact and fully integrated.

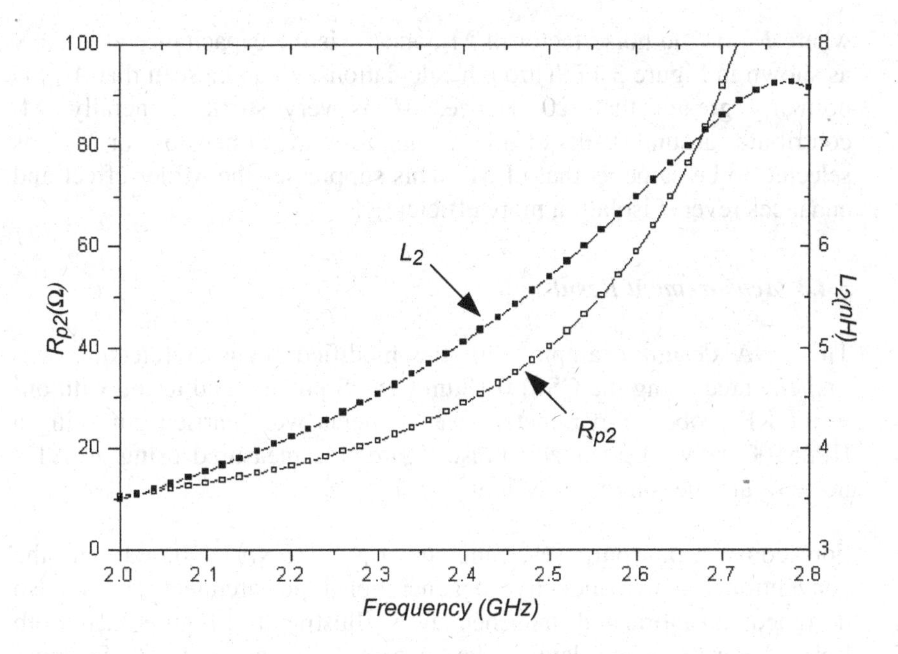

Figure 3.20 Relation of the generated L_2 and R_{p2} with frequency (L_1=1.9nH, C_1=1.4pF, Q=7.5 and ω_{01}=3GHz).

Once the input stage is determined, the resonant load L_dC_d can be decided according to the gain and output matching requirements. In Figure 3.17, transistor M_2 was used to reduce the Miller effect of C_{gd1}. It is known that the Miller effect can degrade power gain, reduce input impedance and increase input referred noise [3]. Although M_2 produces noise that is proportional to its size, in reality the input referred noise generated is not so significant. As shown in [6], the input referred noise of M_2, ΔF can be obtained by

$$\Delta F = \frac{F_{M2} - 1}{A_{M1}^2} \tag{3.40a}$$

$$A_{M1} \approx -\left|\frac{g_{m1}}{j\omega C_X}\right| \tag{3.40b}$$

$$C_X = C_{gs2} + C_{dB1} + (1 + |1/A_{vM1}|)C_{gd1} \approx C_{gs2} + C_{dB1} + C_{gd1} \tag{3.40c}$$

where F_{M2} is the noise factor of M_2, and C_X is the capacitance at node X as shown in Figure 3.17. Through calculations, it can be seen that A^2_{M1} is normally greater than 20. Hence, ΔF is very small. Generally, M_2 contributes around 0.1dB of noise. Therefore, the dimension for M_2 was selected to be same as that of M_1. This suppresses the Miller effect and enhances reverse isolation more efficiently.

3.4.3 Measurement Results

The LNA design example with the modified input architecture was implemented using the CSM 0.18μm CMOS process and tested with on-wafer RF probes. S-parameter measurements were carried out using a HP8510C network analyzer; noise figure was measured using an ATN noise figure measurement system.

For comparison purposes, another LNA (LNA2) based on the conventional source inductive degeneration input architecture was also designed and fabricated. Its schematic is illustrated in Figure 3.21. Both LNAs have the same device sizes except L_g, L_s and L_1C_1 of the input network. Parameters were measured under the same conditions.

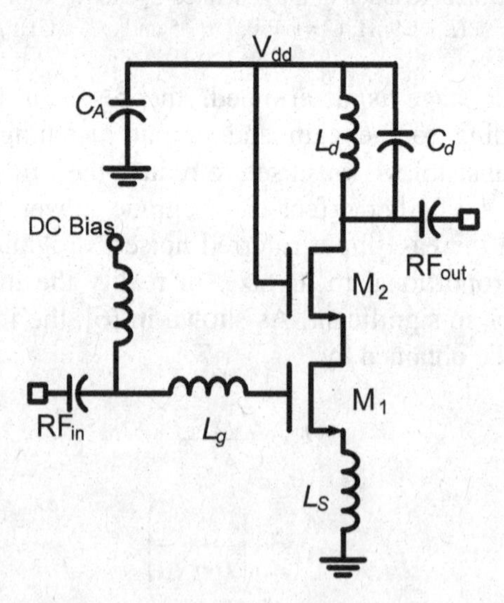

Figure 3.21 A 2.4GHz LNA2 with traditional source inductive degeneration input stage.

Measurement results including power gain, input reflection coefficient, output reflection coefficient, reverse isolation coefficient and noise figure are illustrated and compared in Figure 3.22 and Figure 3.23. It can be seen that with the removal of L_s, LNA1 still achieves an S_{11} of less than -14.4dB over the frequency range of 2.4GHz to 2.5GHz.

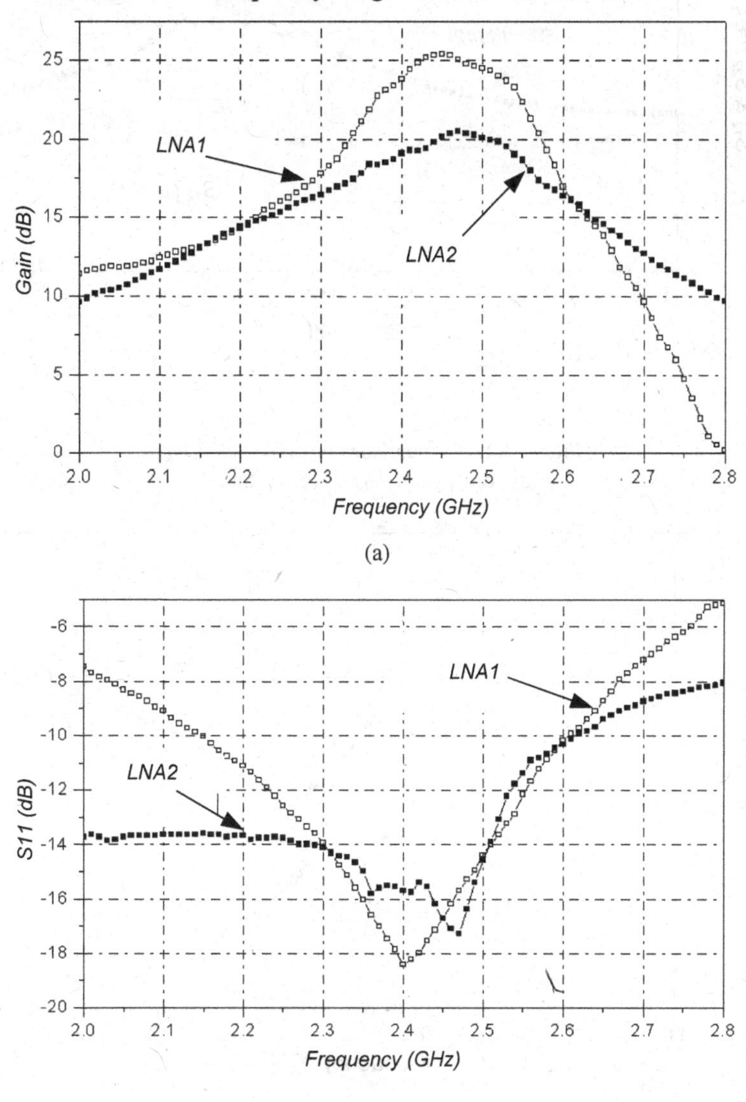

(a)

(b)

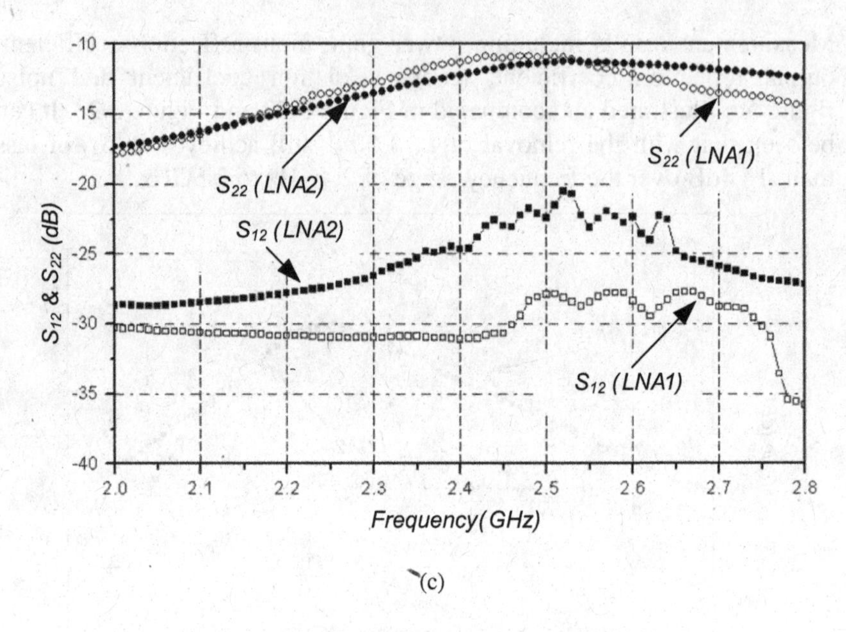

(c)

Figure 3.22 Comparisons of LNA1 and LNA2:
(a) power gain; (b) S_{11}; (c) S_{12} and S_{22}.

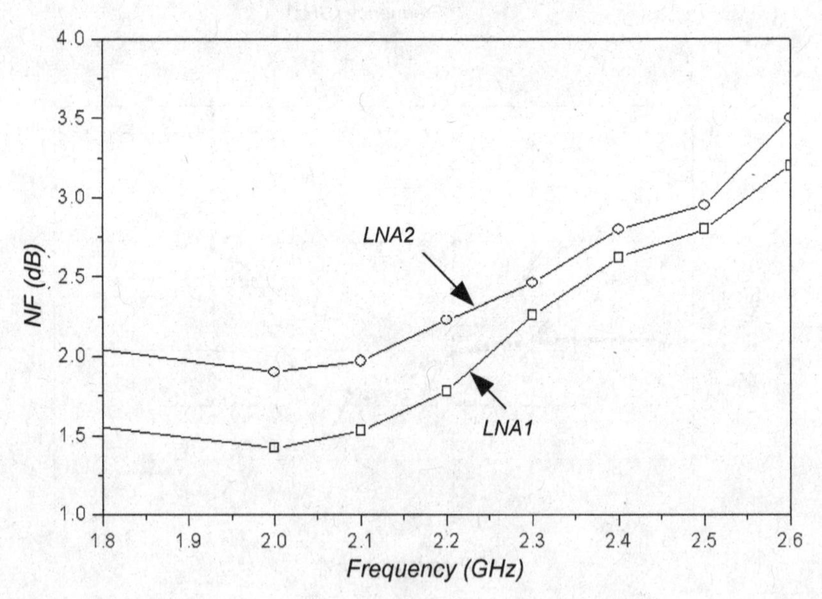

Figure 3.23 *NF* comparisons of LNA1 and LNA2.

The results also indicate that LNA1's power gain and noise figure performance have been improved compared to LNA2. It is suspected that this is the result of the removal of the lossy degeneration inductor L_s and the decrement of the lossy gate inductor L_g. L_s can significantly degrade LNA's gain in a normal topology [15]. From Figure 3.22(c), it can be seen that LNA1 has a better reverse isolation (S_{12}) compared to that of LNA2. There is a compatible output reflection coefficient because both LNAs adopt the same output structure and devices. The better performance for S_{12} of LNA1 is resulted from the removal of L_s.

Figure 3.24 shows the corresponding die photos of LNA1 and LNA2 respectively, where port vb is the bias for the on-chip testing buffer.

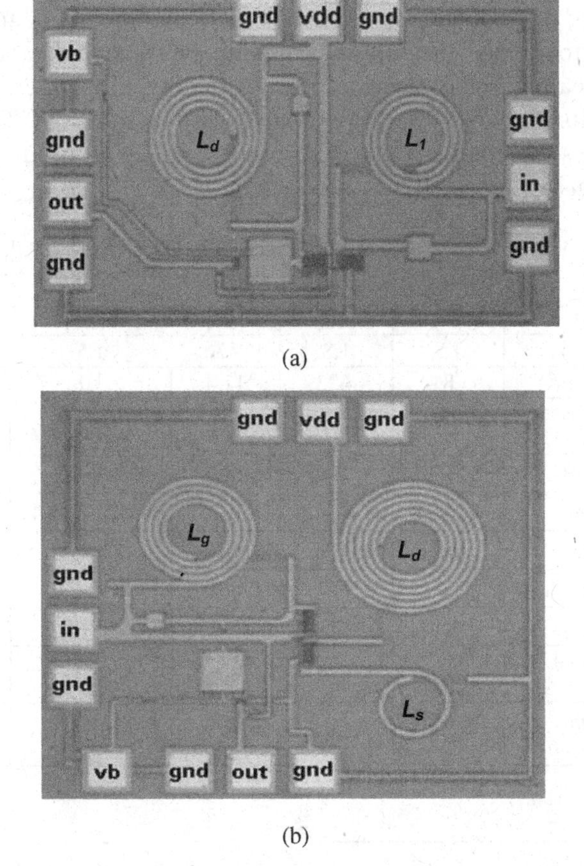

Figure 3.24 Die photos of (a) LNA1; (b) LNA2.

The detailed chip area consumption is illustrated in Table 3.3.

Table 3.3 Chip area consumption comparisons of LNA1 and LNA2.

Parameter	LNA1		LNA2	
	L_1	C_1	L_g	L_s
Value	1.9nH	1.4pF	6nH	1nH
Size	125μm	35μm	170μm	110μm
Area (mm^2)	0.0415	0.0012	0.0908	0.038
Sum (mm^2)	0.0427		0.1288	
Final Chip area (mm^2)[1]	0.35×0.6=0.21		0.53×0.63=0.3339	
Saved area (%)	37		0	

The size of the inductors is the radius including the 50μm space outside, which is to reduce the mutual coupling between inductors and other devices nearby, so as to guarantee its quality [5]. Other performance specifications of LNA1 and LNA2 are shown in Table 3.4. The table also summarizes the comparisons of several recent LNAs used for Bluetooth applications.

Table 3.4 Measured results of recent CMOS LNAs at 2.4GHz.

Parameter	[7]	[16]	[17]	LNA1	LNA2
Vdd (V)	1.8	1.8	N.A.	1.5	1.5
Current (mA)	8	4.23	11.4	10	10
Operating Band (GHz)	2.4	2.4	2.4	2.4	2.4
Gain (dB)	7.5	7	14.7	24 – 25	19 – 20.5
S_{11} (dB)	-12 – -8	N.A.	<-7	-18.5 – -14.4	-17.2 – -14.5
S_{22} (dB)	-19	N.A.	<-25	-12 – -11	-11
S_{12} (dB)	N.A.	N.A.	<-10	-31 – -27.5	-25 – -22
NF (dB)	4.5	3	2.88	2.62 – 2.8	2.8 – 2.95
Chip Area (mm^2)	N.A.	0.79	N.A.	0.21	0.334
CMOS Process (μm)	0.18	0.18	0.25	0.18	0.18

[1] This is the effective chip area without the pad region. With padding, the chip area is 0.4mm^2 and 0.56mm^2 respectively.

The two tables indicate that LNA1 with the parallel *LC* input topology consumes less chip area while keeping an equivalent or better performance compared to other fully integrated 2.4GHz LNA designs. The test results verified the advantages of the parallel *LC* input architecture in the single-band LNA design.

3.5 Summary

In this chapter, the basic concepts and definitions of parameters in LNA design were explained. Subsequently, four commonly used input architectures in LNA designs were classified and examined. This was followed by the input matching analysis of single-band LNA. In addition, the parallel *LC* network was used to replace the large gate inductor, which is normally required for conventional input architecture LNA. A narrow-band CMOS LNA (LNA1) was designed based on the parallel *LC* network input architecture. The results show that the LNA achieves acceptable input reflection coefficients and presents very good gain performance at relatively low power consumption.

References

[1] W. Sheng, A. Emira, E. Sanchez-Sinencio, "CMOS RF Receiver System Design: A Systematic Approach", *IEEE Transactions on Circuits and Systems – I: Regular Papers*, vol. 53, pp. 1023-1034, 2006.

[2] F. Gatta, E. Sacchi, F. Svelto, P. Vilmercati, and R. Castelloo, "A 2-dB noise figure 900-MHz differential CMOS LNA," *IEEE Journal of Solid-State Circuits*, vol.36, No.10, pp.1444-1452, October 2001.

[3] D.K. Shaeffer and T. H. Lee, "A 1.5-V, 1.5-GHz CMOS low noise amplifier," *IEEE Journal of Solid-State Circuits*, vol. 32, No.5, pp.745-759, May 1997.

[4] Y. Tsividis, *Operation and Modeling of the MOS Transistor*, 2nd Edition, McGraw-Hill press, 1999.

[5] 0.18μm CMOS Process Design Manual, Chartered Semiconductor Manufacturing, 2001.

[6] S. X. Mou, J. G. Ma, K. S. Yeo and M. A. Do, "A Modified Architecture Used for Input Matching in CMOS Low-Noise Amplifiers," *IEEE Transactions on Circuits and Systems – II,* vol. 52, No. 11, pp. 784–788, November 2005.

[7] K. Yamamoto, T. Heima, and A. Furukawa, "A 2.4-GHz-band 1.8-V operation single-chip Si-CMOS T/R-MMIC front-end with a low insertion loss switch," *IEEE Journal of Solid-State Circuits,* vol. 36, pp. 1186–1197, August 2001.

[8] P. Andreani and H. Sjolland, "Noise optimization of an inductively degenerated CMOS low noise amplifier," *IEEE Transactions on Circuits and Systems-II: Analog and Digital Signal Processing*, vol.48, No.9, pp.835-841, September 2001.

[9] T. Manku, "Microwave CMOS-device physics and design," *IEEE Journal of Solid-State Circuits,* vol. 34, pp. 277–285, March 1999.

[10] R.E. Lehmann and D.D. Heston, "X-band monolithic series feedback LNA," *IEEE Transactions on Electron Devices*, vol.ED-32, No.12, December 1985.

[11] P. Leroux, J. Janssens, and M. Steyaert, "A 0.8-dB NF ESD-protected 9-mW CMOS LNA operating at 1.23GHz," *IEEE Journal of Solid-State Circuits*, vol.37, No.6, pp. 760-765, June 2002.

[12] F. Bruccoleri, A.M.K. Eric, and B. Nauta, "Wide-band CMOS low-noise-amplifier exploiting thermal noise cancelling," *IEEE Journal of Solid-State Circuits*, vol.39, No.2, pp. 275–282, February 2004.

[13] S. Hyvonen, K. Bhatia, and E. Rosenbaum, "An ESD-protected, 2.45/5.25-GHz dual-band CMOS LNA with series LC loads and a 0.5-V supply," IEEE RFIC Symposium Technique Digest, pp. 43–46, 2005.

[14] A. van der Ziel, *Noise in Solid State Devices and Circuits*, John Wiley & Sons, Inc, New York, 1986.

[15] K. Beom Kyu, and L. Kwyro, "A Comparative study on the various monolithic low noise amplifier circuit topologies for RF and microwave applications," *IEEE Journal of Solid-State Circuits,* vol. 31, pp. 1220-1225, August 1996.

[16] F. Krug, P. Russer, F. Beffa, W. Bachtold, and U. Lott, "A switched-LNA in 0.18μm CMOS for Bluetooth applications," *Proceedings of Silicon Monolithic Integrated Circuits in RF Systems*, pp.80–83, April 2003.

[17] R. Point, M. Mendes, and W. Foley, "A differential 2.4GHz switched-gain CMOS LNA for 802.11b and Bluetooth," *Proceedings of IEEE Radio and Wireless Conference*, pp.221–224, August 2002.

Read to learn more.

Learn to think better.

Think to understand further.

Kiat Seng YEO

CHAPTER 4

RF Mixers

4.1 Introduction

The mixers in the transmitter and the receiver perform the frequency up and down conversions respectively by multiplying the informative signal either in the baseband, or the intermediate frequency (IF) band, or the radio frequency (RF) band with the signals from the local oscillator (LO). Ideally the mixer shifts the input frequency to another frequency according to equation (4.1):

$$f_{out} = \pm f_{LO} \pm f_{in} \tag{4.1}$$

Equation (4.1) can be easily obtained by the multiplication of two sine waves, i.e. $(sin2\pi f_{LO}t)$ x $(sin2\pi f_{in}t)$, hence the term multiplying mixer is used. Multiplying mixers are commonly used in CMOS implementation. Similar mixing products given by equation (4.1) can also be obtained by adding the two sine waves then applying the sum through a non-linear device, e.g. a diode, with a strong content of the second-order distortion.

The second-order term, $(sin2\pi f_{LO}t + sin2\pi f_{in}t)^2$, will also generate the product term, $2\ sin2\pi f_{LO}t\ sin2\pi f_{in}t$ and create the frequency translation as described by equation (4.1). The non-linear mixer is commonly used at much higher frequencies of tens of GHz when it is more difficult to implement the multiplying mixer, which has less spurious output frequencies.

In equation (4.1), f_{in} is the centre frequency of the informative signal. The spectrum of the informative signal can be shifted up or down by $\pm f_{LO}$ to the wanted frequency band or channel with a suitable selection of the LO frequency. The up-conversion is used in the transmitter while the down-conversion is commonly used in the receiver. The following example demonstrates effects of equation (4.1) in four different scenarios of up and down frequency conversion.

Example

The mixer of a Direct-Conversion transmitter is designed to shift the baseband signal of the bandwidth B = 500kHz to an RF signal in the ISM (Industrial, Scientific and Medical) band from 2400 to 2483MHz. As the centre frequency of the baseband signal is zero, we need f_{LO} to be selectable over the range from 2400 to 2483MHz. The RF signal at the mixer output will have a bandwidth of $W = 2B$ = 1MHz due to the addition and subtraction of f_{in} to and from f_{LO}.

If this transmitter is now designed according to a heterodyne topology with an IF of 70MHz, then the IF signal would have a centre frequency of f_{in} = 70MHz and a bandwidth of $W = 2B$ = 1MHz. To translate the IF signal from 70MHz to the RF signal in the (2400-2483) MHz band with a mixer, we have two possible choices of selecting f_{LO}. The first method is called low-side injection and f_{LO} operates in the range from 2330 to 2413MHz, which is 70MHz below the RF signal range. Image frequencies are also generated in this case over the frequency range from 2260 to 2343MHz, or 140MHz below the RF signal range. The second method is called high-side injection and f_{LO} operates in the range from 2470 to 2553MHz, which is 70MHz above the RF signal range. Image frequencies are also generated in this case over the frequency range from 2540 to 2623MHz, or 140MHz above the RF signal range.

For the Direct Conversion receiver, we need f_{LO} to be selectable over the range from 2400 to 2483MHz to down-covert the RF signals to the baseband signals. A low pass filter with the bandwidth of B = 500kHz is used to select the channel.

For the heterodyne receiver with the IF of 70MHz, we have the choices of using either low-side injection or high-side injection as described in 2 to down-convert the RF signals to the IF frequencies. An image reject filter before the mixer is required to suppress the corresponding image frequencies.

In practice, due to the non-linearity of the device or large signals used in mixers, there are also intermodulation products determined by:

$$f_{out} = mf_{LO} + nf_{in} \tag{4.2}$$

where m and n are ± integers. The sum of the absolute values of m and n is the order of the output cross product. For example, ($2f_{LO} - f_{in}$) is a third order cross product. A feed-through of f_{LO} occurs for $n = 0$ and a feed-through of f_{in} for $m = 0$. It is necessary that these intermodulation products do not fall in the same frequency band as that of the wanted f_{out}.

Figure 4.1 shows a simple mixer implemented with a single transistor, where the received signal from the LNA is multiplied with a switching waveform, $S(t) = [0, 1]$ @ f_{LO}. To simplify the discussion we will assume $S(t)$ has a 50% duty cycle. As the fundamental frequency component of $S(t)$ is $(2/\pi)\sin 2\pi f_{LO}t$, and the mixer output is split into two components, $f_{LO} \pm f_{in}$, the voltage gain of this mixer for one frequency component is $1/\pi$, assuming the voltage dropped across the switch is negligible. The multiplying mixer is also called switching mixer in this case. If the LNA and the channel select filter in Figure 4.1 are differential circuits, four transistors can be arranged as two differential toggle switches implement the differential switching mixer. The switching waveform becomes $S(t) = [1, -1]$ @ f_{LO}, and the voltage gain is $2/\pi$ assuming the switches are ideal. Usually, the MOS device acts as a switch which is biased to turn ON and OFF as abruptly as possible. In the switching mixer, the ON resistance of the MOS switch contributes noise. As the gate-source voltage changes, the ON resistance of the switch also changes. This introduces non-linearity in the voltage division between the ON resistance and the load resistance of the filter.

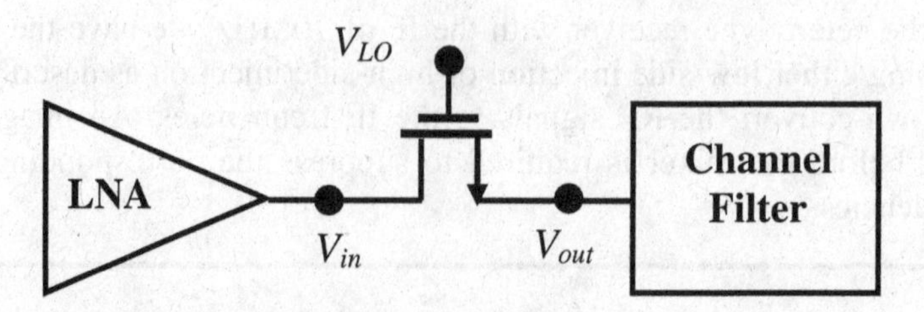

Figure 4.1 A simple mixer.

In RF system-on-chip design, power matching at the input and output of the mixer is no longer required as in conventional RF board level design. Hence the power gain is replaced by the voltage gain of the mixer, which is the ratio of the output voltage of the wanted component of the up or down converted signal to the input voltage, e.g. For the first down-conversion mixer of a receiver, the input voltage is the RF signal voltage while the output voltage is the IF or base-band signal voltage. The amplitude of the LO signal is usually between 200 and 400mV_p and current loading is not a problem as in the diode and bipolar mixers. The voltage gain of the CMOS mixer is usually proportional to the amplitude of the LO signal. The effects of its noise and linearity performance in the voltage mode calculation are described in Chapter 1.

The noise figure of a mixer is usually considered as dependent on the receiver architect. If a heterodyne architect is employed, noise in both the desired channel and its image are superimposed at the intermediate frequency, so noise at the mixer output is increased by at least 3dB even if its circuit is noiseless. The corresponding noise figure is referred to as single-side-band (SSB) noise figure. In a homodyne receiver, the image problem is avoided, so the output noise is not increased if the circuit is noiseless. The corresponding noise figure is referred to as double-side-band (DSB) noise figure and is considered as 3dB lower than the former value. This simplistic view however, does not take into account the effect of the image rejection circuit employed in the heterodyne receiver, and the problem of low frequency noise in the homodyne receiver, as mentioned earlier in Chapter 1.

4.2 Common Configurations of Active Mixers

Active mixers reduce the contribution of the noise factors in subsequent stages of the receiver, so they are widely used when the receiver sensitivity is critical. Figure 4.2 shows the basic configurations of (a) a single-balanced active mixer, and (b) a double- balanced active mixer, which is also referred to as the Gilbert cell. One obvious problem of the single-balanced mixer is the direct feed-through of the LO signal to the output port by the differential pair M_1 and M_2. The large LO signal will desensitize the converted output signal. Thus the LO signal has to be suppressed at the output port. For the double-balance mixer, the two differential pairs M_2-M_3 and M_4-M_5 operate at opposite phases of the LO signal and cancel each other out at the output port. The additional advantage of the double-balanced mixer in reducing the even order distortion has made it the popular choice in many designs.

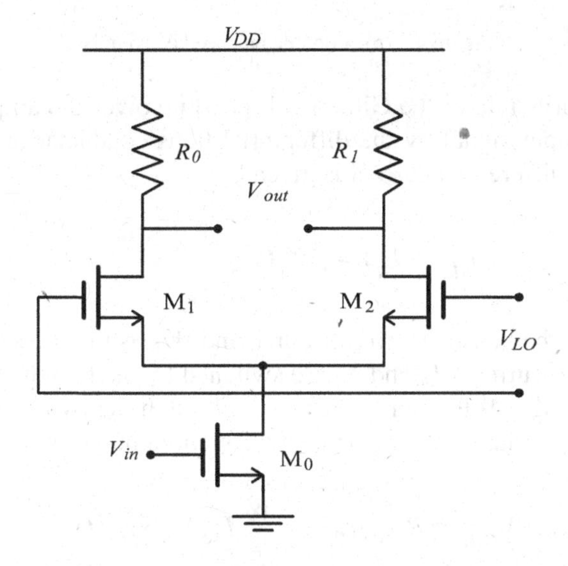

Figure 4.2 (a) A single balanced active mixer.

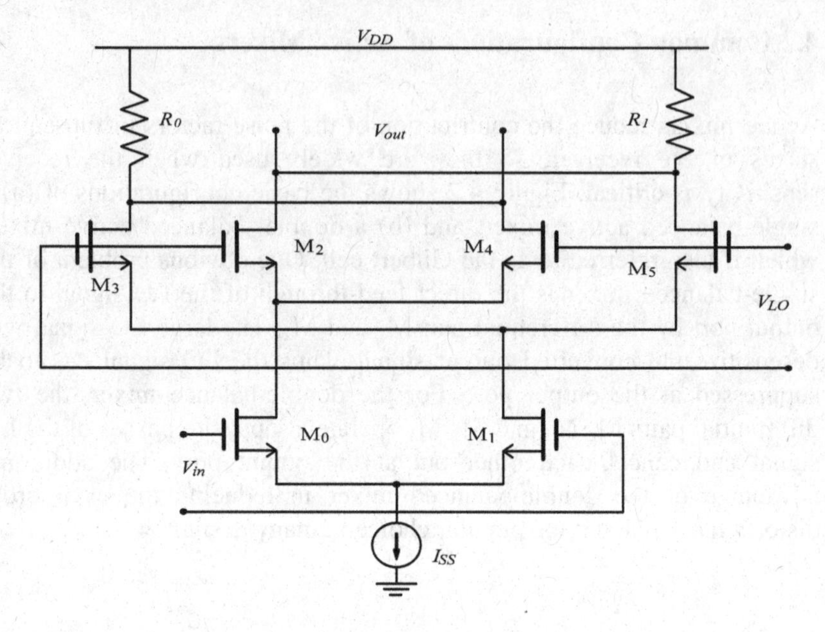

Figure 4.2 (b) A double balanced active mixer.

The basic principle of the Gilbert cell [1-3] involves the amplification of the small input signal by the differential trans-conductance pair M_0-M_1. The output differential current is given by:

$$I_{d0} - I_{d1} = \sqrt{\mu_n C_{ox} \frac{W}{L} I_{SS}} V_{in} \tag{4.3}$$

where I_{SS} is the constant current source and W/L is the aspect ratio of M_0-M_1. The two currents I_{d0} and I_{d1} are switched by the two differential pairs M_2-M_3 and M_4-M_5 to alternatively pass through the two resistors R_0 and R_1 in opposite phases and generate a differential output voltage given by:

$$V_{out} = R_L \sqrt{\mu_n C_{ox} \frac{W}{L} I_{SS}} V_{in} S_{LO}(t) \tag{4.4}$$

where $R_L = R_0 = R_1$, M_2-M_5 are assumed as ideal switches and $S_{LO}(t)$ is the switching function equal to ± 1. Equation (4.4) shows the multiplication of the input voltage with a square wave of unit amplitude. If we consider only the fundamental frequency of the LO with the amplitude of 4/π then the single-sided voltage gain of the mixer is given by:

$$A_v = \frac{2}{\pi} R_L \sqrt{\mu_n C_{ox} \frac{W}{L} I_{SS}} \tag{4.5}$$

Equation (4.5) can be simplified as $A_v = \frac{2}{\pi} R_L G_m$, where G_m is the trans-conductance of the differential amplifier for small V_{in}. Equation (4.5) is also valid for the mixer without the constant current source, but in this case I_{SS} is replaced by the total bias current in M_0 and M_1, i.e. $I_{SS} = 2I_D$, where I_D is the bias current in M_0 and M_1.

Assuming that only two out of four switching transistors are conducting in the saturation region at anytime, the input referred noise spectral density and the noise figure of the mixer are given by equations (4.6) and (4.7) respectively [3].

$$\overline{v_{ni}^2} = 8kT \frac{1}{g_{mRF}^2 R_L} + \frac{16}{3} kT \frac{1}{g_{mRF}} + \frac{8}{3} kT \frac{g_{mlo}}{g_{mRF}^2} + \frac{4K_N}{g_{mRF}^2 W_{lo} L_{lo} C_{ox} f} \tag{4.6}$$

$$NF = 10\log(1 + \frac{\overline{v_{ni}^2}}{4kTR_s}) \tag{4.7}$$

where the *RF*, *lo* and *L* subscripts denote that the parameters belong to the input common source transistors (M_0-M_1), the switching pair transistors (M_2-M_5), and the source and load resistors respectively. K_N is the $1/f$ noise coefficient of NMOS transistor. The gain and the noise equations are derived based on long channel MOS model, hence the non-ideality caused by short channel devices are ignored.

As the output signal of the direct conversion mixer is at low frequencies, the flicker noise is dominant. In choosing the transistor size, the aspect ratio of the switching pair transistors should be increased to reduce noise and to increase the conversion gain. Increasing the bias current also helps in reducing noise. The LO bias voltage is carefully chosen so that v_{lo} can exactly turn on/off the M_2-M_5. In this way, the conversion gain is maximized. The RF bias voltage is used to obtain the drain currents of M_0 and M_1.

Example

A CMOS double-balanced Gilbert cell mixer operating at the center frequency of 2.45GHz is designed by Cadence SpectreRF using the Chartered 0.18um CMOS Digital/Analog/RF Technology file as in Figure 4.3. The switching transistors are biased at the edge of saturation. The circuit parameters and biasing conditions are tabulated in Table 4.1. Apply equation (4.5) with $\mu_n C_{ox}$ = 130μA/V^2 for this technology, I_{SS} = $2I_D$ = 4.46mA, W/L = 60/0.18 = 333.33, and R_L = 600Ω, we have A_v = 5.3 or 14.5dB.

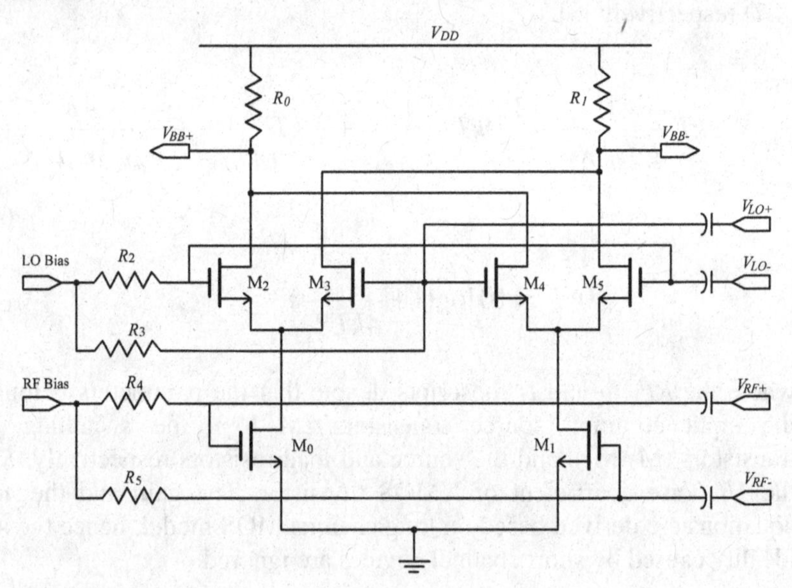

Figure 4.3 A design example of a Gilbert cell without current source.

Table 4.1 Mixer circuit parameters and biasing conditions.

Parameters	Values
W/L (M0-M1)	60μm/0.18μm
W/L (M2-M5)	100μm/0.18μm
R0-R1	600Ω
R2-R5	5kΩ
Voltage source	2.5V
DC bias at LO input	1.35V
DC bias at RF input	650.2mV
M0 drain current	2.23mA

Figure 4.4(a) shows the conversion gain of the mixer when the LO is fixed at 2.45GHz while varying the RF frequency from 2.4GHz to 2.5GHz. The mixer has a flat gain of 15.3dB across the simulated frequency range. Figure 4.4(b) shows the gain when RF is swept from 1.45GHz to 3.45GHz, we can see that at point A and point B, the gain is 3dB lower than the peak value.

In order to investigate the mixer bandwidth for homodyne receiver applications, the RF frequency is set at the same value as the LO frequency and both are swept together. In this case, IF frequency is always 0Hz. In Figure 4.4(c), when the frequency increases to 9GHz, the gain decreases to 12.35dB, which is about 3dB lower than the gain at 2.45GHz.

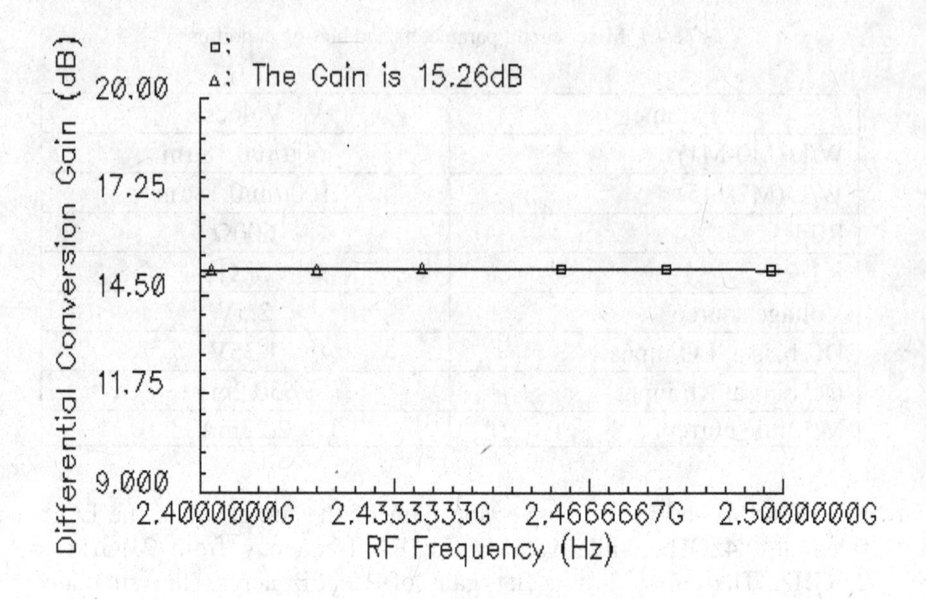

Fig. 4.4 (a) Differential conversion gain.

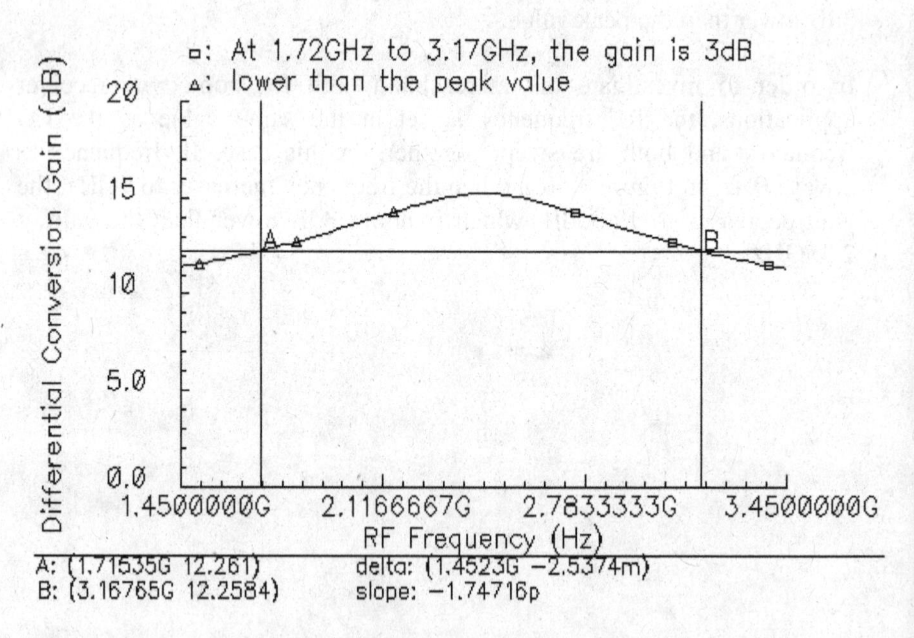

Fig. 4.4 (b) Bandwidth at f_{LO} = 2.45GHz.

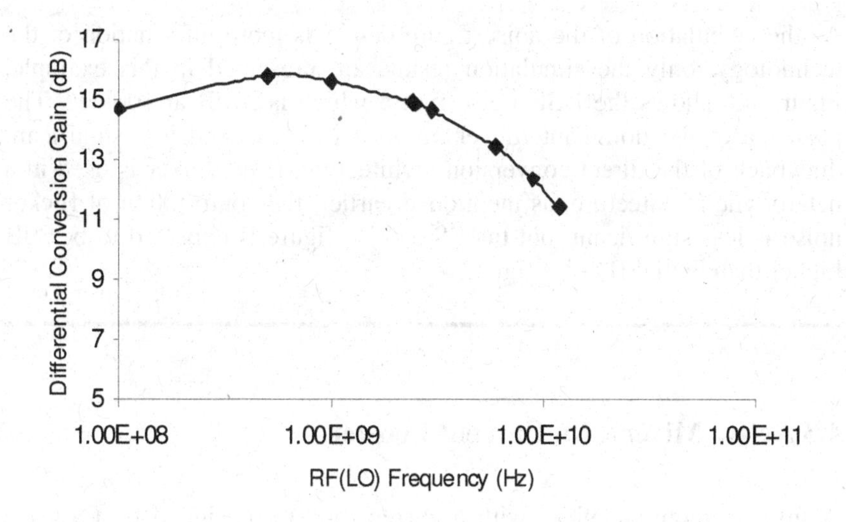

Fig. 4.4 (c) Conversion gain in the homodyne receiver.

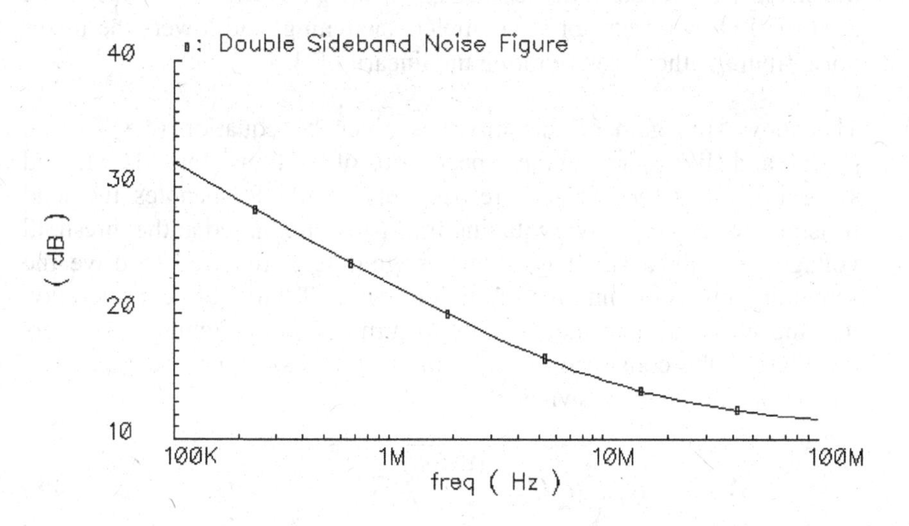

Fig. 4.5 Double sideband noise figure.

As the calculation of the noise figure demands more information of the technology, only the simulation results are presented in this example. Figure 4.5 shows the DSB noise figure which is 32dB at 100kHz. The presence of the dominant flicker noise at the baseband is a significant drawback of the direct conversion architecture. If the mixer is used in a heterodyne architecture, as mentioned earlier, the contribution of flicker noise is less significant, but the SSB noise figure is expected to be 3dB higher than the DSB noise figure.

4.3 Active Mixer with Current Booster

A fully differential mixer with a current booster designed by Cadence SpectreRF as in Figure 4.6 allows independent controls of the currents in the RF transistors and switching transistors. The current boosting PMOS transistor pair relaxes the low supply voltage condition, reduces the required LO overdrive for the complete switching, and lowers the mixer noise figure without deteriorating the linearity [3].

The conversion gain of this mixer is given by equation (4.8), where $(W/L)_{rf}$ and $(W/L)_{lo}$ denote the aspect ratios of RF transistors M1-M2 and switching transistors M3-M6 respectively; while R_L denotes the load resistors, R1 and R2. The switching transistors are biased at the threshold voltage, so only a small v_{lo} swing magnitude is required to drive the switching transistors into the saturation region. Taking into consideration that the v_{lo} swing may not be able to turn off the switching transistors completely, the conversion gain of the mixer is shown to be dependent on the magnitude of v_{lo} swing.

$$A_v = \frac{2}{\pi} R_L \sqrt{\mu_n C_{ox} (\frac{W}{L})_{lo} \mu_n C_{ox} (\frac{W}{L})_{rf}} \, v_{lo} \tag{4.8}$$

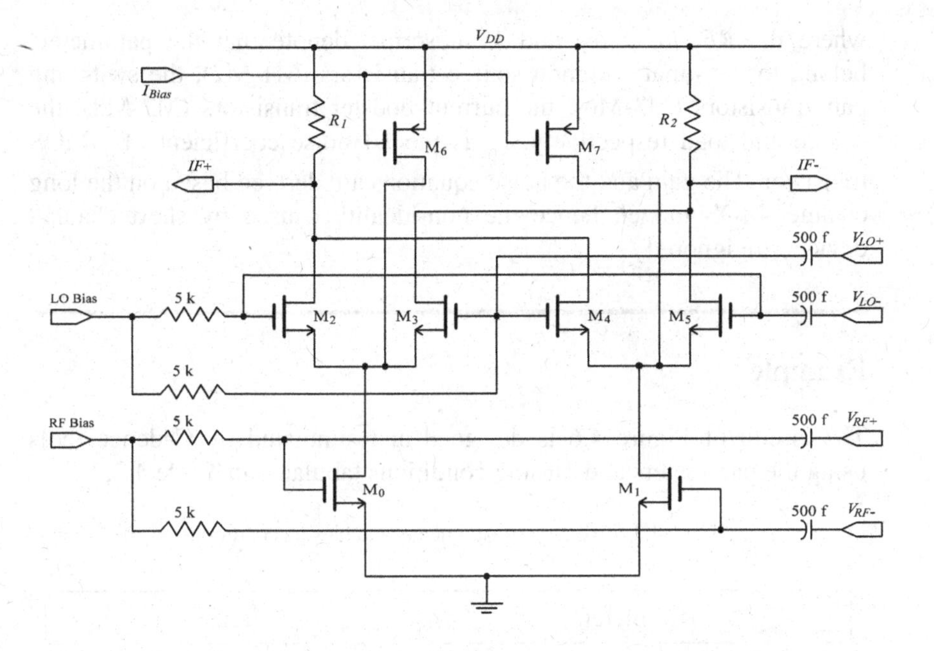

Figure 4.6 Double balanced differential mixer with current booster.

Assuming that only two out of four switching transistors are conducting in the saturation region at anytime, the input referred noise spectral density of the mixer is given by equation (4.9) while the noise figure is given by equation (4.7).

$$\overline{v_{ni}^2} = 8kT\frac{1}{g_{mRF}^2 R_L} + 2.467\left(\frac{16}{3}kT\frac{g_{mboost}}{g_{mRF}^2} + \frac{16}{3}kT\frac{1}{g_{mRF}}\right) + \frac{8}{3}kT\frac{g_{mlo}}{g_{mRF}^2} + \frac{4K_N}{g_{mRF}^2 W_{lo} L_{lo} C_{ox} f} \quad (4.9)$$

where the *RF*, *lo*, *boost* and *L* subscripts denote that the parameters belong to the input common source transistors (M1-M2), the switching pair transistors (M3-M6), the current booster transistors (M7-M8), the source and load respectively. K_N is the 1/f noise coefficient of NMOS transistor. The gain and the noise equations are derived based on the long channel MOS model, hence the non-ideality caused by short channel devices are ignored.

Example

The circuit of Figure 4.6 is designed and simulated by Cadence tools using the parameters and biasing conditions tabulated in Table 4.2.

Table 4.2 Mixer circuit parameters and biasing conditions.

Parameters	Values
W/L (M_1-M_2)	60μm/0.18μm
W/L (M_3-M_6)	100μm/0.18μm
W/L (M_7-M_8)	60μm/0.18μm
R_1-R_2	1kΩ
DC bias at RF input	650mV
DC bias at LO input	1.35V
DC bias for current booster	1.78V
Voltage source	2.5V
M_1 drain current (tail current)	2.5mA
M_7 drain current (injected current)	1.4mA

Apply equation (4.8) with $\mu_n C_{ox}$ = 130μA/V^2, $(W/L)_{rf}$ = 60/0.18, $(W/L)_{lo}$ = 100/0.18, v_{lo} = 0.2V, and R_L = 1000Ω, we have A_v = 7.1 or 17.0dB.

Figure 4.7 shows the conversion gain of a mixer when the LO is fixed at 1.775GHz and the input frequency is varied from 1.725GHz to 1.825GHz. It shows that the mixer has a flat gain across the simulated frequency range. This may suggest the mixer is suitable for use in the wideband image reject down-converter. Table 4.3 shows the mixer performance specifications obtained from the simulation.

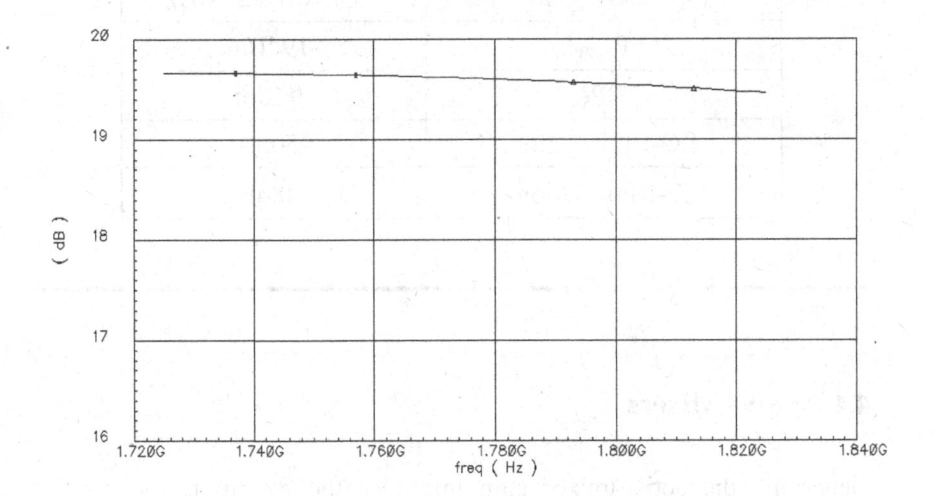

Figure 4.7 Mixer Conversion gain at around 1.8GHz.

The current injection technique is applied to keep the switching current in the cross-coupled transistors low, hence reducing the flicker noise at the baseband or low IF outputs. If the DC currents in the switching transistors are reduced to zero then we approach the condition of the passive mixers.

Table 4.3 Mixer specifications.

Parameters	Simulation
Conversion gain	19.5dB
Power Consumption	11.5825mW
Noise Figure	13.43 dB@25MHz
$P_{1\text{-dB}}$	-19dBm
IIP_3	-9dBm
LO-RF isolation	-50dB
IF-RF isolation	-48dB

4.4 Passive Mixers

Generally, the active-mixer gain improves the overall receiver noise figure at the expense of linearity and speed obtained in passive mixers. The distinct advantages of CMOS passive mixers are the zero power consumption and low noise figure. MOS transistors can turn on without a bias current as long as $V_{gs} > V_{th}$. In recent years, double-balanced passive mixers have become popular in low IF and direct conversion receivers due to their better linearity performance and lower 1/f noise in the absence of dc currents through the switches transistors [4-6]. Hence a higher gain is required for the low noise amplifier.

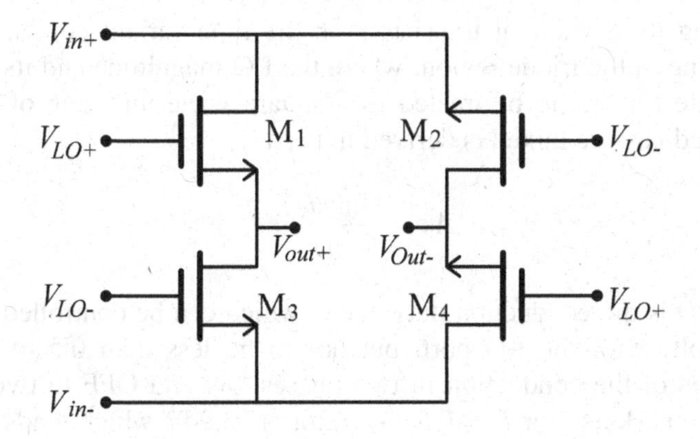

Figure 4.8 (a) Common forms of double balanced passive mixers.

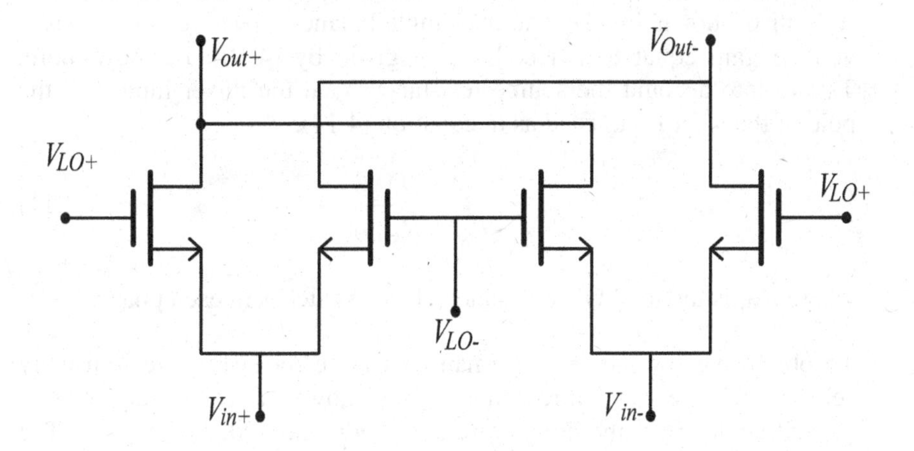

Figure 4.8 (b) Common forms of double balanced passive mixers.

The configuration of the double-balanced passive mixer can be expressed in different forms as shown in Figure 4.8(a)-(b). Note that these forms are identical to the cross connection switching transistor core of the active double balanced mixers.

Viewing the switching transistors as the time-variant conductance *g(t)* operating in the triode region, where the LO magnitude and its dc bias is adequate for *g(t)* to be treated as a square wave, the gain of a double-balanced passive mixed is derived in [4] as:

$$A_v = \frac{\sin(\pi D)}{\pi D} \tag{4.10}$$

where D is the conduction duty cycle, which can be controlled by the dc bias voltage of the LO port, but has to be less than 0.5 to avoid the overlaps of the conduction in two phases ON and OFF in two paths of the mixer. Hence for $D = 0.5$, $A_v = 2/\pi$ or -3.9dB which tends to 1 as D becomes small. The mixer is terminated by a capacitive load C_L. The equivalent conductance of the switches, Dg_{max} and C_L form a first order low pass filter with $\omega_{3dB} = Dg_{max}/C_L$, which should be much larger than the down-converted frequencies of interest. Hence D cannot be too small. It is also shown in [4] that the single-balanced passive mixer has a voltage gain equal to twice the gain given by (4.10), i.e. 6dB more. Taking into account the source resistance R_s at the mixer input [6], the pole of the filter is adjusted as in equation (4.11).

$$\omega_{3dB} = \frac{2D}{(R_S + R_{SW})C_L} \tag{4.11}$$

where R_{SW} is the total ON resistance of two switches in each phase.

To obtain the low noise performance, a large transistor size is usually selected for the low on-resistance. This, however, can create a large capacitive load to the low noise amplifier and reduce its gain. The practical approach has been to increase the transistor size until the gain begins to drop. In [6], it is shown that the total noise density at the mixer output and the mixer noise factor are respectively given by:

$$\overline{v_{no}^2} = 4kT\,\frac{R_S + R_{SW}}{2D} \tag{4.12}$$

$$F = \frac{R_S + R_{SW}}{2DR_S}\left(\frac{\pi D}{\sin \pi D}\right)^2 \tag{4.13}$$

Hence the optimum D can be found by setting the derivative of F to zero. The optimum noise factor can be obtained at $D = 0.375$.

A major advantage of the passive mixer is the inherent linearity of MOS devices operated in the triode region. To obtain a good linearity for the mixer, it is important to keep the current through the switches and the voltage across the device small. Thus the small capacitive load must present a high impedance at the mixer output. The second source of distortion is due to the phase modulation of the mixing of the RF signals. A fast switching of the LO signal, e.g. square wave, or a large LO drive will improve the mixer linearity.

Example

A double balanced passive mixer is designed by Cadence SpectreRF according to Figure 4.8 using the Chartered 0.18μm CMOS process. The widths of all four transistors are 100μm. The threshold voltage of the device is 0.494V. The bias voltage is set at $V_{GS} = 0.45$V. The RF input signal is swept from 0.5 to 10GHz, while the LO signal is set at 1MHz higher than the frequency of the RF signal to produce an output signal of 1MHz. The LO signal voltage is 300mV_p.

A capacitive load of 10pF is connected across the mixer output to suppress all RF frequencies while presenting high impedance to the IF signal. Figure 4.9 shows the simulated results of the voltage gain and noise figure of the mixer. The results show a superior noise performance of the passive mixer as compared to the active mixer. The circuit also has the wideband characteristics suitable for the multiple band operation of the receiver.

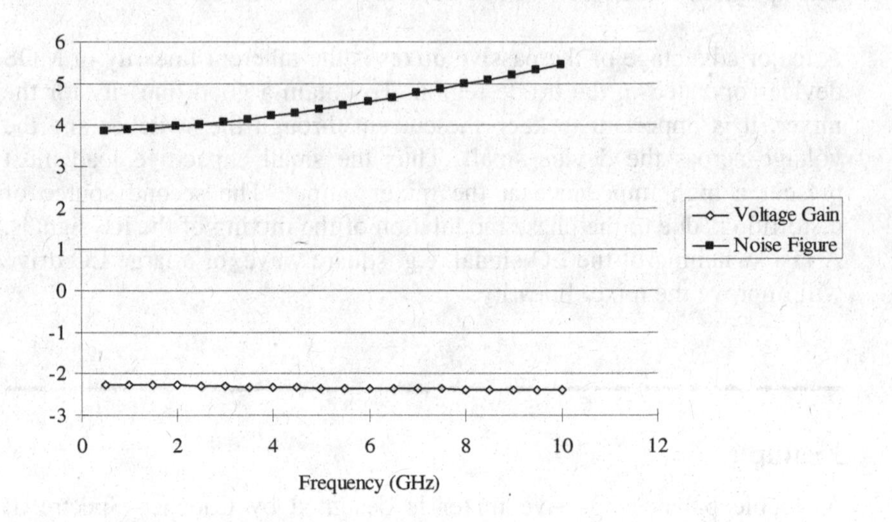

Figure 4.9 Voltage gain and noise figure of a passive mixer.

4.5 Port Isolation and DC Offset in Direct Conversion Mixers

Leakages of high frequency signals from one port to another are unavoidable due to one or more causes such as magnetic and capacitive couplings, conduction through substrate, and unbalanced structures of the circuits. The leakages of the LO signal to the input and output ports of the mixer are most dominant as the LO signal level could be about 60dB higher than that of the RF and down-converted signals. For a double-balanced down-conversion mixer, the leakage of the LO signal to the output port is mostly cancelled out in the mixing process, and attenuated by the low-pass filter characterized by equation (4.11). The channel select filter of the receiver subsequently suppress the LO leakage below the noise floor, thus the LO leakage to the output port seldom poses any problem. The LO leakage to the input port is, however, mixed with the LO input signal to produce a DC offset at the output port as illustrated by Figure 4.10. This DC offset is easily removed by the channel select filter in a heterodyne receiver. However, in a homodyne or direct conversion receiver, the DC offset could be much larger than the DC value of the down-converted base-band signal, causing saturation or heavy distortion of the base-band signal in the subsequent amplification stages. The problem could become more serious if the LO leakage is coupled to the antenna causing the variation of the DC offset with time [7].

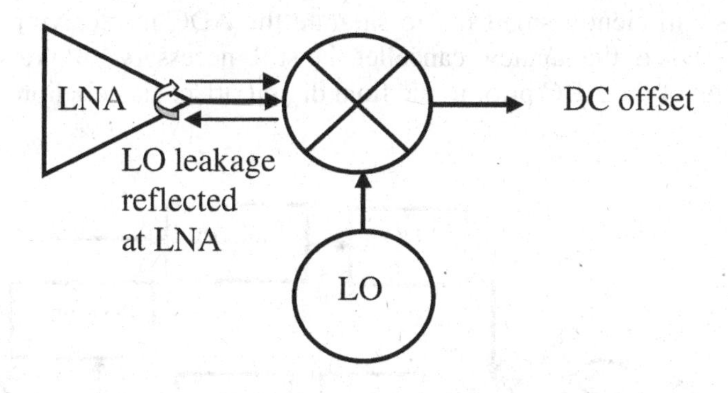

Figure 4.10 LO self mixing process.

Despite the DC offset problem in the direct conversion receiver, the advantage of implementing the fully integrated receiver without an off-chip image reject filter has made this architect popular in the recent years. Several techniques have been used to minimize or remove the DC offset:

A good balance of the circuit structure and a good shielding of the LO and input ports can minimize the LO leakage.

The AC coupling can remove the DC offset if the modulated base-band signal is DC free. However, to track the lower frequency spectrum of the signal, the AC coupling will require large capacitances that are not realizable on-chip [8]. The AC coupling using a large capacitor often fails to track fast variations in the offset voltage.

The offset can be detected and removed digitally by time averaging or by using more complex methods for differentiating the received signal [9, 10]. However, the digital cancellation technique requires analog baseband stages with large spurious-free dynamic ranges (*SFDR*) to tolerate the DC offset. It also requires more bits (as many as 5 or 6 more bits) in the ADC to achieve the same sensitivity and bit error rate as when the offset is not present.

The DC offset can be removed by the digital method only if the total DC value is sufficiently small not to saturate the ADC in receiving a large signal. Hence the analog canceller is still necessary to prevent the saturation of the ADC prior to the final digital offset cancellation.

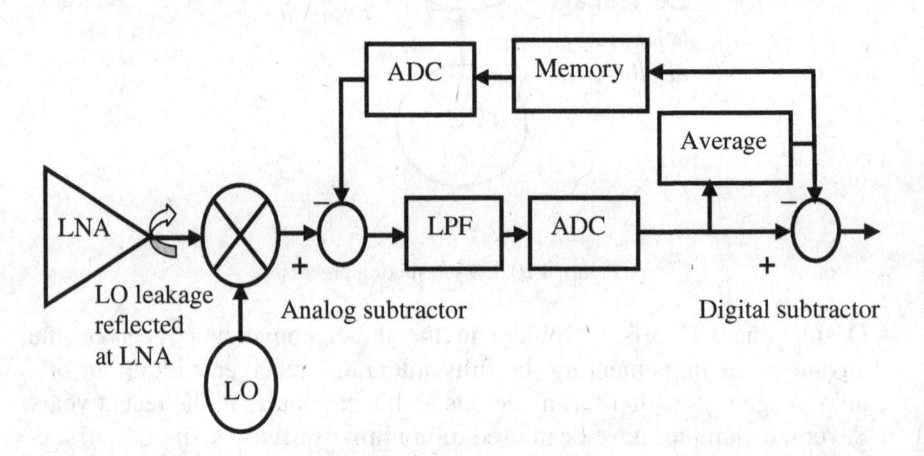

Figure 4.11 A DC offset canceller [11].

Figure 4.11 illustrates the DC offset canceller which includes both analog and digital subtractors [11]. If the DC offset is mainly time-invariant, an analog implementation of this technique should be sufficient to remove the DC offset component.

The DC offset arises because the RF carrier and the LO signal operate at exactly the same frequency. In an even harmonic mixer, the LO frequency is equal to half of the RF carrier, and it is the second harmonic of the LO signal that takes part in the mixing process, translating the RF signal to the base-band. As a result, the LO leakage does not generate a DC component but an output which is still situated at the LO frequency and can be easily filtered out [12, 13]. As the flicker noise is still the remaining problem for this technique, and complicated poly-phase LO waveforms are required, the even harmonic mixer is yet to be a practical technique. Figure 4.12 shows the design of an even harmonic mixer. Its principle is described in [13].

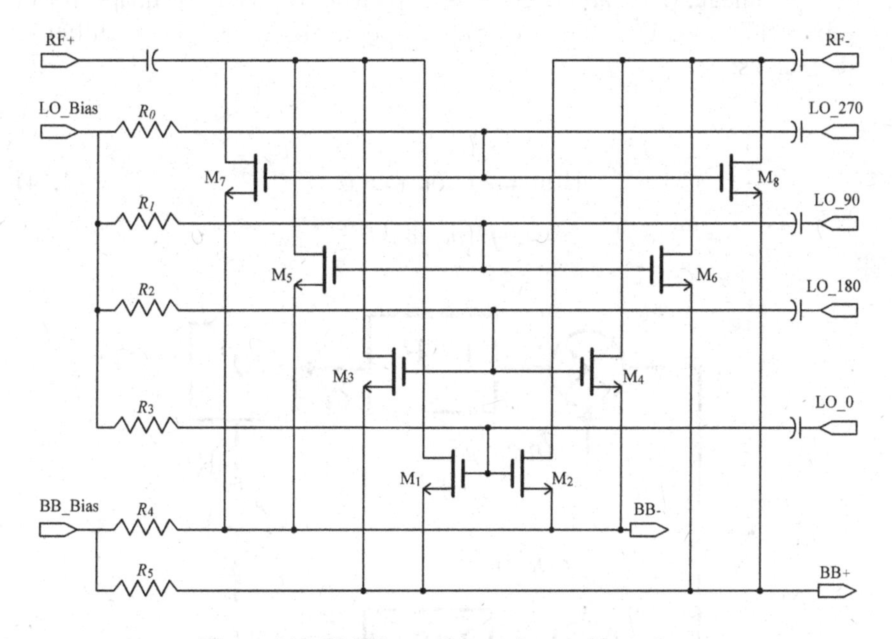

Figure 4.12 CMOS even-harmonic switching mixer.

4.6 Image Reject Mixers for Low IF Architectures

In applications where the requirement for image rejection is not too stringent, and the channel bandwidth is not too wide, the low IF technique is preferred. One of such example is the receivers designed for the Zig-bee standards or IEEE 802.15.4 for applications in the 2.4GHz band [5, 14, 18]. The low IF is selected is selected between 1 and 2MHz so that the DC offset is removed by the channel select filter, the IF circuit consumes less power, and the low IF output signal can be easily digitized. The RF signal is down-converted into two low IF signal paths, namely I and Q, by the quadrature LO signals. The I and Q signals are then phase-shifted by 90° relatively to each other and combined to remove the replica image. This architecture is known as the Hartley image reject mixer (Figure 4.13). The frequency response of the 90° phase shifter is defined as:

$$H(\omega) = -j \text{ for } \omega > 0 \qquad (4.14)$$
$$= +j \text{ for } \omega < 0$$

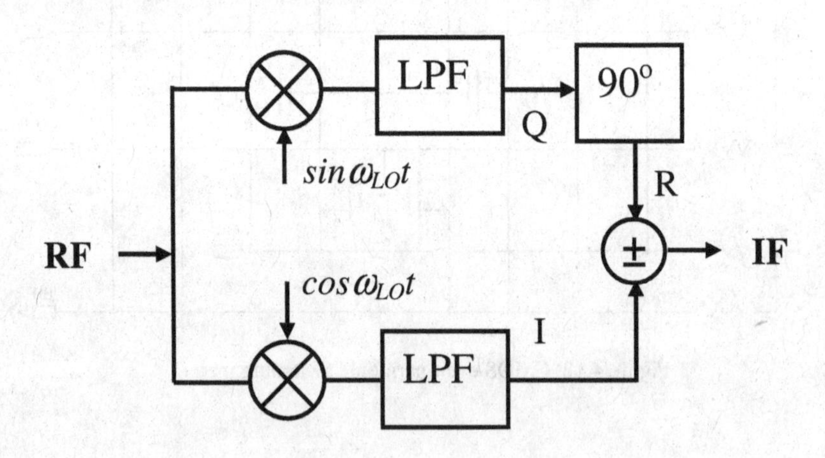

Figure 4.13 Hartley image reject mixer.

Assuming now that the RF signal is accompanied by an image signal as illustrated in Figure 4.14, and the RF signal is down-converted to the IF signal by the low injection of the local oscillator.

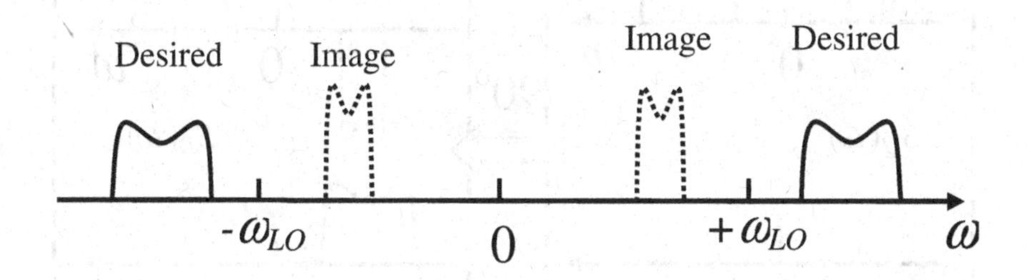

Figure 4.14 RF Signal with an image on lower sideband.

Figure 4.15 illustrates the image cancellation process of the Hartley mixer. The spectra of the signals in Figure 4.14 are shifted up and down by the injection of the local oscillator frequency. After the low-pass filtering, the signal spectra in the I and Q paths are shown as $S_I(\omega)$ and $S_Q(\omega)$. The Q path signal is then 90° shifted to produce the spectra $S_R(\omega)$ which have the image spectra inverted. Thus, the desire upper sideband spectra can be isolated and the image spectra cancelled at the output after the summation of the signals in the two paths. This principle also applies to the case when a high injection of the local oscillator is employed, the desired signal is at the lower sideband and the upper sideband image signal has to be cancelled. The subtraction of the signals in the two paths will be used in this case.

The simplest method to implement the 90° phase shift in I and Q paths is to insert an RC network for one path and a CR network for the other path. The corner frequencies of the low-pass and high-pass networks are set to provide the phase shifts of –45° and +45° respectively at the down-converted IF. This method is applicable only to narrow band signals.

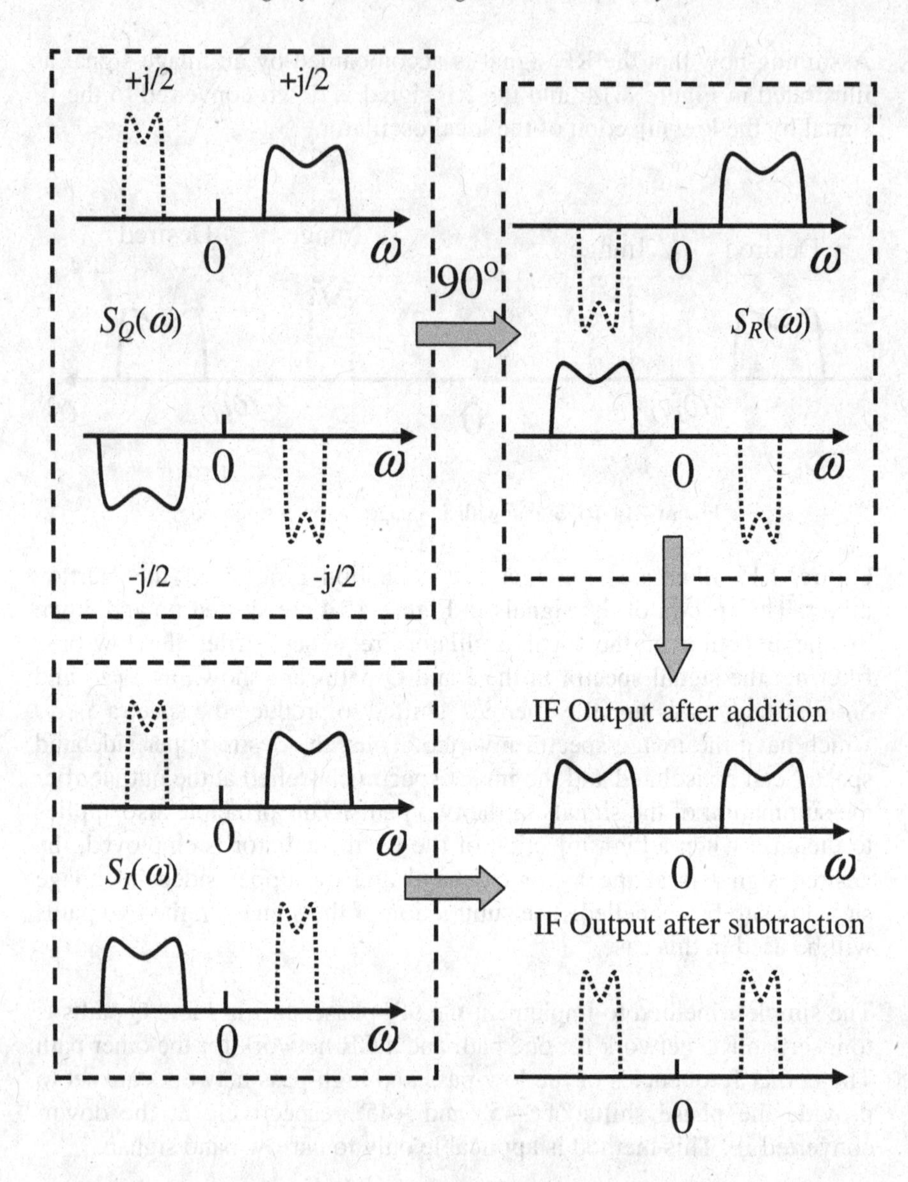

Figure 4.15 Image cancellation process.

The well-known drawback of the Hartley architecture is its sensitivity to mismatches of phase and gain in the two paths, consequently the achievable image rejection is quite limited [16]. Difficulties in implementing an accurate wideband 90° phase shifter lead to the replacement of the phase shifter with a polyphase filter as shown in Figure 4.16, which can accept differential inputs from the outputs from the mixers and can provide the 90° phase shift over a wider bandwidth by cascading two or three filter stages. The differential I_{out} and Q_{out} can be added or subtracted to obtain the sideband signal above or below the LO frequency as described above. The values of R and C are selected so that $1/2\pi RC$ is equal to the intermediate frequency for narrow band applications. For wideband applications, the values of R_i and C_i of multiple stages are chosen so that $1/2\pi R_iC_i$ of individual stages are spread over the selected IF band width. This, however, leads to the high loss of the signal power. The theory of passive polyphase filters is elaborated extensively in [17].

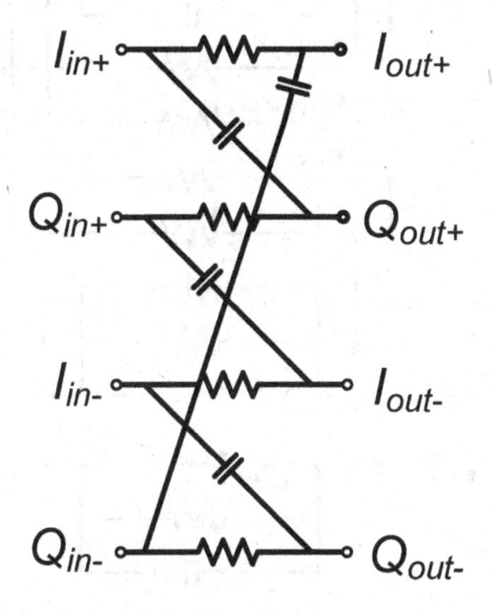

Figure 4.16 A single stage passive poly phase filter.

Active polyphase filters have also been designed to overcome the signal power loss in passive filters [18, 19]. Figure 4.17 shows the first order active polyphase band pass filter. As this building block is a band pass filter, multiple stages are usually implemented to obtain both the necessary band width as well as to perform the channel selection task. However, as the complexity of these circuits' increases, the power consumption is increased and the mismatch problem is naturally worsened, thus the effectiveness of these methods depends heavily on the good layout.

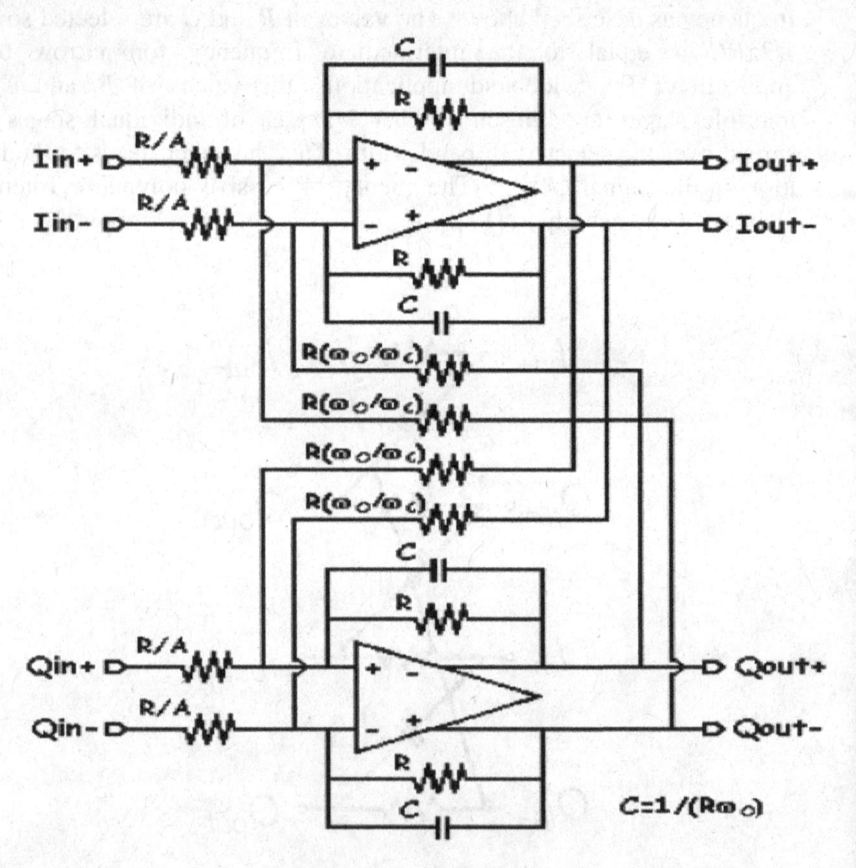

Figure 4.17 First order active polyphase filter building block [17].

An image rejection of more than 35dB is achievable by this method in recent publications [20, 21]. An alternative and more practical approach to this problem has been carried out in many commercial single chip solutions, where the low IF I and Q signals are digitized and processed digitally for the image cancellation.

The Weaver image reject down-converter has a similar operation as that of the Hartley. However, it uses an additional quadrature mixing stage to avoid the requirement of the 90° phase shifter. The work was first introduced by Weaver in 1956 [22] and well described in [1]. The additional mixing operation, however, creates the problem of secondary image. This technique has not been seen in either recent research publications or in commercial chip solutions.

References

[1] B. Razavi, "RF Microelectronics", Prentice Hall, 1998.

[2] K.L. Fong, R.G. Meyer, "Monolithic RF Active Mixer Design", *IEEE Transaction on Circuits and Systems II: Analog and Digital Signal Processing,* Vol. 46, No. 3, pp. 231-239, March. 1999.

[3] K. Kivekas, A. Parssinen, J. Jussila, and K. Halonen, "Design of Low Voltage Active Mixer for Direct Conversion Receivers," *The 2001 IEEE International Symposium Circuits and Systems (ISCAS), 2001.* Vol. 4, 6-9 May 2001.

[4] S. Zhou and M.C.F. Chang, "A CMOS Passive Mixer with Low Flicker Noise for Low-Power Direct Conversion Receiver", *IEEE Journal of Solid State Circuits,* Vol. 40, No. 5, pp. 1084-1093, May-2005.

[5] T.K. Nguyen, V. Krizhanovskii, J. Lee, S.K. Han, S.G. Lee, N.S. Kim, and C.S. Pyo, "A Low-Power RF Direct Conversion Receiver/Transmitter for 2.4GHz-Band IEEE 802.15.4 Standard in 0.18 μm CMOS Technology", *IEEE Transaction on Microwave Theory and Techniques*, Vol. 54, No. 12, pp. 4062-4071, Dec-2006.

[6] B.W. Cook, A. Berny, A. Molnar, S. Lanzisera, and K.S.J. Pister, "Low Power 2.4GHz Transceiver with Passive RF front-end and 400mV Supply", *IEEE Journal of Solid State Circuits*, Jan-2007.

[7] B. Razavi, "Design Considerations For Direct Conversion Receivers", *IEEE Trans. On Circuits and Systems II,* Vol. 44, No. 6, pp. 428-435, June-1997.

[8] A. Parssinen et al., "A 2GHz Wide-Band Direct Conversion Receiver for WCDMA Applications", *IEEE J. of Solide-State Circuits*, vol. 34, no. 12, pp. 1893-1903, 1999.

[9] B. Lindquist et al., "A New Approach to Eliminate the DC Offset in TDMA Direct Conversion Receiver", *IEEE Vehicular Technology Conference*, pp. 754-757, 1993.

[10] J. K. Cavers, M. W. Liao, "Adaptive Compensation for Imbalance and Offset Losses in Direct Conversion Transceivers", *IEEE Transaction on Vehicular Technology*, Vol. 42, No. 4, pp. 581-588, 1993.

[11] H. Yoshida, H. Tsurumi, Y. Suzuki, "DC Offset Canceller in a Direct Conversion Receiver for QPSK Signal Reception", *The Ninth IEEE International Symposium on Personal, Indoor and Mobile Radio Communications*, 1998., Vol. 3, pp. 1314 – 1318, 8-11 Sept. 1998.

[12] M. Goldfarb, E. Balboni, J. Cavey, "Even harmonic double-balanced active mixer for use in direct conversion receivers", *IEEE J. Solid-State Circuits,* Vol. 38, No. 10, pp. 1762 – 1766, 2003.

[13] J.J. Liu, M.A. Do, X.P. Yu, J.G. Ma, K.S. Yeo, and S. Jiang "CMOS Even Harmonic Switching Mixer for Direct Conversion Receivers", *Journal of Circuits, Systems, and Computers*, Vol. 15, No. 2, pp. 183-196, 2006.

[14] Wolfram Kluge, Frank Poegel, Hendrik Roler, Matthias Lange, Tilo Ferchland, Lutz Dathe, and Dietmar Eggert, "A fully Integrated 2.4GHz IEEE 802.15.4-Compliant Transceiver for ZigBee™ Applications", *IEEE Journal of Solid-State Circuits*, Vol. 41, No.12, pp. 2767-2775, Dec-2006.

[15] R. Hartley "Modulation Systems", U.S. Patent 1,666,206, April-1928.

[16] Razavi, B., "Challenges and trends in RF design", *ASIC Conference and Exhibit, 1996. Proceedings, Ninth Annual IEEE International*, 1996, Page(s): 81 –86.

[17] Behbahani, F.; Kishigami, Y.; Leete, J.; Abidi, A.A., "CMOS mixers and polyphase filters for large image rejection", *IEEE Journal of Solid-State Circuits*, Volume: 36 Issue: 6, Page(s): 873 –887, June 2001.

[18] Crols, J.; Steyaert, M, "An analog integrated Polyphase Filter for a High Performance Low-IF Receiver", *1995 Symposium on VLSI Circuit Digest of Technical Papers*, Kyoto, Japan, Pages: 87-88, 1995.

[19] Hornak, T.; Knudsen, K.L.; Grzegorek, A.Z.; Nishimura, K.A.; McFarland, W.J., "An image-rejecting mixer and vector filter with 55-dB image rejection over process, temperature, and transistor mismatch", *IEEE Journal of Solid-State Circuits*, Volume: 36 Issue: 1, Page(s): 23 –33, Jan. 2001.

[20] Wolfram Kluge, Frank Poegel, Hendrik Roller, Matthias Lange, Tilo Ferchland, Lutz Dathe, and Dietmar Eggert, "A Fully-Integrated 2.4GHz IEEE 802.15.4-CompliantTransceiver For ZigBee™ Applications", *IEEE Journal of Solid-State Circuits*, Vol. 41, No. 12, pp. 2767-2775.

[21] Ilku Nam, Kyudon Choi, Joonhee Lee, Hyouk-Kyu Cha, Bo-Ik Seo, Kuduck Kwon, Kwyro Lee, "A 2.4-GHz Low-Power Low-IF Receiver and Direct-Conversion Transmitter in 0.18-μm CMOS for IEEE 802.15.4 WPAN Applications", *IEEE Trans. on Microwave Theory and Techniques*, Vol. 55, No. 4, pp. 682-689, April-2007.

[22] Weaver, D.K., "A Third Method of Generation and Detection of Single-Sideband Signals", Proc. IRE, vol.44, no.12, pp. 1703-1705, 1956.

We get more when sharing knowledge but less when sharing things.

Kiat Seng YEO

CHAPTER 5

RF CMOS Oscillators

Oscillator is a critical component in a frequency synthesizer which is used to generate certain frequencies. A common application of the oscillator is in the receiver, where the periodic signal generated from the oscillator is used to drive the mixer.

In this chapter, firstly, the basic concepts and terminologies used in oscillator will be discussed. Secondly, various LC oscillators will be shown. Subsequently, detailed descriptions of a methodology for the design of an LC voltage-controlled oscillator (VCO) will be shown. Then two LC VCOs will be presented as design examples.

5.1 Introduction

An oscillator can be described as a positive feedback system and it amplifies its own noise at a selected frequency ω_0, as shown in Figure 5.1. The transfer function of the oscillator is

$$A = \frac{V_{out}(s)}{V_{in}(s)} = \frac{G(s)}{1 - G(s)} \tag{5.1}$$

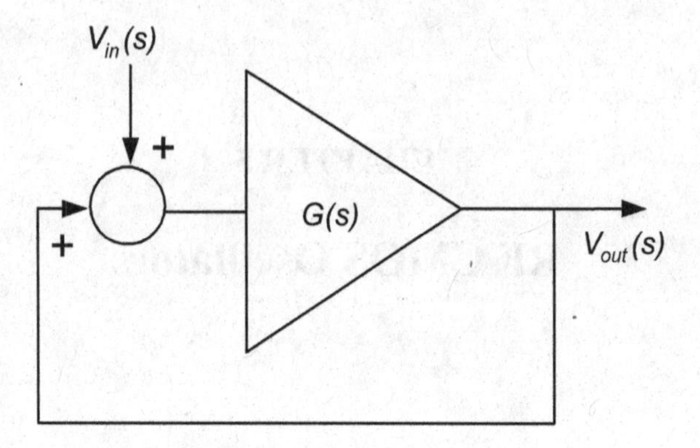

Figure 5.1 Feedback diagram of an oscillator.

From equation (5.1), it can be concluded that the closed loop gain will approach infinity under the following conditions, which are called the Barkhausen's Criteria: (1) the open loop gain is equal to unity, i.e. $|G(S)|=1$, and (2) the total phase shift of the loop is equal to 0°, i.e. $\angle G(S)=0^{\circ}$.

In an environment with the existence of noise at all frequencies, the Barkhausen's Criteria can only be satisfied with the noise at a specific frequency ω_0. When the oscillation is properly started, the noise signal at frequency ω_0 is amplified and increased till the amplifying devices are saturated. Hence, the stable oscillation is maintained. In order to ensure the startup of the oscillation in presence of temperature and process variations, the small signal loop gain is typically chosen to be 2-3 times of the required value.

In most RF applications, it is required that the oscillator to be tunable, where the output frequency is a function of a control input, mostly in the form of voltage. This circuit is commonly referred to as voltage-controlled oscillator (VCO). A VCO can be described by

$$\omega_{out} = \omega_{fr} + K_{vco} \times v_c \tag{5.2}$$

where ω_{fr} is the free running frequency of the VCO, v_c is the control voltage of the VCO and K_{vco} is the gain of the VCO specified in rad/s/V. A voltage signal with magnitude of V_m can be described as

$$V_{out}(t) = V_m \cos(\theta(t)) \tag{5.3}$$

where $\theta(t) = \int \omega_{out} \cdot dt + \theta_0$ and θ_0 is the phase offset. Substituting equation (5.2) into equation (5.3), the sinusoidal voltage output signal of a VCO is given by

$$V_{out}(t) = V_m \cos(\omega_{fr} t + K_{vco} \int v_c dt + \theta_0) \tag{5.4}$$

where K_{vco} is assumed to be linear. Commonly, the CMOS oscillators or VCOs in today's technology are implemented as a ring oscillator or an LC oscillator. The basic concepts of the ring oscillator will be studied first followed by that of an LC oscillator. Due to the superior performance of an LC oscillator over ring oscillator, LC oscillator is more popular and will be the focus of this chapter.

5.1.1 Ring Oscillator

Ring oscillators are widely used in the PLL frequency synthesizers and the clock recovery circuits. A ring oscillator usually employs three or more stages of inverters in a loop configuration. If there is an even number of inversion stages in the loop, the overall DC phase shift around the loop is 0°. Consequently, the positive feedback at DC will finally drive the transistors permanently to either the cut-off state or the triode state. Therefore, the number of inversion stages in the loop must be odd so that the circuit can oscillate, rather than latch up. For an odd number of inversion stages, the total phase shift of the loop is dependent on the frequency. Therefore, the phase condition of 0° of the Barkhausen's Criteria is satisfied at only a specific frequency.

The oscillation frequency of a ring oscillator is determined by either the 3-*dB* bandwidth of each stage or the time delay of each stage, depending on whether the circuit is working in a small signal condition with an output resistance or in a large signal condition with the rail-to-rail swing. Hence, the oscillation frequency of an N-stage ring oscillator can be tuned by varying the time delay of each stage.

Figure 5.2 shows an example of a three-stage ring oscillator, where each inverter can be a common-source stage, or a CMOS inverting stage, or a differential stage.

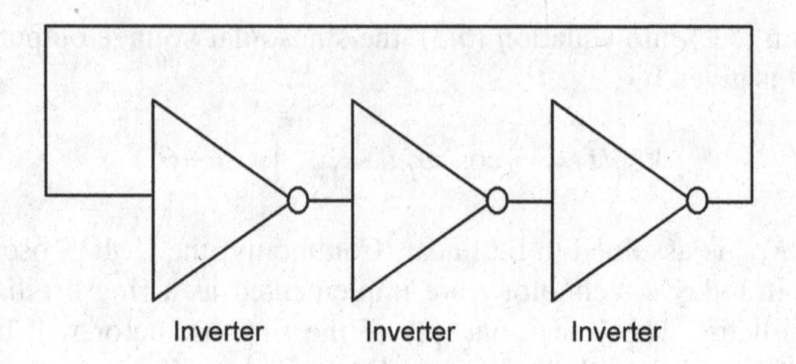

Figure 5.2 Three-stage ring oscillator.

In the absence of a frequency selective network, the ring oscillator is easy to implement with the current integration technology, but it suffers from a relatively high phase noise level, compared to that of the LC oscillator, which will be discussed in the next section. The switching activities in a ring oscillator introduce a lot of disturbances in the oscillator. In addition, the multiple-stage design also increases the noise level, making the ring oscillator unpopular in RF systems. A general discussion of the ring oscillator phase noise is given in [1] and the resulting phase noise is given by

$$L(\Delta\omega) = A_k \cdot \frac{kTR_n}{{V_A}^2} \cdot \left(\frac{\omega_0}{\Delta\omega}\right)^2 \tag{5.5}$$

where A_k is a factor depending on the noise generation mechanism studied, k is the Boltzmann's constant, T is the temperature in Kelvin, R_n is an equivalent noisy resistor, V_A is the voltage amplitude of the signal, ω_0 is the oscillation frequency, and $\Delta\omega$ is the frequency offset. The only way to lower the phase noise is to lower the equivalent resistance R_n, but this inherently implies larger power consumption. Typical noise value is –94dBc/Hz at 1MHz offset from a 2.2GHz carrier [1] or –83dBc/Hz at 100kHz offset from a 900MHz carrier [2].

5.1.2 LC Oscillator

A resonator–based oscillator has an LC resonator tank (LC tank) which acts as a frequency selective element. The energy loss in the tank has to be compensated by active devices. A parallel LC tank is modeled in Figure 5.3. The positive resistor R_P models the resistive loss in the tank, and the negative resistor $-R_a$ models the active device, which provides the energy into the tank. Once the energy loss is equal to the energy provided by the active device, a stable oscillation can be sustained [3]. The effective impedance of the LC tank can be expressed as

$$Z(j\omega) = \frac{1}{\frac{1}{j\omega L} + j\omega C + \frac{1}{R_P}} \tag{5.6}$$

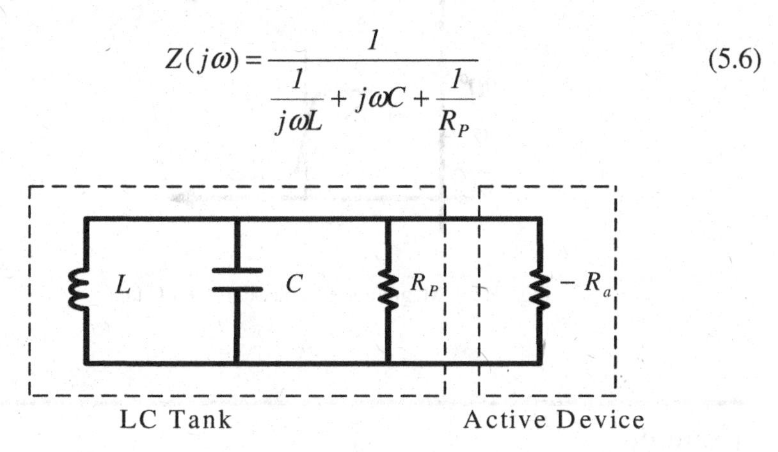

Figure 5.3 Typical schematic of the LC oscillator.

The resonant characteristics of $Z(j\omega)$ are illustrated in Figure 5.4. At resonance, the effective impedance $Z(j\omega)$ of the LC tank is purely real, which is equal to the effective resistance R_P. This means that the phase of the effective impedance is equal to zero. The frequency f_0 at resonance is given as

$$f_0 = \frac{1}{2\pi\sqrt{L\,C}} \tag{5.7}$$

Therefore, the phase condition of Barkhausen's criteria is satisfied, and the magnitude condition can be achieved by setting $R_P / R_a \geq 1$. In general, R_P is selected to be 2 to 3 times of R_a so that proper start-up for an LC oscillator can be guaranteed [4].

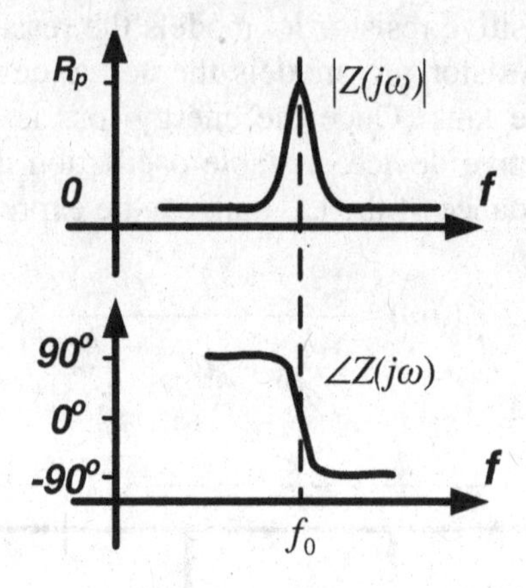

Figure 5.4 Magnitude and phase of the LC tank.

Example

Design an LC resonator to be operating at 3GHz using a 1nH inductor with a series resistor of 2Ω. The capacitor's series resistor is negligible.

Solution

Series to parallel transformation can be performed to convert series resistance R_s and series inductor L_s to parallel resistance R_P and parallel inductor L_p respectively as illustrated in Figure 5.5.

If the impedance of the parallel circuit is equal to that of the series circuit, then the circuits are equivalent and can be written as

$$Z_{par} = Z_{ser} \tag{5.8}$$

Or
$$\frac{R_P sL_p}{R_P + sL_p} = R_s + sL_s \tag{5.9}$$

Taking the imaginary part of equation (5.9)

$$R_s sL_p + R_P sL_s = R_P sL_P \tag{5.10}$$

Assuming $R_P >> R_s$, from equation (5.10), $L_s \approx L_p$. Let $L_s = L_P = L$, and substitute this into the real part of equation (5.9)

$$R_P = \frac{\omega^2 L_P L_s}{R_s} = \frac{(\omega L)^2}{R_s} \tag{5.11}$$

Since L = 1nH, $R_P = (2\pi \cdot 3x10^9 \cdot 1x10^{-9})^2 / 2$ = 177.65Ω. In order to satisfy the magnitude condition of the Barkhausen's criteria $R_P \geq R_a$ must be satisfied. By choosing $R_P = 2.5R_a$, the magnitude of the negative resistor R_a is chosen to be 177.65/2.5 = 71.06Ω. From equation (5.7), C=$1/[(2 \cdot \pi \cdot 3x10^9)^2 (1x10^{-9})$ = 2.8pF.

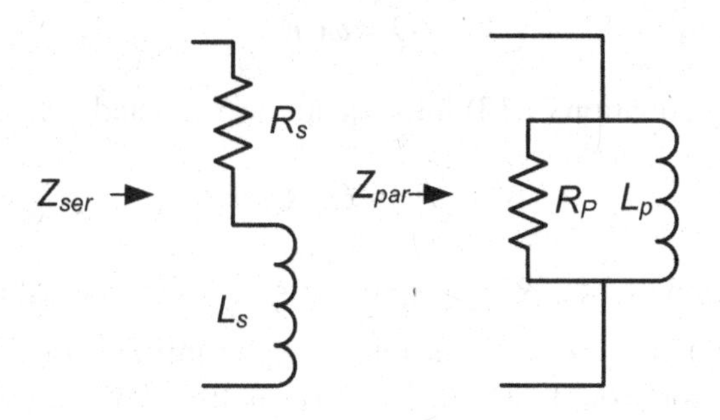

Figure 5.5 Series and parallel LR circuits.

In an LC tank shown in Figure 5.3, the quality or the energy efficiency of the tank is often specified in terms of the quality factor Q. The quality factor Q for a system under sinusoidal excitation at a frequency ω is fundamentally defined as

$$Q = \omega(\text{Energy Stored})/(\text{Average Power Dissipated}) \tag{5.12}$$

If a voltage V_E was applied across the LC tank in Figure 5.3, the energy stored across the inductor and capacitor can be obtained to be

$$\text{Energy Stored} = \frac{1}{2}\frac{V_E^{\ 2}}{\omega^2 L}$$

Or
$$\text{Energy Stored} = \frac{1}{2}CV_E^{\ 2}$$

The average power dissipated for the LC tank is

$$\text{Average Power Dissipated} = \frac{1}{2}\frac{V_E^{\ 2}}{R_P}$$

The quality factor for the resonator in Figure 5.3 is thus

$$Q = \frac{R_p}{\omega L} \tag{5.13}$$

Or
$$Q = \omega C R_p$$

Substituting equation (5.13) into equation (5.11) and rearranging yields

$$R_p = Q^2 R_s \tag{5.14}$$

From equation (5.14), $R_p >> R_s$ for large Q. For an ideal resonator, from equation (5.13), Q will be infinite as R_p is infinite. In other words, the ideal resonator would not need to have active devices and no energy is needed to maintain the oscillation once the oscillation starts as no energy is lost per cycle. However, in practical, Q of an on-chip inductor is quite low, typically from 3 to 20. This is mainly due to their parasitic series resistance and the substrate loss.

Example

In a practical resonator, capacitor has finite series resistance. Figure 5.6 shows a resonator consisting of an inductor L, a varactor C_v and a fixed capacitor C_f. The series resistances of L, C_v and C_f are R_L, R_v and R_f respectively. Calculate Q of the resonator given that the values of L, C_v, C_f, R_L, R_v and R_f are 1nH, 2pF, 0.8pF, 2Ω, 0.5Ω and 0.5Ω respectively at the frequency of 3GHz.

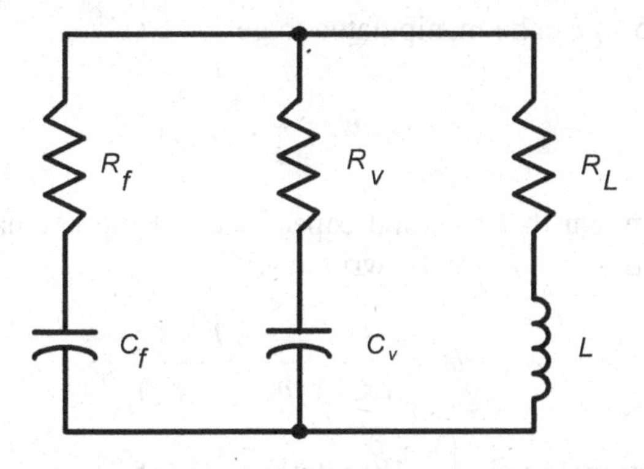

Figure 5.6 A practical LC resonator.

Solution

From equations (5.13) and (5.14), the equivalent series resistance for the inductor $R_S = \omega L / Q$ and the equivalent series resistance for the capacitor $R_S = 1 / Q\omega C$ can be obtained.

The quality factor of the inductor Q_L, the quality factor of the fixed capacitor Q_f and the quality factor of the varactor Q_v can then be modeled by the series resistance $R_L = \omega L/Q_L$, $R_f = 1/(Q_f\omega C_f)$ and $R_v = 1/(Q_v\omega C_v)$, respectively. By applying equation (5.14), the equivalent parallel resistance R_p for the LC resonator in Figure 5.6 can be obtained from

$$\frac{1}{R_p} = \frac{1}{R_f Q_f^2} + \frac{1}{R_v Q_v^2} + \frac{1}{R_L Q_L^2} \tag{5.15}$$

Through substituting equation (5.15) into equation (5.13) and rearranging, the quality factor Q for the LC resonator can be obtained

$$\frac{1}{Q} = \frac{\omega L}{R_p} = \omega L[\frac{1}{R_f Q_f^2} + \frac{1}{R_v Q_v^2} + \frac{1}{R_L Q_L^2}] \tag{5.16}$$

Equation (5.7) can be manipulated to get

$$\omega L = \frac{1}{\omega C} \tag{5.17}$$

It can be shown that the total capacitance of the resonator $C \approx C_f + C_v$. Thus, equation (5.17) can be written as

$$\omega L = \frac{1}{\omega C} = \frac{1}{\omega (C_f + C_v)} \tag{5.18}$$

By substituting equation (5.18) into equation (5.16) and manipulating

$$\begin{aligned} \frac{1}{Q} = \frac{\omega L}{R_p} &= \omega L[\frac{1}{R_f Q_f^2} + \frac{1}{R_v Q_v^2} + \frac{1}{R_L Q_L^2}] \\ &= \frac{1}{C_f + C_v}\left(\frac{C_f}{Q_f} + \frac{C_v}{Q_v}\right) + \frac{1}{Q_L} \end{aligned} \tag{5.19}$$

The quality factors of the inductor, fixed capacitor and varactor can be calculated as shown below

$$Q_L = \frac{\omega L}{R_L} = \frac{2\pi \cdot 3x10^9 \cdot 1x10^{-9}}{2} = 9.425$$

$$Q_f = \frac{1}{\omega C_f R_f} = \frac{1}{2\pi \cdot 3x10^9 \cdot 0.8x10^{-12} \cdot 0.5} = 132.63$$

$$Q_v = \frac{1}{\omega C_v R_v} = \frac{1}{2\pi \cdot 3x10^9 \cdot 2x10^{-12} \cdot 0.5} = 53.05$$

It is interesting to note that though the series resistances of capacitor C_v and C_f are the same, Q_f is much larger than Q_v. The reason is due to the smaller C_f, resulting in larger impedance compared to the larger C_v. Generally, fixed capacitors like metal-insulator-metal (MIM) capacitor usually have the quality factor a few times higher than that of a varactor like MOS-varactor. Substituting all the values into equation (5.19) yields

$$\frac{1}{Q} = \frac{1}{C_f + C_v}\left(\frac{C_f}{Q_f} + \frac{C_v}{Q_v}\right) + \frac{1}{Q_L}$$

$$= \frac{1}{0.8x10^{-12} + 2x10^{-12}}\left(\frac{0.8x10^{-12}}{132.63} + \frac{2x10^{-12}}{53.05}\right) + \frac{1}{9.425}$$

$$= \frac{1}{64.03} + \frac{1}{9.425} = 0.1217$$

$$Q = 8.22$$

An important observation from equation (5.19) and the above calculations is that adding a high Q fixed capacitor to a low Q varactor will improve the total quality factor of the capacitors. The total quality factor of the capacitors is 64.03, which is higher than the quality factor of varactor alone. This technique can be used to improve quality factor of the resonator if the quality factor of the varactor is low. Nevertheless, in this example and in most cases, $Q_v >> Q_L$, thus the quality factor of the resonator is determined mainly by the quality factor of the inductor, which makes $Q \approx Q_L$ in general.

In order to compare the performance between oscillators, figure of merit (*FOM*) can be used. The Leeson [4] heuristic expression for the phase noise $L(\Delta\omega)$ of an LC oscillator in the $1/f^2$ region is given by

$$L(\Delta\omega) = F\frac{kT}{2P_{sig}}\frac{\omega_0^{\ 2}}{Q^2(\Delta\omega)^2} \tag{5.20}$$

where Q is the loaded quality factor of the resonator as defined in equation (5.16), $\Delta\omega = 2\pi\Delta f$ is the angular frequency offset, P_{sig} is the average signal power (in watts), and F is the device noise excess factor. This equation was verified in [5].

The figure of merit (*FOM*) is commonly used to determine the phase noise performance of an oscillator, with respect to the power and oscillation frequency at a certain offset frequency [6],[27]

$$FOM = \left(\frac{\omega_0}{\Delta\omega}\right)^2 \frac{1}{P_{vco} L(\Delta\omega)} \tag{5.21}$$

where P_{vco} is the total power consumption of an oscillator in milli-watts. Therefore, from equations (5.20) and (5.21)

$$FOM = \frac{2}{F} \times \frac{P_{sig}}{P_{vco}} \times \frac{1}{kT}\left(Q^2\right) \tag{5.22}$$

Equation (5.22) shows that the *FOM* is proportional to the squared quality factor of the resonator. This suggests that an oscillator using a good inductor with high quality factor will be able to attain better phase noise with the same power consumption, resulting in a better *FOM*.

In order to compare between oscillators with different Q to reflect a change in performance that is independent of Q, for example, due to a change in circuit topology, equation (5.22) must be normalized. An arbitrary value of Q = 10 is taken as the nominal value, the normalized *FOM* is [7]

$$FOM_{NORM} = FOM\left(\frac{10}{Q}\right)^2 \tag{5.23}$$

5.2 Various LC VCO Topologies

5.2.1 Colpitts and Hartley LC VCOs

Colpitts and Hartley are two types of LC VCOs, with different methods of impedance transformation. A Colpitts LC VCO utilizes a capacitance divider to supply the loop gain while forming the positive feedback as depicted in Figure 5.7(a). An inductance divider is employed in Hartley LC VCO as shown in Figure 5.7(b). Although both are theoretically feasible, in reality the Colpitts structure is more attractive in integrated circuit implementation as it requires fewer inductors. In the Colpitts structure, a positive feedback is formed by connecting the amplifier output to its positive input, which is the source of the transistor. Without an impedance transformer, the low impedance of the transistor seen from its source will significantly affect the performance of the LC tank by decreasing its *Q* value. By employing the capacitance divider, the source impedance of the transistor is up–converted by a factor of $(1+\frac{C_2}{C_1})^2$. It can be deduced that the ratio of C_2 and C_1 must be large enough to ensure high impedance.

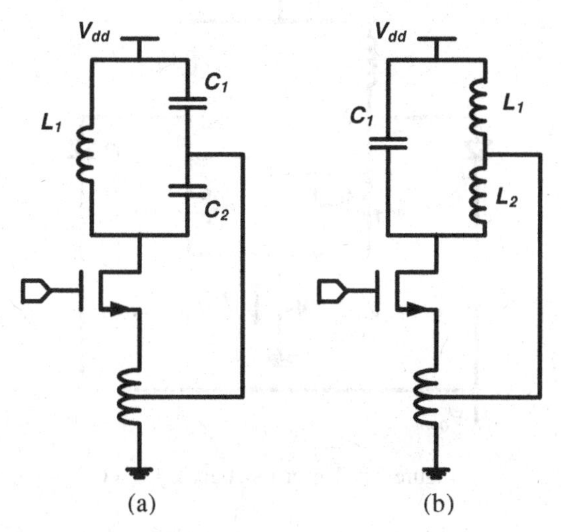

Figure 5.7 (a) Colpitts LC VCO; (b) Hartley LC VCO.

5.2.2 Differential LC VCOs

The RF local oscillator is usually required to have differential outputs. A differential implementation of the Colpitts LC VCO, also known as the cross-coupled structure, is shown in Figure 5.8. A cross-coupled structure has differential outputs. This is an advantage because the most commonly used mixer in RF systems is the double balanced Gilbert-cell mixer where differential inputs are needed.

The operation of the cross-coupled LC VCO can be viewed as two identical Colpitts oscillators in a cross-coupled configuration. Each of them serves the other as an active impedance transformer and as part of its feedback loop. For example, MOSFET M_2 detects the signal at node X. Being a common source amplifier, the signal at the gate of M_2 will be passed to the drain, which is the gate of M_1. M_1 will then serves as the feedback path, which completes the positive feedback loop by passing the signals from its gate back to the node X. Similarly, M_1 can be viewed as the oscillation amplifier and M_2 as the feedback circuit. The cross-coupled pair in Figure 5.8 can also be shown to have negative impedance of $-2/g_m$, which will be discussed in next page.

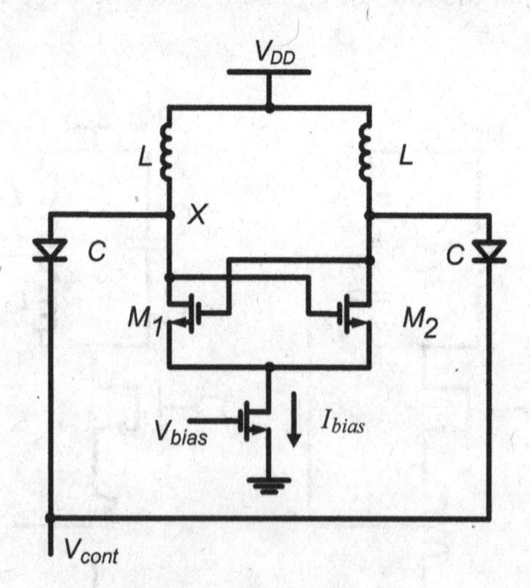

Figure 5.8 The cross-coupled LC VCO.

Hence, with enough negative impedance, the cross-coupled pair can compensate for the loss of the LC tank and maintain stable oscillation. Consequently, the cross-coupled LC oscillator is also called the *negative-Gm* oscillator.

At resonance, the impedance of the capacitor and the impedance of the inductor are equal in magnitude and opposite in polarity.

$$\frac{1}{2\pi \cdot f_0 \cdot C} = 2\pi \cdot f_0 \cdot L \tag{5.24}$$

where f_0 is the resonant frequency, C and L are the capacitance and inductance of the cross-coupled LC oscillator, respectively. Simplifying equation (5.24), the resonant frequency f_0 is obtained to be

$$f_0 = \frac{1}{2\pi\sqrt{LC}} \tag{5.25}$$

In practice, due to the parasitic capacitance, C in the above equation should be replaced with the total tank capacitance C_{tank}. C_{tank} includes the variable capacitance of the varactor, the parasitic capacitance of the inductors if they are implemented on chip, and the parasitic capacitance at the gate and drain of the MOSFET transistors M_1 and M_2. Hence, the resonant frequency can be tuned by varying the varactor capacitance through a control voltage.

Example

For the VCO shown in Figure 5.8, calculate the minimum inductor size if its quality factor Q_L is 10. Given that the transconductance g_m of the transistor is 5mS. Assume ideal capacitor.

Solution

Figure 5.8 can be redrawn as Figure 5.9 to calculate the negative resistance $-R_{da}$ provided by the cross-coupled transistor pair. Let the transconductance for transistor M_1 and M_2 to be g_{m1} and g_{m2}. Due to symmetry, $g_{m1} = g_{m2} = g_m$. From Figure 5.9

$$-R_{da} = V_{do} / I_{do} \tag{5.26}$$

$$V_{do} = V_{DS1} - V_{DS2} = V_{GS2} - V_{GS1} = -\frac{I_{do}}{g_{m2}} - \frac{I_{do}}{g_{m1}} = -\frac{2I_{do}}{g_m} \tag{5.27}$$

By substituting equation (5.27) into equation (5.26)

$$-R_{da} = -\frac{2}{g_m} \tag{5.28}$$

For half circuit of the cross-coupled transistor pair, it follows that the negative resistance is

$$-R_a = -\frac{1}{2}R_{da} = -\frac{1}{g_m} \tag{5.29}$$

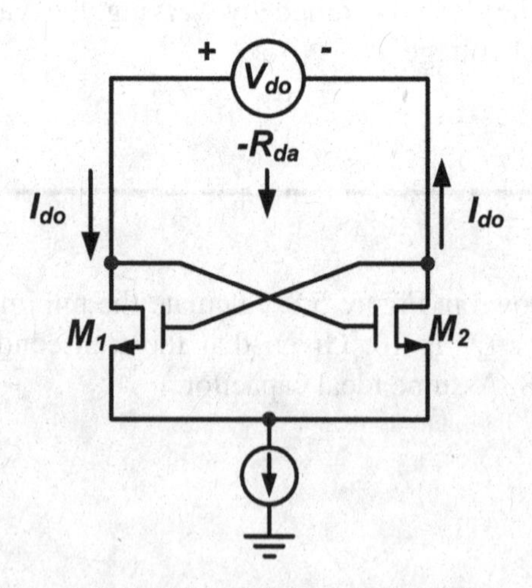

Figure 5.9 Calculation of the negative resistance.

Each half circuit is connected to one LC tank. The load resistance of each LC tank is equal to R_p. For proper oscillation start-up, $R_p = 2.5R_a$ will be chosen. Given that $g_m = 5\text{mS}$

$$R_p = 2.5R_a = 2.5\frac{1}{g_m} = \frac{2.5}{5m} = 500\Omega \tag{5.30}$$

From equation (5.15), the quality factor of the resonator is equal to that of the inductor as the capacitor is ideal. Hence $Q = Q_L$. From equation (5.13), the minimum value of the inductor L in order to guarantee oscillation can be obtained

$$R_p = Q\omega L$$

$$500 = 10 \cdot 2\pi \cdot 3x10^9 \cdot L$$

$$L = 2.65nH$$

From the above calculations, a few insights can be obtained. Firstly, R_p increases with L. Secondly, a larger R_p in turn requires a smaller g_m, which means smaller current consumption. This means that for a fixed Q, larger inductor will result in lower current consumption for the oscillator. Hence, it is generally recommended to have large L for low power design. However, this is done at the cost of smaller frequency tuning range due to the smaller possible varactor size.

5.2.2.1 Complementary LC VCOs

Three various complementary LC VCOs that use cross-coupled PMOS pair in addition to cross-coupled NMOS pair are shown in Figure 5.10.

In each of the LC VCO, the LC tank is not connected to the power supply directly but connected to the output nodes. The inductor is usually formed by 2 inductors in series, each with an inductance of L to ensure better symmetry.

The complementary LC VCO is popular for several reasons. Firstly, by reusing the current of NMOS transistor for PMOS transistor, the complementary LC VCO can achieve a total transconductance g_m of $g_{m,n}+g_{m,p}$ compared to just $g_{m,n}$ of a non-complementary LC VCO with the same current. $g_{m,n}$ and $g_{m,p}$ are the transconductance of NMOS transistor and PMOS transistor respectively. Due to the larger g_m, the oscillation amplitude of this structure is larger than that of a non-complementary type with the same current consumption. This results in better phase noise performance and faster switching for a cross-coupled differential pair.

Secondly, the complementary LC VCO is able to have a symmetric tank by designing such that $g_{m,n} = g_{m,p}$. A symmetric tank results in less *1/f* noise up-conversion thus improving the phase noise performance.

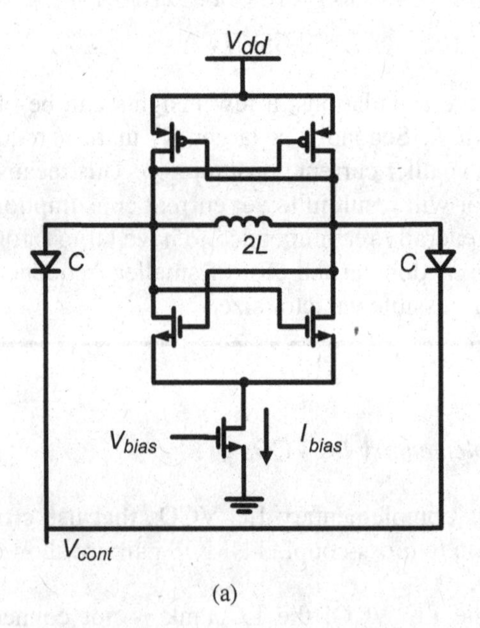

(a)

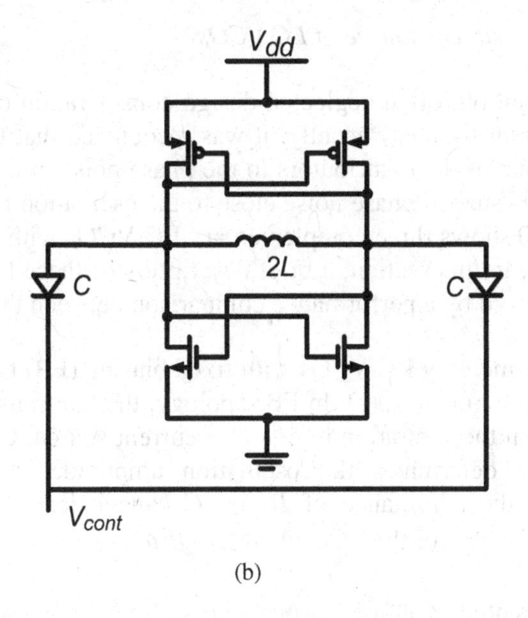

(b)

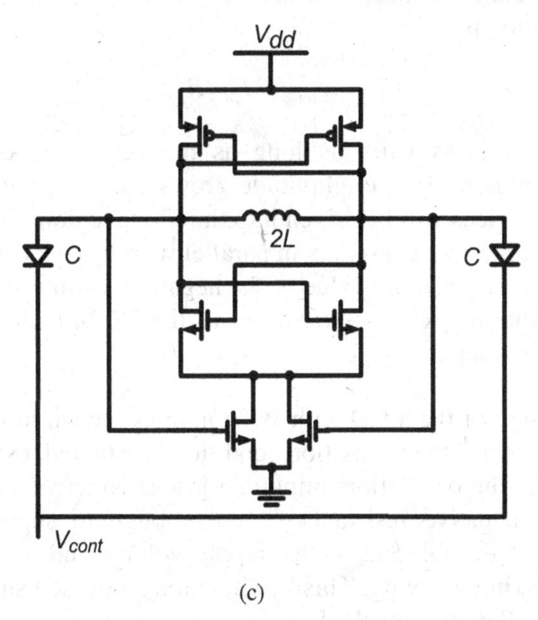

(c)

Figure 5.10 Complementary LC VCOs (a) with fixed biasing (FB) tail current source; (b) without tail (WT) current source; (c) with memory reduction (MR) tail current source.

5.2.2.2 Tail current source of LC VCOs

An important but often neglected design consideration of an LC VCO is its tail current source. Recently, it was recognized that the tail transistor might be one of the contributors to the phase noise in a VCO, especially to the $1/f^3$ - shaped phase noise close to the oscillation frequency [3],[8]. Figure 5.10 shows three complementary LC VCOs with different biasing techniques. In this section, a brief description of these LC VCOs will be given followed by a performance comparison between them.

The complementary LC VCOs with fixed biasing (FB) tail current source is shown in Figure 5.10(a). In FB topology, the tail transistor is designed to operate in the saturation region as a current source. Consequently, the tail current determines the oscillation amplitude. At the resonance frequency, the admittance of L and C cancel, leaving the equivalent parallel resistance of the LC tank $R_P = QL\omega$.

The differential voltage swing across the tank is given in first approximation by

$$V_{tank} = I_{bias} R_P \tag{5.31}$$

Equation (5.31) is valid as long as the active devices work in the saturation region. As the amplitude grows closer to the supply voltage, the active devices will be driven into the triode region. The cross-coupled transistors now act as resistors in parallel with R_P or it can be viewed as a reduction in the absolute value of the negative resistance that balances R_P. Hence, additional loss is introduced to the VCO, which leads to lower VCO quality factor.

The operation of the VCO with WT topology shown in Figure 5.10(b) is as follows. When the oscillation condition is satisfied, oscillation starts to develop. As the oscillation amplitude grows larger, it will reach a point where the negative resistance is not enough to support the positive resistance of the LC tank if the supply voltage and ground do not first clip the maximum swing. This is where the amplitude stops growing and a stable oscillation is reached.

Figure 5.10(c) shows the memory reduction (MR) tail transistor VCO [7]. The operation of the MR VCO is as follows. Initially, when the circuit is balanced, both the output voltages and currents flowing in the two sides are set by the size of the tail transistors. The tail transistors will go into the saturation region first while the cross-coupled NMOS transistors are still in the cut-off region. When both the tail transistors and cross-coupled NMOS transistors are in the triode region, the tail transistors determine the current as the voltages at the source of the cross-coupled NMOS transistors are floating.

The flicker noise is known for its long correlation time and an associated physical process, which has a "long-term memory" [31]. The "carrier trapping in localized oxide states" is a process that plays a significant role in the generation of the flicker noise in a MOSFET. Moreover, the memory involved with the flicker noise is related to the long occupation time constants of the traps. As a switched transistor will force a trap to release its captured electron, rendering the transistor to be memory-less, the flicker noise will be reduced.

Since all the transistors in the MR VCO topology are switched biasing rather than fixed biasing, it is expected to have lower flicker noise [9],[10]. Moreover, as the transistors operate in the triode region for a large portion of the oscillation period, they exhibit lower current flicker noise than the transistors that operate in the saturation region, for example the tail transistor in the FB topology [11].

A comparison of the three topologies will reveal the advantages and disadvantages of these VCOs.

The main advantage of the WT topology and MR topology over the FB topology is that without the tail transistor flicker noise source, the only flicker noise source now is the cross-coupled transistors, which have an inherently lower flicker noise due to the switched biasing, resulting in better phase noise performance.

Another disadvantage of the FB topology compared to the MR topology and WT topology is that the tail transistor in the FB topology reduces the headroom available for oscillation by around 0.2V to 0.4V in the CMOS 0.18μm technology. The effect is not negligible for low voltage design.

A smaller signal power P_{sig} has an adverse effect on the phase noise, as phase noise is essentially the noise to signal ratio of the VCO. The tail transistors of the MR topology mostly work in the triode region, so the headroom requirement is smaller. The WT topology can achieve the largest oscillation amplitude among the three topologies.

For the FB topology, extra circuitry is needed to provide biasing voltage to the tail transistor. This not only increases the power consumption, but also introduces noise sources to the VCO. The noise current coming from the biasing network will be mirrored into the tail transistor. Both the MR topology and the WT topology do not encounter this problem.

The major obstacle in implementing the WT topology is the power consumption. This is especially true in the case of an over-designed loop gain. For the complementary LC oscillator, in order to maintain the oscillation, the loop gain condition is

$$(g_{m,n} + g_{m,p}) = \alpha_g (1/R_p) \tag{5.32}$$

where α_g is the excess gain factor and typically from 2 to 3. $g_{m,n}$ and $g_{m,p}$ are the transconductance of the NMOS and PMOS cross-coupled transistors, respectively. The excess gain factor is a safety margin to guarantee oscillation. However, in the case of the WT topology, the VCO will consume a lot of "short-circuit current" that is useless to the functioning of the oscillator. For the same tank characteristics, which include the excess gain factor α_g and the oscillation frequency, the WT topology has the highest power consumption while both the FB topology and the MR topology have the same power consumption for the main LC tank.

Another disadvantage of the WT topology is the absence of the high tail transistor impedance in series with the cross-coupled transistors to stop the transistors from loading the resonator in the triode region [12]. In a balanced circuit, the odd harmonics circulate in a differential path, while even harmonics flow in a common-mode path. The even harmonics that are usually dominated by the second harmonic components travel through the resonator capacitors and the cross-coupled transistors to ground. The high impedance acts to suppress the noise in the tail transistor by making it appear noiseless to the VCO, thus improving the phase noise performance. Compared to the WT topology, the FB topology and the MR topology suppress the second harmonic noise more and prevent the cross-coupled transistors from loading the resonator. Thus, an improvement of phase noise in the $1/f^3$ region is expected from the MR topology and the FB topology over the WT topology. However, the improvement on the phase noise performance of the VCO with the FB topology is masked by the up-converted flicker noise of the tail transistor.

Finally, the FB topology is less susceptible to the frequency pushing effect, which is the frequency sensitivity to the voltage supply. Both the WT topology and the MR topology are affected by the frequency pushing effect.

For comparison, three VCOs with the same tank characteristics and oscillation frequency were designed and simulated using the three topologies. The excess gain factor α_g is made to be 2.5 and the tank quality factor is about 9.

The VCOs were designed for GSM-1800 applications where the oscillation frequency is at 1.88GHz and they were optimized for *FOM*. The post-layout simulation performance of the VCOs for three topologies is summarized in Table 5.1.

Table 5.1 Summary of the performance of the three VCOs.

	Without Tail Transistor Topology (WT)	Fixed Biasing Tail Transistor Topology (FB)	Memory Reduced Tail Transistor Topology (MR)
Power Consumption	1.7mA * 2V = 3.5mW	1.4mA * 2V = 2.8mW	1.4mA * 2V = 2.8mW
Phase Noise	-84dBc/Hz @10kHz Offset -126.6dBc/Hz @600kHz Offset	-81dBc/Hz @10kHz Offset -126dBc/Hz @600kHz Offset	-87dBc/Hz @10kHz Offset -127.6dBc/Hz @600kHz Offset
FOM @600kHz Offset	191.3dB	191.5dB	193dB

From Table 5.1, it can be seen that the MR topology gives the maximum improvement of the phase noise of 6dB from the FB topology and 3*dB* from the WT topology while consuming lower power consumption than the WT topology. The MR topology also shows the best *FOM* of 193dB, which corresponds to a normalized *FOM* of 194dB using equation (5.23). Nevertheless, in the following discussions, the FB topology will be our focus as it is the most generic and popular LC VCO topology. Some other methods to reduce the effect of tail current noise in the FB topology include adding a large capacitor or a resonator in parallel with the tail transistor [7]. In addition, by maintaining a good symmetry in the rise and fall time of the output voltage swing as well as proper sizing of the tail transistor can also reduce the effect of tail transistor noise on the VCO [3].

5.3 LC VCO Design Methodology

Due to the ever increasing demand for the bandwidth, stringent requirements are placed on the spectral purity of oscillators. Efforts to improve the phase noise performance of integrated LC VCOs have resulted in many research works [14],[16],[17]. The reported methods are summarized as: to achieve low phase noise, the oscillation amplitudes must be maximized. This means either increasing the biasing current or increasing the tank inductance (assuming a given tank quality factor). The former increases the power consumption, while the latter reduces the frequency tuning range due to a lower total tank capacitance [13-15]. Even with these endeavors, design and optimization of integrated LC tank VCOs still pose many challenges to circuit designers as simultaneous optimization of multiple variables is required.

Although many phase noise models have been developed for different types of oscillators, each of these models makes restrictive assumptions applicable only to a limited class of oscillators. Most of these models are based on a linear time invariant (LTI) system assumption, which assumes all noise sources are stationary [1,4]. Therefore, those models are incapable of making accurate predictions about phase noise in oscillators because oscillator is a periodically time varying system. Even though the phase noise model reported in [15] takes LTI into account, it still cannot evaluate the phase noise of the LC tank VCO accurately due to the incompletely device noise mechanism and the complicated simulation.

A methodology to design the LC tank VCO with lower phase noise will be illustrated and verified in this section. A fundamental relationship between the channel lengths of MOSFET in LC tank VCOs and the phase noise will be derived. From this relationship, an optimum channel length can be obtained for the LC tank VCO with the lowest phase noise performance. The phase noise model based on the linear time variant phase noise analysis to be presented subsequently can properly assess the effects of both stationary and cyclostationary noise sources on phase noise. Stationary noise sources are noise sources in the oscillator whose statistical properties do not depend on time and the operation point of the circuit, for example, resistor's thermal noise. Cyclostationary noise sources are noise sources with periodic changes in the currents and voltages. This model will explain the exact mechanism by which the gate induced noise, the drain noise, the tail transistor noise, and random noise, are converted into phase noise and amplitude noise.

5.3.1 Topology

5.3.1.1 Operation theory

The circuit diagram and output waveform of the complementary LC VCO is shown in Figure 5.11. A fully integrated complementary cross-coupled configuration is chosen as a vehicle of the methodology.

It is already known that the active devices ($NMOS_1$, $NMOS_2$, $PMOS_1$, and $PMOS_2$) serve as the negative resistor to compensate for the energy loss from the tank due to the tank effective resistance.

The operation of the complementary VCO is described below. Firstly, it is noted that the oscillator forces V_{gd} of both NMOS transistors to change in equal magnitudes but in opposite polarity to generate a differential voltage across the resonator. At the differential zero voltage, four switching transistors ($NMOS_1$, $NMOS_2$, $PMOS_1$ and $PMOS_2$) are all in saturation region and form a small-signal negative conductance that breeds the startup of the oscillation. As the differential oscillation voltage crosses $V_{th,n}$ (the threshold voltage of NMOS), V_{gd} of $NMOS_1$ exceeds $+V_{th,n}$, forcing it into the triode region, while V_{gd} of $NMOS_2$ falls below $+V_{th,n}$, driving the device into deeper saturation region, and then $NMOS_2$ turns off.

Simultaneously, as the falling differential oscillation voltage crosses $-V_{th,p}$, V_{gd} of $PMOS_2$ exceeds $-V_{th,p}$, forcing it into the triode region. At the same time, V_{gd} of $PMOS_1$ forces itself into deeper saturation, and then $PMOS_1$ turns off. Thus, both NMOS and PMOS pairs are in the saturation region for a short period of time, followed by $NMOS_1$ and $PMOS_2$ in the off state, as $NMOS_2$ and $PMOS_1$ are on, or vice versa.

Such switching process is periodical throughout the operation of the VCO as depicted in Figure 5.11(b). The current I_{bias} as indicated in Figure 5.11(a) drives the LC tank VCO into stable oscillation.

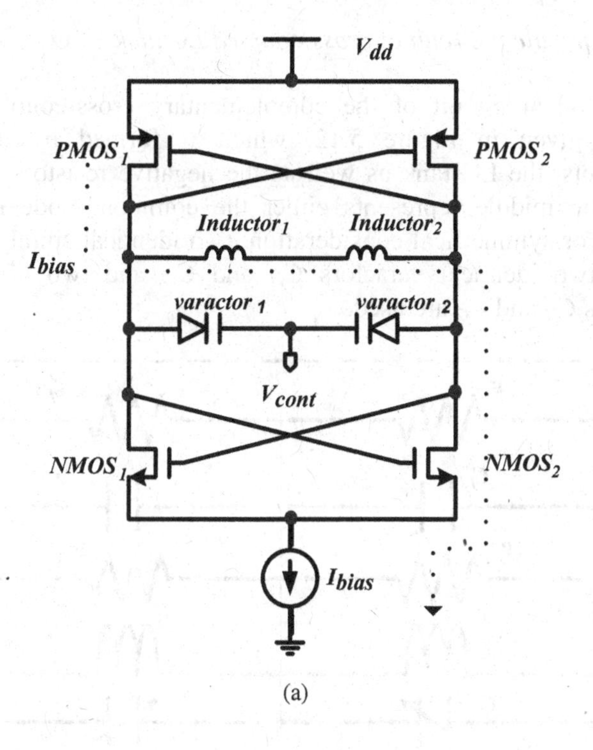

(a)

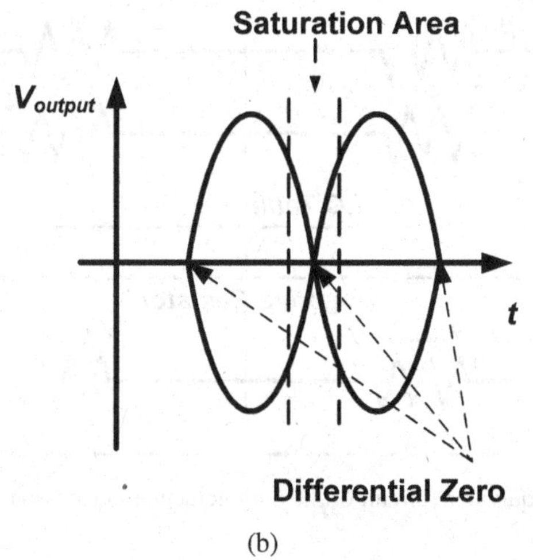

(b)

Figure 5.11 (a) Schematic of the cross-coupled complementary LC VCO with the parasitic elements; (b) differential zero.

5.3.1.2 Equivalent circuit of cross-coupled LC tank VCO

The equivalent circuit of the complementary cross-coupled LC tank VCO is given in Figure 5.12, which is formed by the parasitic components, the LC tank as well as the negative resistors. The broken line in the middle represents either the common mode reference or ground. For symmetrical consideration, two identical spiral inductors L_1 and L_2, two identical varactors C_{v1} and C_{v2} and two identical fixed capacitors C_{f1} and C_{f2} are used.

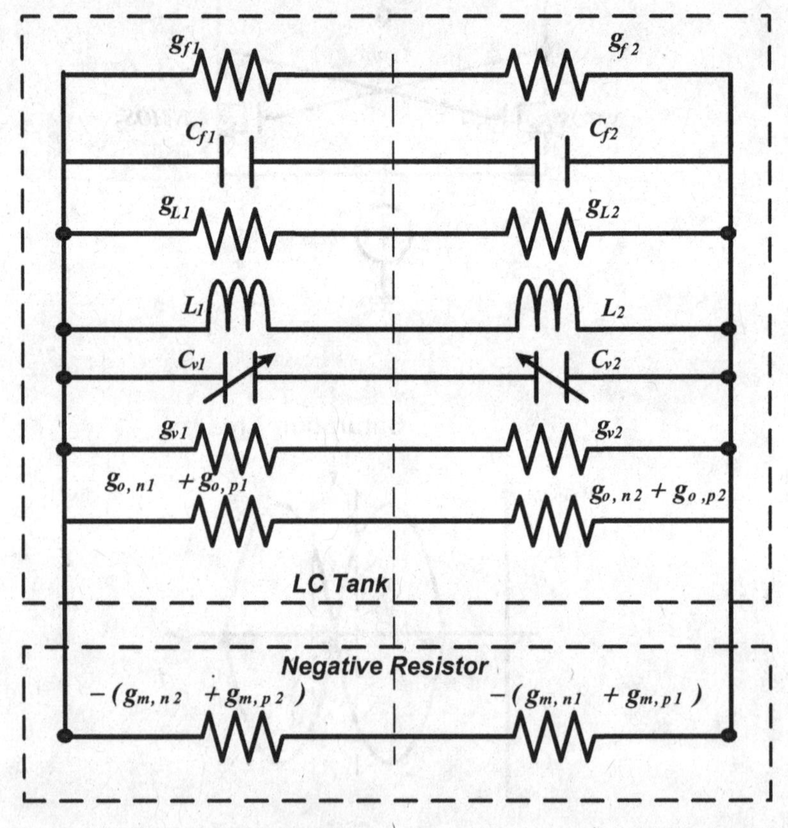

Figure 5.12 Equivalent circuit of the complementary cross-coupled LC tank VCO.

Each inductor has an inductance of L ($L_1 = L_2 = L$), and conductance of g_L ($g_{L1} = g_{L2} = g_L$). Similarly, the varactors are modeled as the parallel of a capacitance C_v ($C_{v1} = C_{v2} = C_v$) and conductance g_v ($g_{v1} = g_{v2} = g_v$). For each fixed capacitor, the capacitance is C_f ($C_{f1} = C_{f1} = C_f$) and conductance is g_f ($g_{f1} = g_{f2} = g_f$). Note that the parasitic capacitance of NMOS (C_{NMOS}) and PMOS (C_{PMOS}) are absorbed into the value of C_{f1} and C_{f2}. C_{NMOS} and C_{PMOS} are capacitances associated to the gate of NMOS and PMOS ($C_{NMOS} = C_{gs,n} + C_{db,n} + 2C_{gd,n}, C_{PMOS} = C_{gs,p} + C_{db,p} + 2C_{gd,p}$) [14], where $g_{m,n}$ and $g_{m,p}$ are small signal transconductances of the cross-coupled NMOS and PMOS pair to form the negative resistors, $g_{o,n}$ and $g_{o,p}$ are the output conductance of the NMOS and PMOS pair.

5.3.2 Associated Noise Sources of Complementary LC Tank VCO

Figure 5.13 depicts the total noise sources existing in an LC tank VCO. Of the three main noise sources, two will be discussed in this section, namely noise contribution from the LC tank and noise contribution from the tail transistor. The noise contribution from the active devices will be discussed in the next section.

5.3.2.1 Noise sources of the LC tank

The noise contributions from the LC tank comprise of the noise due to the loss of inductor, varactor and fixed capacitor. $\frac{\overline{i_{ind}^2}}{\Delta f}$, $\frac{\overline{i_v^2}}{\Delta f}$ and $\frac{\overline{i_f^2}}{\Delta f}$ are the current noise power spectral density (PSD) of the inductor's loss, varactor's loss and fixed capacitor's loss respectively. The total current noise PSD due to the LC tank can be thus represented by

$$\frac{\overline{i_{n,\text{tank}}^2}}{\Delta f} = \frac{\overline{i_{ind}^2}}{\Delta f} + \frac{\overline{i_v^2}}{\Delta f} + \frac{\overline{i_f^2}}{\Delta f} \quad (5.33)$$

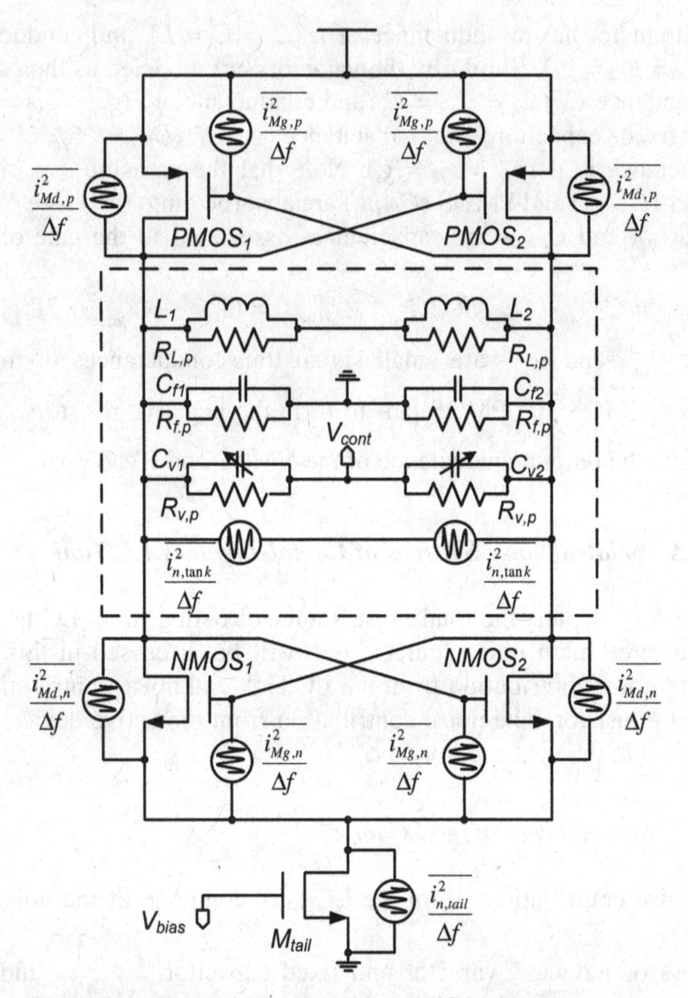

Figure 5.13 The complementary LC tank VCO with noise sources.

The current noise spectral density of each component is given by [15]

$$\frac{\overline{i_{ind}^2}}{\Delta f} = 2kTg_L \tag{5.34}$$

$$\frac{\overline{i_v^2}}{\Delta f} = 2kTg_v \tag{5.35a}$$

$$\frac{\overline{i_f^2}}{\Delta f} = 2kTg_f \tag{5.35b}$$

where the conductance of the inductor g_L is given in equation (5.36); the conductance of varactors g_v and the conductance of fixed capacitors g_f are given in equation (5.37a) and (5.37b), respectively. The quality factor Q_L of the inductors in equation (5.36), the quality factor Q_v of the varactors and the quality factor Q_f of the fixed capacitors in equation (5.37) can be extracted via S-parameter simulations. The noise sources of the tank are dependent on design and process technology and will be treated as constant in the following analyses.

$$g_L = \frac{1}{R_{L,p}} = \frac{1}{Q_L \cdot \omega \cdot L} \tag{5.36}$$

$$g_v = \frac{1}{R_{v,p}} = \frac{\omega \cdot C_v}{Q_v} \tag{5.37a}$$

$$g_f = \frac{1}{R_{f,p}} = \frac{\omega \cdot C_f}{Q_f} \tag{5.37b}$$

where $R_{L,,p}$, and $R_{v,,p}$ $R_{f,,p}$ are the equivalent parallel resistances of inductor, varactor and fixed capacitor respectively as shown in Figure 5.13.

5.3.2.2 Upconversion of 1/f noise in the tail transistor

The mechanism of the flicker noise upconversion was explained by Rael [5]. The common mode node of the current source oscillates at twice the oscillation frequency $2\omega_0$, because the current source is pulled every time as one of the differential transistors switches on. Through the channel length modulation, the noise of the tail current source is upconverted to $2\omega_0$.

The upconverted noise enters the LC tank and is mixed with the fundamental frequency, resulting in the phase noise sidebands at the oscillator frequency and 3rd harmonics. Therefore, to minimize the upconversion of the flicker noise and the white noise from the tail transistor, balance must be preserved, meaning that all even harmonics must be suppressed. Odd harmonics have little importance on the flicker noise upconversion because they do not affect the symmetry. It is known that the flicker noise contribution from the differential transistors is reduced due to the high frequency switching operation. The flicker noise is a correlated noise and can only exist in the systems with memory. When transistors are ideally switched, all memories and the consequent flicker noise will be removed [10]. When the switching is not ideal, a small amount of the differential MOSFET flicker noise is upconverted. Therefore, the symmetry of the cross-coupled LC tank VCO is very important and critical for suppression of the *1/f* noise upconversion.

The cross-coupled LC tank VCOs should be designed to be as symmetrical as possible in order to suppress the *1/f* noise upconversion. Additionally, a large capacitor in parallel with the tail transistor can be added (not shown in the schematic) to reduce the tail current noise at twice the oscillation frequency [3]. This will make the noise contribution from the tail transistor to be small. Thus for simplicity, its effect will be neglected in the following analyses.

5.3.3 Noise Sources in Active Devices

5.3.3.1 High frequency noise

It has been known that Non-Quasi-Static (NQS) effects influence the power spectral density of the drain current noise at very high frequencies. This is because the thermal noise in the strong inversion saturation region is the result of random potential fluctuations in the channel. These fluctuations are coupled to the gate terminal through the oxide capacitance, and cause a gate noise current to flow even if all terminal voltages are fixed [18,19]. This current is called induced gate noise.

At high frequencies, the impedance of the gate capacitance becomes smaller, and this effect becomes more pronounced. Thus, using channel noise is not enough to model the noise characteristics of the MOS transistors when it operates at high frequency. Figure 5.14 shows an equivalent small signal circuit model for high frequency noise sources of a MOS device, in which a common source configuration is used as an example.

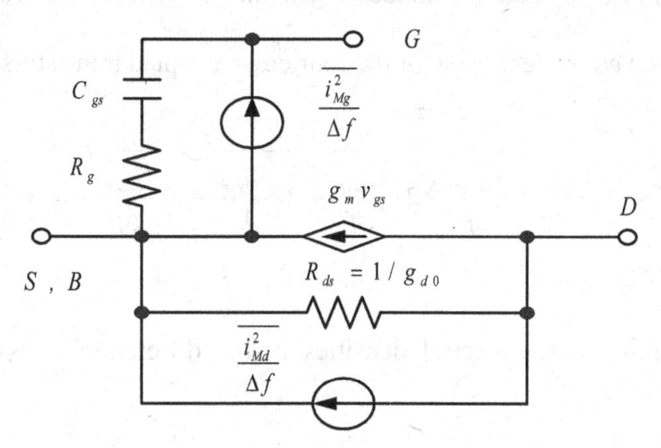

Figure 5.14 High frequency noise model for MOSFET.

C_{gs} is the capacitance between the gate and source. R_g is an equivalent resistor which produces the same amount of the thermal noise as the induced gate current, and its value is equal to $1/g_g$ (g_g is caused by distributed nature of channel). $g_m v_{gs}$ is the small signal current existing in the channel. R_{ds} is the channel resistance, which is equal to $1/g_{d0}$ (g_{d0} is defined as channel resistance at $V_{ds}=0$). The current noises $\frac{\overline{i_{Mg}^2}}{\Delta f} = 4kT\delta g_g$ and $\frac{\overline{i_{Mg}^2}}{\Delta f} = 4kT\gamma g_{d0}$ are the power spectral densities of the induced gate current noise and drain current noise, respectively. γ is the drain current noise empirical constant and δ is the gate current noise empirical constant. The cross correlation coefficient of the above current noises can be approximated as a constant in saturation region [20,21]. Therefore, the induced gate noise and drain noise can be considered as two independent noise sources in the following analyses.

5.3.3.2 Noise sources in cross-coupled transistors

$\frac{\overline{i_{n,active}^2}}{\Delta f}$ represents the total current noise power spectral density of the active device in an LC tank VCO, which is the sum of the channel current noise $\frac{\overline{i_{Md}^2}}{\Delta f}$ and the induced gate noise $\frac{\overline{i_{Mg}^2}}{\Delta f}$ [15,16]. Taking the differential equivalent noise of the four cross-coupled transistors [16]

$$\sum \frac{\overline{i_{n,active}^2}}{\Delta f} = \frac{1}{4}\left(2 \cdot \frac{\overline{i_{Md,n}^2}}{\Delta f} + 2 \cdot \frac{\overline{i_{Md,p}^2}}{\Delta f} + 2 \cdot \frac{\overline{i_{Mg,n}^2}}{\Delta f} + 2 \cdot \frac{\overline{i_{Mg,p}^2}}{\Delta f}\right) = \frac{1}{2}\left(\frac{\overline{i_{Md}^2}}{\Delta f} + \frac{\overline{i_{Mg}^2}}{\Delta f}\right) \tag{5.38}$$

The current power spectral densities in the differential cross-coupled VCO are

$$\frac{\overline{i_{Md}^2}}{\Delta f} = \frac{\overline{i_{Md,n}^2}}{\Delta f} + \frac{\overline{i_{Md,p}^2}}{\Delta f} = \gamma \cdot 4kT(g_{d0,n} + g_{d0,p}) \tag{5.39}$$

$$\frac{\overline{i_{Mg}^2}}{\Delta f} = \frac{\overline{i_{Mg,n}^2}}{\Delta f} + \frac{\overline{i_{Mg,p}^2}}{\Delta f} = \delta \cdot 4kT(g_{g,n} + g_{g,p}) \tag{5.40}$$

where, γ is the drain current noise empirical constant, which varies from 2/3 for long channel devices to 2.5 for short channel devices. The gate current noise empirical constant δ typically ranges from 0.3 to 0.35 for short channel devices and 0.395 for long channel devices [22]. In practice, δ will rise to 2 or even as large as 5 due to hot electron effects at high frequencies [20]. In order to simplify the evaluation, $\delta \approx 2\gamma$ will be adopted in the methodology.

5.3.3.3 Optimization of channel length L_{ch}

The noises from the cross-coupled NMOS and PMOS are the main contributors to the phase noise of a cross-coupled LC VCO. Reducing the channel noise and the induced gate noise is still a stringent challenge for RFIC designs. Hence, a methodology to suppress the noise sources from the NMOS and PMOS pairs is necessary to improve the phase noise performance of VCOs.

The channel conductance g_{d0} and the gate transconductance g_g are given by equation (5.41) and equation (5.42) for a MOSFET respectively [20]

$$g_{d0} = \frac{I_{dc}}{E_{sat} \cdot L_{ch}} = \frac{\alpha_1}{L_{ch}}, \text{ where } \alpha_1 = \frac{I_{dc}}{E_{sat}} \tag{5.41}$$

$$g_g = \frac{\omega^2 C_{gs}^2}{5 g_{d0}} = \frac{\omega^2 C_{gs}^2 E_{sat} \cdot L_{ch}}{5 I_{dc}} \tag{5.42}$$

where E_{sat} is the electric field at which the carrier velocity reaches its saturation velocity. Although the values of g_g and g_{d0} vary with the change of the transistors' operating points during oscillation, the bias condition for g_g and g_{d0} is $V_{gs} = V_{dd}/2$ when the voltage across the LC tank is zero (i.e., differential zero crossing point, which also means that the active devices are in saturation region). From Figure 5.11(a), it can be observed that the drain current of NMOS and PMOS I_{dc} is equal to $0.5I_{bias}$ when the oscillator is working close to differential zero. C_{gs} can be calculated as [20]

$$C_{gs} = \frac{2}{3} C_{ox} \cdot W_{ch} \cdot L_{ch} = \frac{2}{3} \cdot S \cdot C_{ox} \cdot L_{ch}^2. \tag{5.43}$$

Here, the aspect ratio of the MOS transistor $S = \frac{W_{ch}}{L_{ch}}$ is used to simplify the derivation. W_{ch} is the channel width of the MOSFET, and L_{ch} is the channel length of the MOSFET. C_{ox} is the gate oxide capacitance.

Substituting equation (5.43) into equation (5.42), the following equation can be obtained

$$g_g = \frac{4\omega^2 \cdot S^2 \cdot C_{ox}^2 \cdot E_{sat} \cdot L_{ch}^5}{45 I_{dc}} = \alpha_2 \cdot L_{ch}^5 \tag{5.44}$$

where $\alpha_2 = \frac{4\omega^2 \cdot S^2 \cdot C_{ox}^2 \cdot E_{sat}}{45 I_{dc}}$.

It is assumed that NMOS and PMOS have the same channel length L_{ch} but different aspect ratios for the analyses. The current noise spectral density of the cross-coupled LC tank VCO $\frac{\overline{i_{n,active}^2}}{\Delta f}$ in equation (5.38) is rewritten as

$$\begin{aligned} \frac{\overline{i_{n,active}^2}}{\Delta f} &= \frac{1}{2}\left(\frac{\overline{i_{Md}^2}}{\Delta f} + \frac{\overline{i_{Mg}^2}}{\Delta f}\right) \\ &= 2kT\gamma\left[2(g_{g,n} + g_{g,p}) + (g_{d0,n} + g_{d0,p})\right] \\ &= 2kT\gamma\left[2(\alpha_{2,n} + \alpha_{2,p})L_{ch}^5 + (\alpha_{1,n} + \alpha_{1,p})/L_{ch}\right] \end{aligned} \tag{5.45}$$

where $\alpha_{2,n}, \alpha_{2,p}, \alpha_{1,n}$ and $\alpha_{1,p}$ are the factors of transconductances in equation (5.44) and conductance in equation (5.41) for the NMOS and PMOS, respectively. These factors are all constants because NMOS and PMOS pairs are operating in the saturation region. The channel length, which is used to minimize the phase noise, is obtained by differential equation (5.45) with respect to the channel length as follows

$$L_{opt} = \left(\frac{\alpha_{1,n} + \alpha_{1,p}}{10(\alpha_{2,n} + \alpha_{2,p})} \right)^{1/6} \tag{5.46}$$

This means that an optimization gate length exists for any cross-coupled differential VCOs to minimize the noise contributions from active devices. Therefore, it breaks the conventional rule for RFIC designs to adopt the minimum technology size for the gate length of the active device to improve the phase noise performance.

The phase noise minimization of the cross-coupled LC tank VCO is achieved when the gate length of the active device is equal to the optimum gate length L_{opt}. This insight can be very valuable in the efforts to improve the phase noise performance of VCOs for the design engineers. The total thermal noise of the LC tank VCO can be expressed by the following equation

$$\sum \frac{\overline{i_n^2}}{\Delta f} = \frac{1}{2}\left(\frac{\overline{i_{n,active}^2}}{\Delta f} + \frac{\overline{i_{n,tank}^2}}{\Delta f}\right)$$
$$= (2kT\gamma g_{d0,n} + 2kT\delta g_{g,n} + 2kT\gamma g_{d0,p} + 2kT\delta g_{g,p}) + (2kTg_L + 2kTg_v + 2kTg_f) \tag{5.47}$$

5.3.4 Linear Time Variant (LTV) Phase Noise Analysis

Linear time variant analysis method introduces an impulse sensitivity function (ISF) for the excess phase of the LC tank VCO and it is a general method to calculate the total phase noise of an oscillator with multiple nodes and multiple noise sources [15].

5.3.4.1 Definition of Impulse Sensitivity Function (ISF) $-\Gamma(\omega_0 t)$

Impulse Sensitivity Function (ISF) is utilized to analyze the impact of the current noise on the amplitude and phase through their associated impulse responses, i.e., its impact on the phase and amplitude changes if the time at which the current impulse is injected is varied. Therefore, the impact of the current noise is time-varying as shown in Figure 5.15.

It can be observed that when the impulse occurs close to the differential zero of the VCO output, the impact on the phase is at the maximum, as marked by $\Delta\Phi_{max}$ in Figure 5.15. Furthermore, when the impulse occurs at peak of the output, it causes small impact on the phase. On the other hand, conversely is true for the amplitude deviations.

However, the amplitude deviations are suppressed to zero by an automatic gain control (AGC) circuit or the intrinsic nonlinearity of the device. The phase deviations are accumulated in the whole system, and it continues indefinitely. Consequently, the effect of the phase will be the focus of the following analyses.

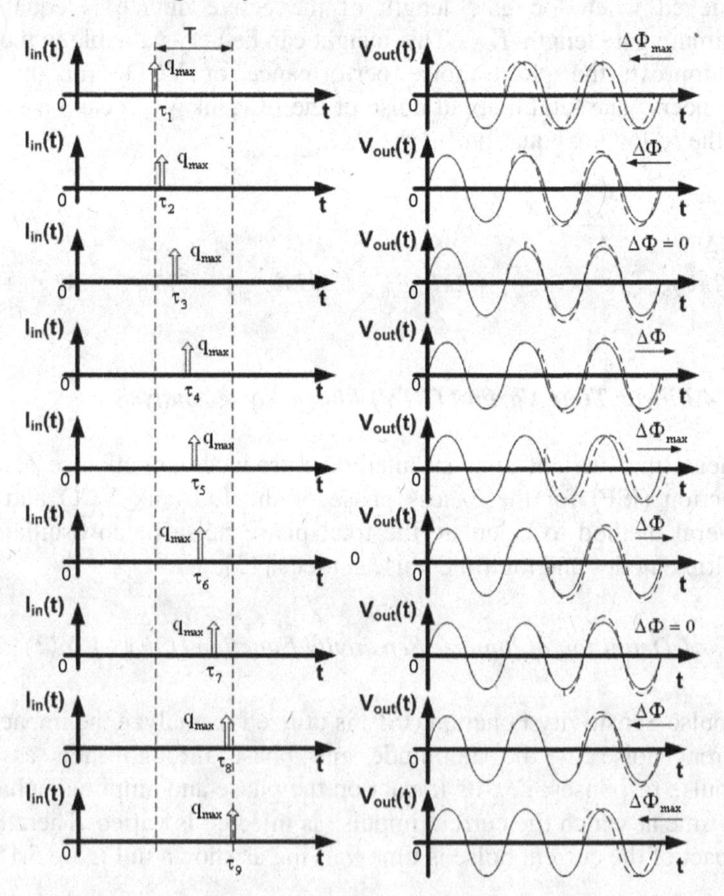

Figure 5.15 Impulse response of an ideal LC oscillator when the current noise is injected at the varying time.

ISF is constructed by calculating the phase deviations as the impulse position is varied, thus it is periodic function with the same period as the VCO output as shown in Figure 5.16.

For the differential output, ISF is approximately proportional to the derivative of the VCO output waveform, and the unit phase impulse response $h_\Phi(t,\tau)$.

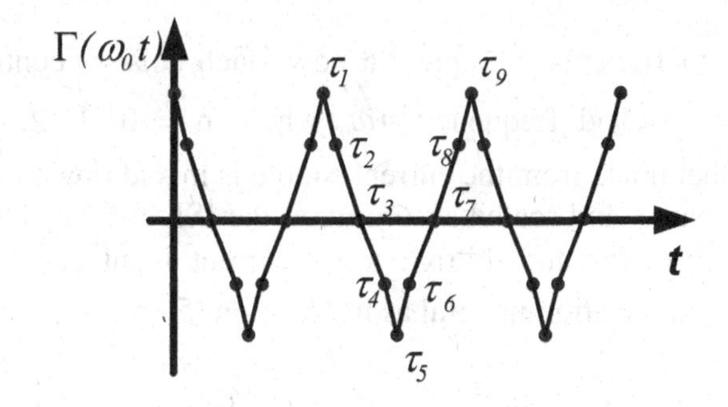

Figure 5.16 ISF Function.

5.3.4.2 Parameterized phase impulse response $h_\Phi(t, \tau)$ using ISF

$h_\Phi(t,\tau)$ is used to characterize the phase deviations, which is defined in equation (5.48). Here $u(t)$ is the unit step and τ is the time when the impulse is injected. The equivalent system of the phase noise is modeled in Figure 5.17.

$$h_\Phi(t,\tau)=\frac{\Gamma(\omega_0\tau)}{q_{\max}}u(t-\tau) \tag{5.48}$$

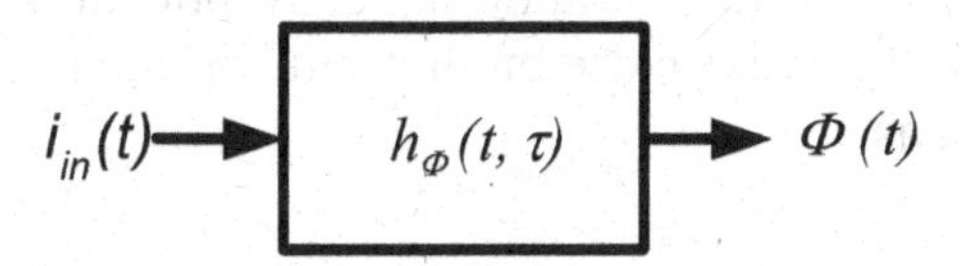

Figure 5.17 Equivalent system of the phase noise.

Since ISF is a periodic function at frequency ω_0, only noise around DC, ω_0 and its harmonics will result in non–zero excess phase. Noise at all other frequencies will average out over time. ISF can be expanded in Fourier series as

$$\Gamma(\omega_0\tau) = \frac{c_0}{2} + \sum_{n=1}^{\infty} c_n \cos(n\omega_0\tau + \theta_n) \tag{5.49}$$

where the coefficients c_n represent how much noise is contributed from the vicinity around frequency $n\omega_0$ where n = 0, 1, 2.... It can be observed that noise from the current source is mixed down from different frequency bands and scaled according to the ISF coefficients. The phase deviation $\Phi(t)$ for an arbitrary noise current input can be calculated using the superposition integral as in equation (5.50).

$$\Phi(t) = \int_{-\infty}^{\infty} h_\Phi(t,\tau) i(\tau) d\tau = \frac{1}{q_{max}} \int_{-\infty}^{t} \Gamma(\omega_0\tau) i(\tau) d\tau \tag{5.50}$$

where $i(\tau)$ is the current impulse injected at the node of the circuit, and $\Gamma(\omega_0\tau)$ is ISF as expressed by equation (5.49). ISF is essentially a transfer function between an arbitrary noise source and an excess phase at the output of the oscillator. q_{max} is the maximum charge swing, and $q_{max} = C_{node} V_{max}$, where V_{max} is the voltage swing across the capacitor at the output node of the circuit.

5.3.4.3 Phase noise calculation

Substituting equation (5.47) into the conventional SSB phase noise expression [15], the phase noise of an oscillator due to the thermal noise can be rewritten as

$$L(\Delta\omega) = 10\log\left(\frac{\Gamma_{rms}^2}{q_{\max}^2} \cdot \frac{\sum \frac{\overline{i_n^2}}{\Delta f}}{2 \cdot (\Delta\omega)^2}\right) \tag{5.51}$$

and the single-side bandwidth (SSB) phase-noise of an oscillator due to the flicker noise can be rewritten as

$$L(\Delta\omega) = 10\log\left(\frac{c_0^2}{q_{\max}^2} \cdot \frac{\sum \frac{\overline{i_n^2}}{\Delta f}}{8 \cdot (\Delta\omega)^2} \cdot \frac{\omega_{1/f}}{\Delta\omega}\right) \tag{5.52}$$

The quantity Γ_{rms} is the *rms* value of ISF. In this chapter, all ISFs were obtained using Spectre Time Simulations (Periodic Steady-State) by injecting a small current pulse into an oscillator node over one oscillation cycle, and observing the output phase shifts several cycles later. $\omega_{1/f}$ is the corner frequency of the device *1/f* noise, and c_0 is the DC value of ISF, which can be calculated by $c_0 = \frac{1}{2\pi}\int_0^{2\pi}\Gamma(x)dx$.

In an oscillator, some of the noise sources may change cyclically, or in other words, change with time in a periodic manner. These noise sources, for example, the channel noise of the active devices, are referred to as cyclostationary. On the other hand, stationary noises are noise sources in the oscillator whose statistical properties do not depend on time and the operation point of the circuit, for example, resistor's thermal noise.

ISF contains only the sensitivity to noise as a function of the time. It does not include the information of the noise amplitude modulation due to cyclostationary noise. $\alpha(x)$ is used to represent the deterministic periodic function describing the noise amplitude modulation, it is defined to be a normalized function with a maximum value of 1. Hence, the effective ISF is given by [15]

$$\Gamma_{rms,eff}(x) = \Gamma_{rms}(x) \times \alpha(x) \tag{5.53}$$

where $\alpha(x)$ represents the Noise Modulation Function (NMF). $\alpha(x)$ can be derived from the device noise characteristics.

5.3.4.4 Steps to achieve minimal phase noise

The steps of the methodology to achieve low phase noise are summarized as:

Step 1: Calculate the optimized gate length of the active device using equation (5.46) to minimize the spectral density of each oscillator noise source.

Step 2: Derive the impulse sensitivity function of each oscillator noise source from the transient simulation with a current noise injected at the noise source node of the oscillator.

Step 3: Combine the above results to obtain $\Gamma_{rms,eff}(\omega_0 t)$ for each oscillator noise source.

Step 4: Calculate the Fourier series coefficients for each $\Gamma_{rms,eff}(\omega_0 t)$.

Step 5: Calculate the overall output phase noise using the results from steps 3 and 4 and the phase noise expressions in equation (5.51) and equation (5.52).

In order to verify the methodology, two LC tank VCO design examples will be presented in the next two sections.

5.3.5 A 2GHz Cross-Coupled LC Tank VCO

5.3.5.1 2GHz cross-coupled LC tank VCO

Figure 5.13 shows the schematic of the 2GHz LC tank VCO design example that employs the methodology described above. Six different cases with different sizes of the cross-coupled NMOS and PMOS are given in Table 5.2. The resonant tank consists of a spiral inductor of 4nH with a quality factor of 8 and two PN varactors with tuning range from 2pF to 2.8pF and a quality factor of 40. In order to ensure the six VCOs operate at the same frequency of 2GHz, capacitors C_{f1} and C_{f2} shown in Figure 5.13 are used to compensate for the capacitance variations of the tank due to the different sizes of the NMOSs and PMOSs. In addition, the aspect ratio S for NMOSs and PMOSs in the six cases are fixed. N is the number of fingers, where each finger has the aspect ratio of W_{ch}/L_{ch}, thus the aspect ratio for each transistor is given by $S = NW_{ch}/L_{ch}$.

Table 5.2 Aspect ratios of NMOS and PMOS devices for the 2GHz VCOs.

Case	NMOS				PMOS				C_1/C_2 (fF)
	L_{ch} (μm)	W_{ch} (μm)	N	S	L_{ch} (μm)	W_{ch} (μm)	N	S	
1	0.18	2.5	8	111	0.18	2.5	20	278	933
2	0.27	3.75	8	111	0.27	3.75	20	278	863
3	0.36	5	8	111	0.36	5	20	278	733
4	0.45	6.25	8	111	0.45	6.25	20	278	582
5	0.54	7.5	8	111	0.54	7.5	20	278	399
6	0.63	8.75	8	111	0.63	8.75	20	278	182

5.3.5.2 Verifications and discussions

Based on the above component sizes and parameters, an optimum channel length of 0.26μm is calculated by using equation (5.46). All the six cases were simulated using Spectre-RF of Cadence (EDA tool), and the simulated phase noise performances at an offset frequency of 600kHz are shown in Figure 5.18.

Note that the VCO with the channel length of 0.27μm in Case 2 has the lowest phase noise of −123dBc/Hz, which confirms the analyses in Section 5.3.3.3.

As an additional note, the default value for γ is 2/3 in BSIM3 model. However, in this design, γ of around 2.0 is extracted when MOSFETs with 0.18μm gate length are working in the saturation region ($V_{ds} = V_{gs} \approx$ 0.9V). In addition, γ can be approximated as $\gamma = \frac{2}{3}\frac{g_m}{g_o}$ when the transistors operating in the saturation region [23].

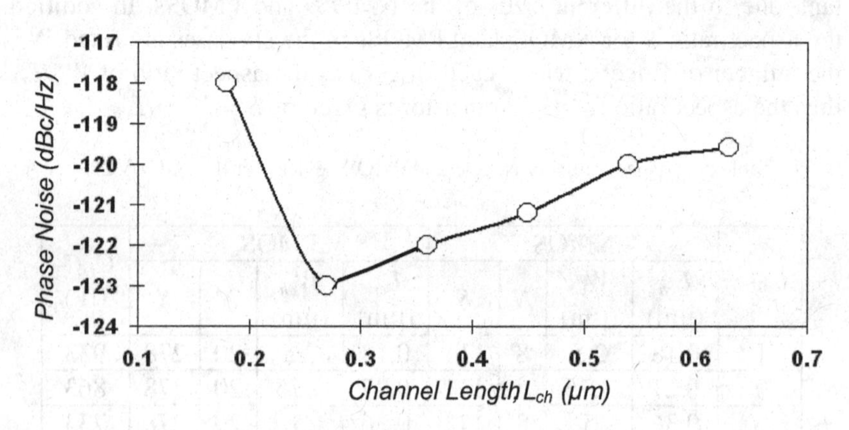

Figure 5.18 Simulated phase noise performances of the six VCOs.

5.3.5.3 Experimental results

The case 2 VCO was fabricated using CSM 0.18μm RF CMOS technology. Figure 5.19 shows the microphotograph of the VCO. The area is approximately 0.5 × 0.6 mm^2.

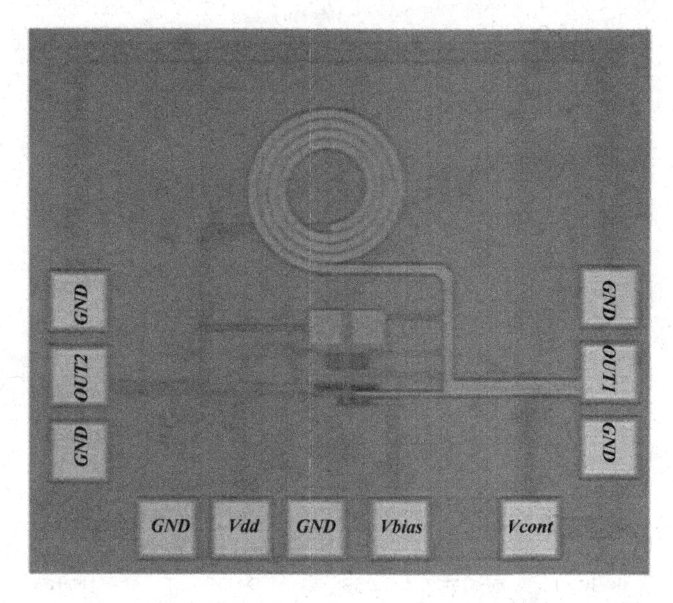

Figure 5.19 Microphotograph of the VCO.

The phase noise performance shown in Figure 5.20 was measured by HP8563E Spectrum Analyzer. In order to measure the phase noise accurately, a G-P-G (Ground-Power-Ground) DC probe and an HP11612B K21 Bias Network were utilized at the power supply V_{dd}, which was at 1.8V. The DC bias of the tail transistor V_{bais} was at 0.8V and the control voltage V_{cont} (N+ terminal of the PN varactor) was at 1.5V. Note that the control voltage between two terminals of the PN varactor was around 0.6V because P+ was connected to the output at the DC level of 0.9V.

A low phase noise of −103dBc/Hz@100kHz, −118dBc/Hz@600kHz and −128dBc/Hz@3MHz away from the carrier frequency of 1.968GHz were achieved. The measured phase noise is around 5dB higher than the simulated phase noise. This 5dB discrepancy can be attributed to the uncertain channel noise factor γ, which was around 2 in the implementation rather than 2/3, and the degradation of tank amplitude due to the parasitic resistors in metal layers. Furthermore, as expected, the oscillation frequency slightly shifted down to 1.968GHz, due to the parasitic capacitance [24-26].

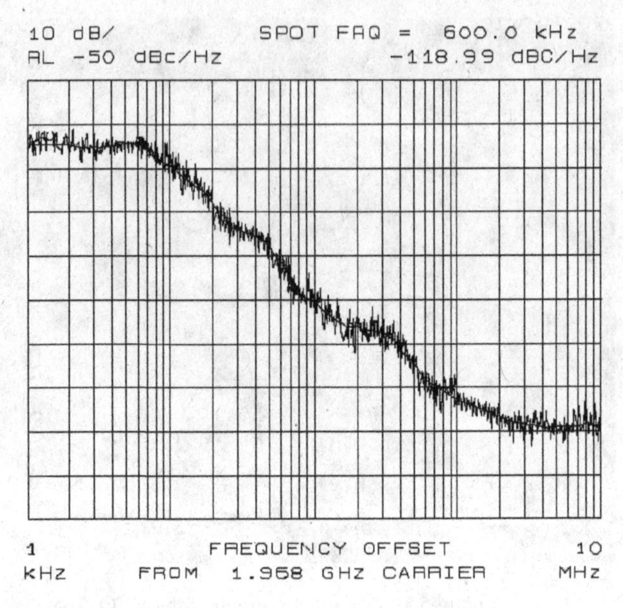

Figure 5.20 Measured phase noise of the 2GHz VCO.

An optimized gate length of 0.27μm for the lowest phase noise performance is obtained by applying the methodology. This VCO has a tuning range from 1.9GHz to 2.18GHz as depicted in Figure 5.21.

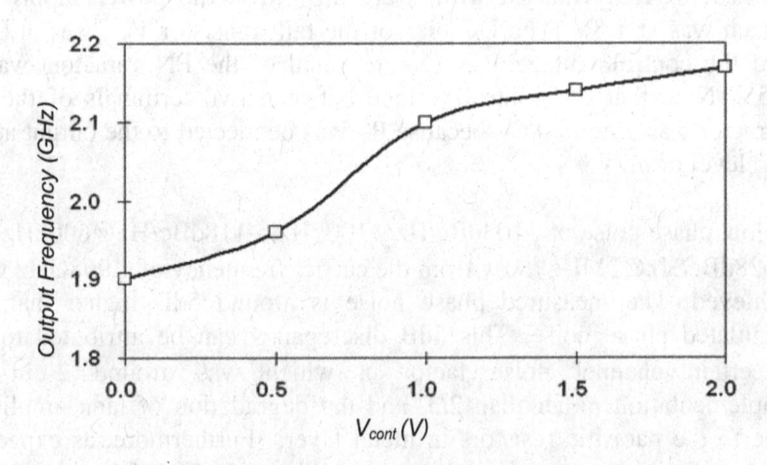

Figure 5.21 Measured tuning range of the LC VCO.

The total current dissipated by the VCO was 1.75mA, and the total power consumption of the VCO *(P_{vco})* was obtained to be 3.15mW. Using equation (5.21), the *FOM* for case 2 is obtained to be 186, which is the highest *FOM* value compared with the reported literatures as shown in Table 5.3.

Table 5.3 Measured results compared with other CMOS VCOs.

Ref.	Tech.	Frequency (GHz)	Phase Noise (dBc/Hz)	Power Consumption (mW)	*FOM*
This work	0.18μm*	2.0	−118.9	3.15	186
[26]	0.25μm*	1.8	−121	6	182.8
[24]	0.65μm**	2.0	−116	6	177.1
[25]	0.35μm*	2.03	−117	10	177.5
[28]	0.65μm**	2.0	−126	34.2	181.1

Note: * - CMOS; ** - BiCMOS; Phase noise was obtained at 600kHz offset.

5.3.6 A 9.3 ~10.4GHz Cross-Coupled Complementary Oscillator

Based on the phase noise optimization methodology, the optimized gate length of the active device for a 10GHz LC VCO can be calculated using Equation (5.46).

The schematic of this design example is given in Figure 5.22, in which an impulse injection current source is connected to the output of the LC tank VCO.

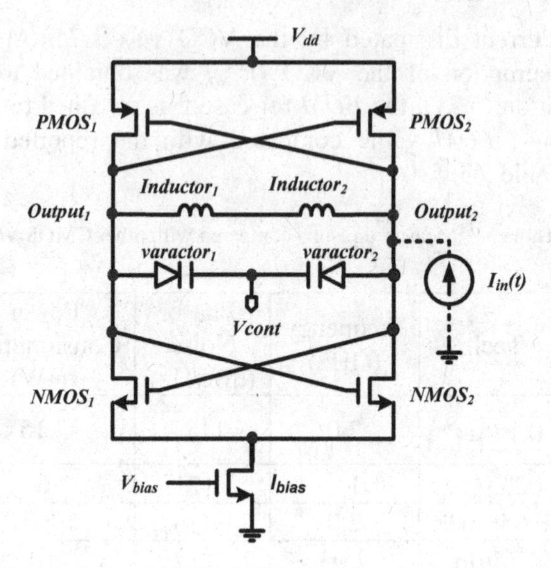

Figure 5.22 Schematic of 10GHz LC VCO with an impulse injection current.

5.3.6.1 Phase noise estimation for 10GHz LC tank VCO

The simulated ISFs with an injected charge of 0.1pC are shown in Figure 5.23(a). The NMOS and PMOS noise modulation functions (NMFs) are presented in Figure 5.23(b), and the effective ISFs are shown in Figure 5.23(c). According to equation (5.53), the effective Γ_{rms} of all active devices' noise sources, $\Gamma_{rms,n,eff} = 0.19$ and $\Gamma_{rms,p,eff} = 0.15$ are obtained when the LC VCO operates at 9.83GHz.

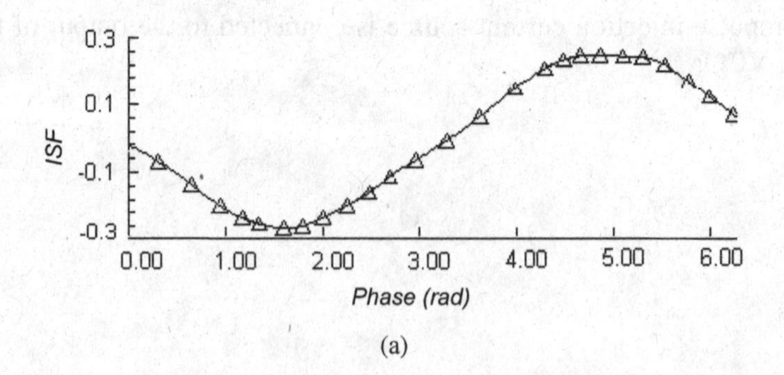

(a)

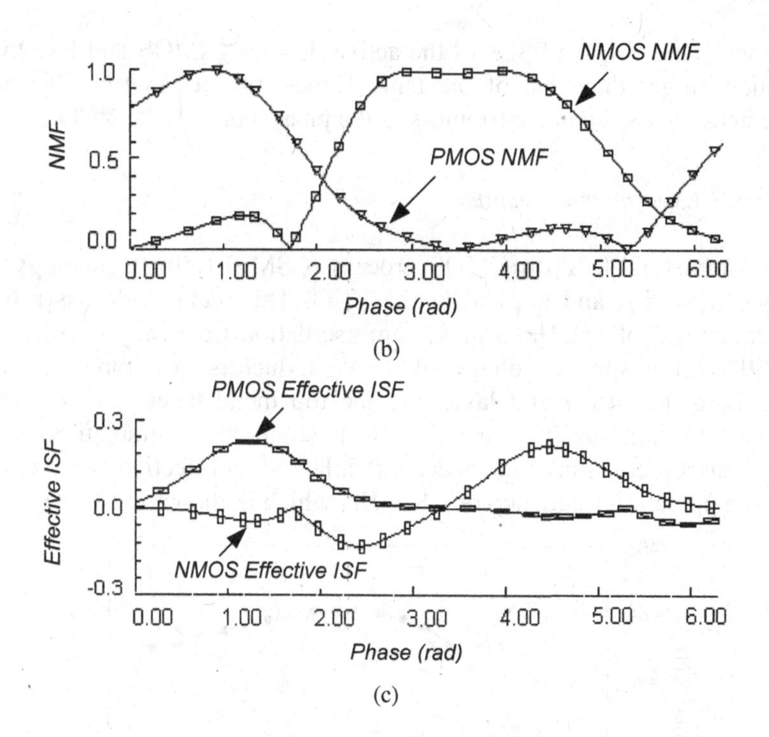

Figure 5.23 (a) ISF (Γ_{rms} (x)) of NMOS and PMOS; (b) NMF ($\alpha(x)$) of NMOS and PMOS; (c) effective ISF ($\Gamma_{rms,eff}$ (x)) of NMOS and PMOS.

The effective ISFs of the active device and LC tank are summarized in Table 5.4. Subsequently, the phase noise of –93dBc/Hz@100kHz away from the carrier of 9.83GHz was calculated using equation (5.51). Similarly, the phase noises of –95dBc/Hz@100kHz and –89dBc/Hz@100kHz were calculated when the VCO operates at 9.3GHz and 10.4GHz respectively.

Table 5.4 PSD of all noise sources and effective ISF.

Noise Source	PSD (A^2/Hz)	Γ_{rms}
NMOS	2.86e-22	$\Gamma_{rms,n,eff} = 0.19$
PMOS	1.41e-22	$\Gamma_{rms,p,eff} = 0.15$
Tank	8.10e-23	$\Gamma_{rms,tank} = 0.28$

From Table 5.4, the PSDs of the active devices (NMOS and PMOS) are much bigger than that of the tank. Hence, the noises from the active devices are the main contributors to the phase noise of the VCO.

5.3.6.2 Experimental results

A commercial 0.18μm CMOS process (CSM 0.18μm technology) was used to design and implement the VCO. This technology has a transit frequency f_t of 58GHz, a maximum oscillation frequency f_{max} of 67GHz [29], and a supply voltage of 1.8V. Inductors are implemented by stacking the fifth metal layer and the top metal layer (the sixth metal layer) to minimize the metal and substrate loss. Simulations of the inductors predict an inductance of 0.5nH and an effective Q value of 10 at the operation frequency of 9.83GHz, which is shown in Figure 5.24.

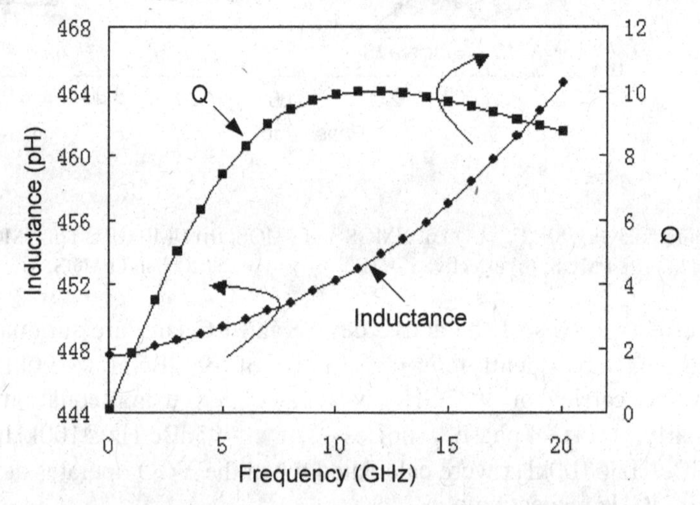

Figure 5.24 Simulated inductance and Q value of an inductor.

The P+/N-well varactor of the design incorporates a capacitance of C_{var} = 0.5pF with the Q value of around 38 at 9.83GHz, and the ratio of maximum varactor capacitance over minimum varactor capacitance (C_{max}/C_{min}) is about 2.5 with 0V < V_{var} < 4V [30], where V_{var} is the voltage applied across the varactor.

The area of the die is approximately 0.5 x 0.6 mm^2 as shown in Figure 5.25. The VCO was measured on wafer. A G-S-G RF probe was used at the outputs of the oscillator (*OUT1* or *OUT2* as shown in Figure 5.22). The G-P-G DC probe connected to the Bias Network was adopted for power supply V_{dd} to filter off noise from the DC supply equipment HP4142.

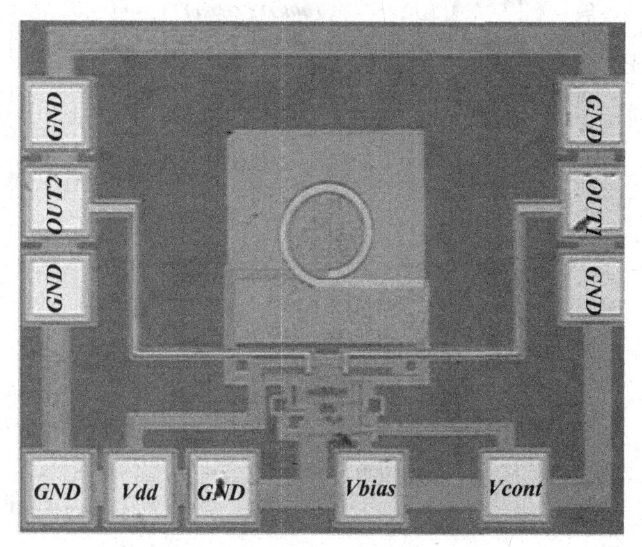

Figure 5.25 Die photo of the LC tank oscillator.

The phase noise and the power spectral density were measured using HP8564E spectrum analyzer and the phase noise characteristic for the 9.83GHz oscillator is shown in Figure 5.26. The phase noises are –91dBc/Hz@100kHz, –89dBc/Hz@100kHz and –84dBc/Hz@100kHz from the oscillation frequency of 9.3GHz, 9.83GHz and 10.4GHz respectively.

The *FOM* can be calculated using equation (5.21). Figure 5.27 depicts the curves of the oscillation frequency and *FOM* versus the controlled voltage (which is the voltage difference between the cathode and anode of the PN varactors) from the measurements. The tuning range of 1.1GHz and *FOM* value of 165 at operation frequency 9.83GHz were achieved.

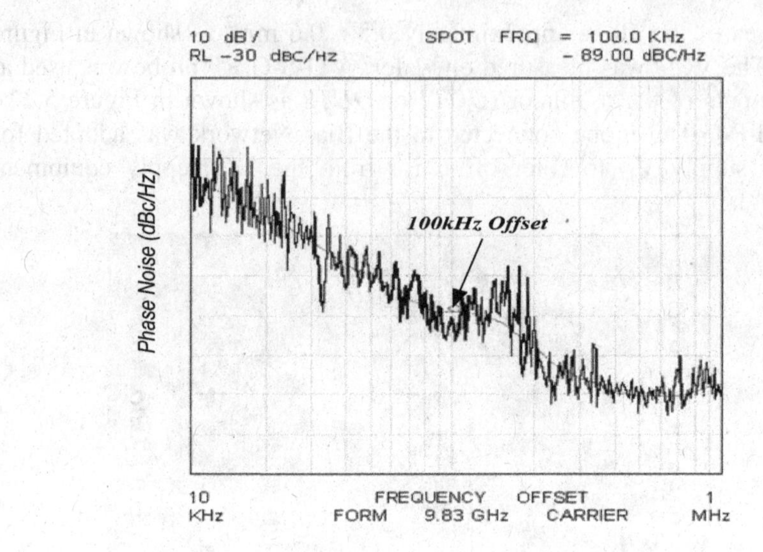

Figure 5.26 Measured phase noise at 9.83GHz.

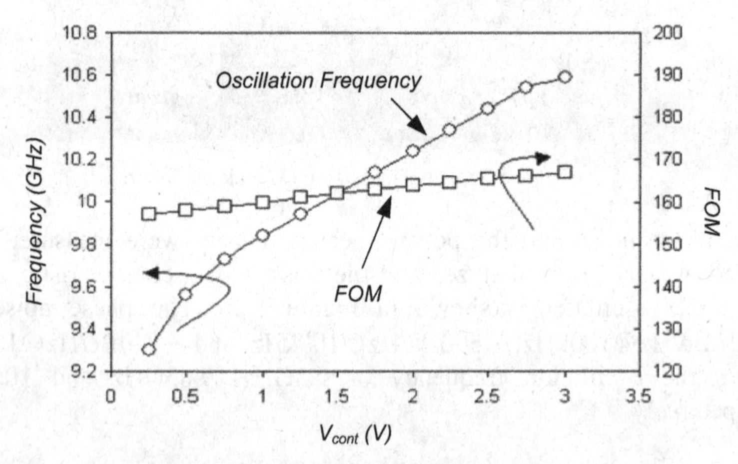

Figure 5.27 Tuning characteristics and *FOM* of the VCO.

The phase noise performance from the measurements and LTV method for 9.83GHz VCO is summarized in Table 5.5. The phase noise calculated using LTV analysis has a good agreement compared with the measured phase noise at a 100kHz offset frequency, and the maximum difference of about 5dB is due to the degradation of tank amplitude due to the parasitic resistors in metal layers. The performance of the 10GHz VCO is one of the best VCOs reported in the literature using 0.18μm CMOS technology.

Table 5.5 Performance of the 10GHz VCO.

Parameters	Measured	LTV method
Supply Voltage	1.8V	1.8V
Core circuit power supply	5.8mW	5.6mW
Center Frequency	9.83GHz	10.2GHz
Tuning Range	11.2%	15.6%
Phase noise (f_0=9.3GHz@100kHz)	–91dBc/Hz	–95dBc/Hz
Phase noise (f_0=9.83GHz@100kHz)	–89dBc/Hz	–93dBc/Hz
Phase noise (f_0=10.4GHz@100kHz)	–84dBc/Hz	–89dBc/Hz

5.4 Summary

This chapter begins with the basic concepts and terminologies used in an oscillator, it is then followed by a brief discussion of various LC oscillators. Then, the detailed methodology of using optimized channel length for the design of LC VCO is shown.

The methodology was verified through two design examples. One is a 2GHz cross-coupled LC tank VCO with power consumption of 3.15mW. The measured phase noises were obtained to be −103.3dBc/Hz@100kHz, −118.9dBc/Hz@600kHz and −128dBc/Hz@3MHz from the carrier frequency of 1.968GHz. The *FOM* is 186. This VCO has one of the best *FOM* performance among all the 2GHz VCOs fabricated in 0.18μm CMOS technology reported in the literature.

The second VCO has a 1.1GHz tuning range from 9.3GHz to 10.4GHz. This VCO has a low phase noise of –89dBc/Hz@100kHz from the center frequency of 9.83GHz. It was also designed and fabricated using CSM 0.18μm technology. The power consumption of the core circuit of VCO is 5.8mW and the output peak-to-peak voltage is around 1.1V.

In conclusion, this methodology has the capability to minimize the phase noise performance of an LC tank VCO and the phase noise predictions made by the LTV phase noise analysis are in good agreement with the measurement results.

References

[1] B. Razavi, "A Study of Phase Noise in CMOS Oscillators", *IEEE Journal of Solid-State Circuits,* vol. 31, no. 3, pp. 331-343, Mar. 1996.

[2] M. Thamsirianunt and T. Kwasniewski, "CMOS VCOs for PLL Frequency Synthesis in GHz Digital Mobile Radio Communications", *Proceedings of the IEEE CICC*, pp. 331-334, May 1995.

[3] Hajimiri and T. H. Lee, *The Design of Low Noise Oscillator*, Kluwer Academic Publishers, 1999.

[4] Leeson, "A Simple Model of Feedback Oscillator Noise Spectrum", *Proceedings of IEEE*, vol. 54, pp. 136-154, Feb. 1966.

[5] J. J. Rael and A. A. Abidi, "Physical Processes of Phase Noise in Differential LC Oscillators", *Proceedings of CICC 2000*, pp. 569-572, 2000.

[6] M. Tiebout, "Low Power Low-Phase-Noise Differentially Tuned Quadrature VCO Design in Standard CMOS", *IEEE Journal of Solid-State Circuits*, vol. 36, no. 7, pp. 1018-1024, Jul. 2001.

[7] C. C. Boon, M. A. Do, K. S. Yeo, J. G. Ma, and X. L. Zhang, "RF CMOS Low-Phase-Noise LC Oscillator Through Memory Reduction Tail Transistor *IEEE Transaction On Circuits and Systems- II: Express Briefs, Analog and Digital Signal Processing*, Vol. 51, No. 2, Feb. 2004.

[8] C. Samori, A. L. Lacaita, F. Villa, and F. Zappa, "Spectrum Folding and Phase Noise in LC Tuned Oscillators", *IEEE Transactions on Circuits and Systems – II*, vol. 45, pp. 781-790, Jul. 1998.

[9] E. A. M. Klumperink, S. L. J. Gierkink, A. P. van der Wel, and B. Nauta, "Reducing MOSFET 1/f Noise and Power Consumption by Switch Biasing", *IEEE Journal of Solid-State Circuits*, vol. 35, no. 7, pp. 994-1001, Jul. 2000.

[10] S. L. J. Gierkink, E. A. M. Klumperink, A. P. van der Wel, G. Hoogzaad, E. van Tuijl, and B. Nauta, "Intrinsic 1/f Device Noise Reduction and Its Effect on Phase Noise in CMOS Ring Oscillators", *IEEE Journal of Solid-State Circuits*, vol. 34, no. 7, pp. 1022-1025, Jul. 1999.

[11] Hung, P. K. Ko, C. Hu, and Y. C. Cheng, "A Unified Model for Flicker Noise in Metal-Oxide-Semiconductor Field-Effect Transistors", *IEEE Trans. Electron Devices*, vol. 37, pp. 654-665, Mar. 1990.

[12] E. Hegazi, H. Sjoland, and A. A. Abidi, "A Filtering Technique to Lower LC Oscillator Phase Noise", *IEEE Journal of Solid-State Circuits*, vol. 36, no. 12, pp. 1921-1930, Dec. 2001.
[13] G. F. Svelto and R. Castello, "A bond–wire inductor–MOS varactor VCO tunable from 1.8GHz to 2.4GHz," *IEEE Trans. Microwave Theory Techn.*, Vol.50, No.1, pp.403–407, Jan. 2002.
[14] Hajimiri and T. H. Lee, "Design issues in CMOS differential LC oscillators," *IEEE J. Solid–State Circuits*, Vol.34, No.5, pp.717–724, May 1999.
[15] Hajimiri and T. H. Lee, " A general theory of phase noise in electrical oscillators," *IEEE J.Solid–State Circuits*, Vol.33, No.2, pp.179–194, Feb. 1998.
[16] D. Ham and A. Hajimiri, "Concepts and methods in the optimization of integrated LC VCOs," *IEEE J. Solid–State Circuits*, Vol.36, No.6, pp.896–909, Jun. 2001.
[17] G. F. Svelto and R. Castello, "A bond–wire inductor–MOS varactor VCO tunable from 1.8GHz to 2.4GHz," *IEEE Trans. Microwave Theory Techn.*, Vol.50, No.1, pp.403–407, Jan. 2002.
[18] V. D. Ziel, "Thermal noise in field effect transistors," *Proc. IEEE*, pp.1801–1812, Aug. 1962.
[19] L. J. Pu and Y. Tsividis, "Small signal parameters and themal noise of four terminal MOSFET in non–quasistatic operation," *Solid–State Electronics*, Vol.33, pp.521, 1990.
[20] Y. Tsividis, "*Operation and modeling of the MOS Transistor*", McGraw–Hill International Editions 1999.
[21] M. Shoji, " Analysis of high frequency thermal noise of enhancement mode MOS field effect transistors," *IEEE Trans. Electron Devices*, Vol.13, No.6, pp.520–524, Jun. 1996.
[22] M. J. Deen and T. A. Fjeldly, "*CMOS RF Modeling, Characterization and Applications, Selected Topics in Electronics and Systems–Vol.24*", World Scientific, pp.192–193, Apr. 2002.
[23] D. P. Triantis, A. N. Birbas, and D. Kondis, " Thermal noise modeling for short–channel MOSFET's," *IEEE Trans. Electron Devices*, Vol.43, No.11, pp. 1950–1955, Nov. 1996.
[24] D. Muer, M. Steyaert, and G. L. Puma, "A 2GHz low–phase noise integrated LC-VCO set with flick–noise upconversion minimization," *IEEE J. Solid–State Circuits*, Vol.35, No.7, pp.1034–1038, Jul. 2000.
[25] J. Craninckx and M. S. J. Steyaert, "A 1.8GHz low–phase–noise CMOS VCO using optimized hollow spiral inductors," *IEEE J. Solid–State Circuits*, Vol.32, No.3, pp.736–744, Mar. 1997.
[26] P. Andreani, "A low–phase–noise low–phase–error 1.8GHz quadrature CMOS VCO," *Proc., ISSCC 2002*, pp.228–229, 2002.
[27] P. Kinget, "*Integrated GHz Voltage Controlled Oscillator*"s, Klower Academic Publishers, pp.353–381, 1999.
[28] J. J. Kucera, " Wideband BiCMOS VCO for GDMUMT direct convention receivers," *Proc., ISSCC 2000*, pp.375–375, 2000.
[29] CSM 0.18 μm process design manual. pp.1–120, 2002.
[30] T. P. Liu, "1.5V 10–12.5GHz integrated CMOS Oscillators," *Proc., Symposium on VLSI Circuit Digest of Technical Papers*, pp.55–56, 1999.
[31] E. A. M. Klumperink, S. L. J. Gierkink, A. P. van der Wel, and B. Nauta, "Reducing MOSFET 1/f Noise and Power Consumption by Switch Biasing", *IEEE Journal of Solid-State Circuits*, vol. 35, no. 7, pp. 994-1001, Jul. 2000.

Success does not come by just finding the perfect solution, but by learning to see and understanding the problem perfectly.

Kiat Seng YEO

CHAPTER 6

RF CMOS Phase-Locked Loops

The local oscillator of a transceiver achieves the stability, accuracy and programmability of its RF output frequency by using a phased-locked loop (PLL). The PLL offers the universally-accepted method for synthesizing an RF frequency equal to an integral or fractional number of a reference frequency generated by a crystal oscillator. The theory of PLL and the design methodology are well documented in many text books [1-4]. Recent attention has been drawn to the CMOS/BiCMOS modern transceiver system-on-chip (SoC), where a larger silicon area is occupied by one or more fully-integrated PLLs [5-7]. The performance of most PLL frequency synthesizers is usually decided by the two most critical RF building blocks, namely, the voltage controlled oscillator (VCO) described in Chapter 5 and the programmable frequency divider in Chapter 7. While the application of the PLL in the demodulation of FM (frequency modulated) signals has never been popular in analog transceivers, its application in modern data recovery circuits (DRC) for data rates of several gigabits/second in optical communications has pushed the design of the phase-frequency detector (PFD) into the RF range, making the implementation of CMOS/BiCMOS PLL one of the most challenging task in RFIC design [8-11].

6.1 Fundamental Principles of a Phase-Locked Loop (PLL)

A PLL is a control system, where phase is the variable of interest. The block diagram of a PLL is shown in Figure 6.1. The circuit is called a phase-locked loop because the feedback operation in the loop automatically adjusts the phase of the output signal F_{out} to follow the phase of the reference signal F_{ref}. Under the locked condition, the phase difference between F_{ref} and F_{out}/N is constant with time, and their frequency difference is zero. Hence, $F_{out} = N.\ F_{ref}$. Figure 6.2 shows the state variable diagram of the PLL, the circuit is constructed by using the phase of the reference signal $\theta_{ref}(t)$ and the phase of the output signal $\theta_{out}(t)$ as loop variables. The frequency divider divides the VCO frequency (and phase) by a division modulus of N. Let $\theta_{div}(t)$ be the phase of the output signal of the divider, the following equation is obtained:

$$\theta_{div}(t) = \theta_{out}(t)/N \tag{6.1}$$

In order to analyze the steady state transfer function of the PLL, assumptions are made that a constant reference frequency is applied and the division modulus is fixed. Now, assuming that the loop is locked and the phase detector gives an output voltage $v_{pd}(t)$ proportional to the phase difference between its inputs $\theta_{ref}(t)$ and $\theta_{div}(t)$.

$$v_{pd}(t) = K_{pd}\left[\theta_{ref}(t) - \theta_{div}(t)\right] \tag{6.2}$$

where K_{pd} is the gain of the phase detector in V/rad.

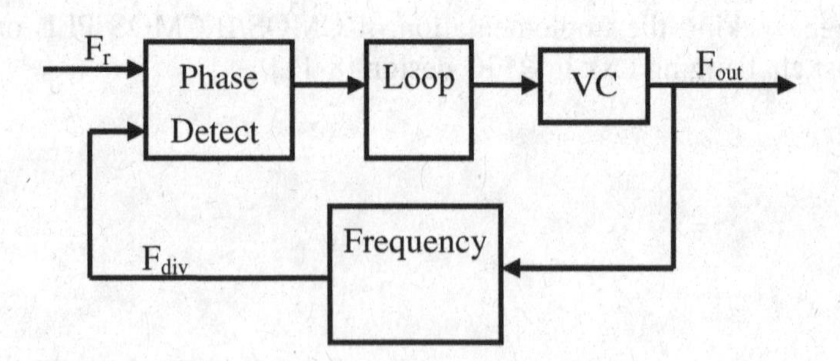

Figure 6.1 Block diagram of a PLL.

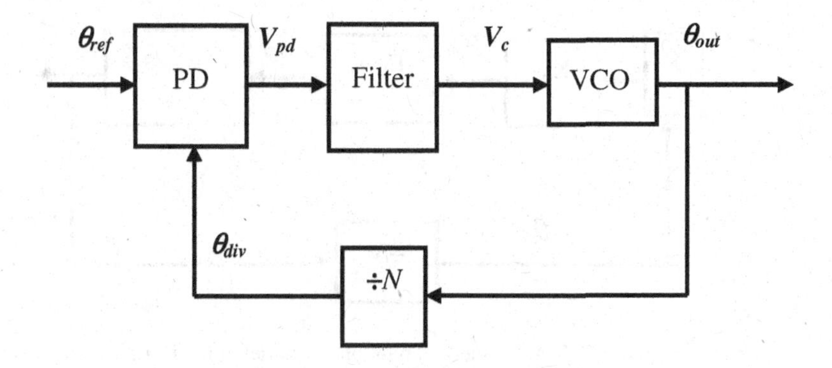

Figure 6.2 State variable diagram of a PLL.

The phase detector output voltage is then passed through the low-pass filter where the high frequency signal components are suppressed. The loop filter determines largely the noise and dynamic performance of the loop. An ideal voltage-controlled oscillator (VCO) generates a periodic output whose frequency is a linear function of the control voltage $v_c(t)$ and is given by:

$$\omega_{out}(t) = \omega_{fr} + K_{vco} \times v_c(t) \tag{6.3}$$

where ω_{fr} is the free running frequency of the VCO and K_{vco} is the gain of the VCO specified in rad/s/V. Since frequency is the derivative of phase, the VCO operation can be described by equation (6.4) where the constant ω_{fr} is excluded from the phase model.

$$\frac{d\theta_{out}(t)}{dt} = K_{vco} \cdot v_c(t) \tag{6.4}$$

Taking the Laplace transform of (4), we have

$$\theta_{out}(s) = \frac{K_{vco} \cdot V_c(s)}{s} \tag{6.5}$$

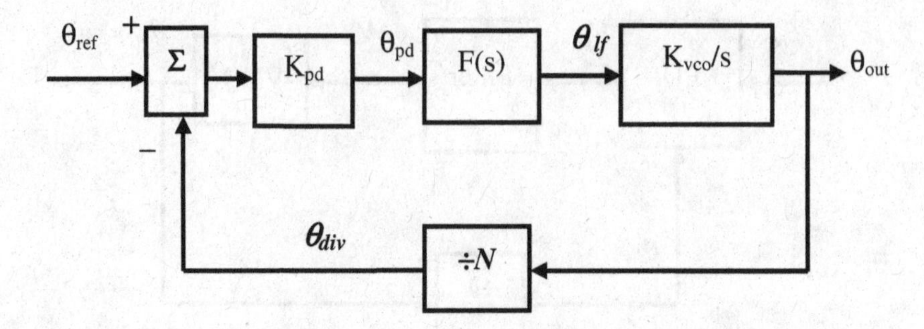

Figure 6.3 Linear time-invariant phase-model of the PLL.

If the signals around the loop are interpreted by their phases, the small-signal and noise behavior of the loop can be explored by linearizing the components and evaluating the transfer functions. Figure 6.3 shows the linear time-invariant (LTI) phase-model of the PLL. The forward gain is equal to $G(s) = K_{pd} \times F(s) \times K_{vco}/s$, while the feedback gain is equal to $H(s) = 1/N$. Hence, the open loop gain $OL(s)$ is obtained as:

$$OL(s) = G(s) \cdot H(s) = \frac{K_{pd} \cdot F(s) \cdot K_{vco}}{N \cdot s} \tag{6.6}$$

Equation (6.6) is useful to study the operation of the PLL, such as the step response and the stability of the system. The closed loop transfer function and the phase error transfer function are respectively given by:

$$\frac{\theta_{out}(s)}{\theta_{ref}(s)} = \frac{G(s)}{1+G(s)H(s)} = \frac{K_{pd}F(s)K_{vco}}{s+K_{pd}F(s)K_{vco}/N} \tag{6.7}$$

and

$$\frac{\theta_{err}(s)}{\theta_{ref}(s)} = \frac{1}{1+G(s)H(s)} = \frac{s}{s+K_{pd}F(s)K_{vco}/N} \tag{6.8}$$

6.2 Transient Characteristics - Tracking

In this section, the dynamic behavior of the loop when it is subjected to a phase step or a frequency step in the reference frequency will be examined. Then the startup behavior of the loop will be discussed. To simplify the following concepts, we assume $F(s) = 1$, and a first-order PLL is obtained.

Equation (6.9) represents a phase step with a magnitude of $\Delta\theta$ being applied to the reference signal. The assumptions are made that the loop is in lock state and the phase error is sufficiently small to justify an assumption of linearity.

$$\theta_{ref}(t) = u(t)\Delta\theta \tag{6.9}$$

where $u(t)$ is the unit step function. Applying the Laplace transform, equation (6.9) becomes

$$\theta_{ref}(s) = \Delta\theta/s \tag{6.10}$$

For t approaching infinity, the final value theorem of the Laplace transforms states

$$\lim_{t\to\infty}\theta(t) = \lim_{s\to 0} s\theta(s) \tag{6.11}$$

Therefore, the resulting steady state phase error for a first-order loop is

$$\theta_{err,ss} = \lim_{s\to 0}[s\cdot\theta_{ref}(s)\cdot\frac{1}{1+G(s)H(s)}] = \lim_{s\to 0}[s\cdot\frac{\Delta\theta}{s}\cdot\frac{1}{1+K_{pd}K_{vco}/(Ns)}] = 0 \tag{6.12}$$

It can be seen from the above equation that the PLL will reduce any phase error to zero if given sufficient time.

If the input frequency changes with a step size of $\Delta\omega$, the input phase equals $\theta_{ref}(t) = t\times\Delta\omega$. This situation appears when the division modulus in synthesizer is changed. The resulting steady-state phase error is:

$$\theta_{err,ss} = \lim_{s\to 0}[s\cdot\frac{\Delta\omega}{s^2}\cdot\frac{1}{1+K_{pd}K_{vco}/(Ns)}] = \frac{\Delta\omega\cdot N}{K_{pd}K_{vco}} = \frac{\Delta\omega}{\omega_p} \tag{6.13}$$

where $\omega_p = K_{pd}K_{vco}/N$ is the crossover frequency in radians/sec, which is defined as the frequency where the open loop gain magnitude is equal to one. Equation (6.7) shows that for the first-order PLL, ω_p is also equal to the closed-loop bandwidth.

In a system such as GSM, the local oscillator (LO) has to switch from the receive channel to the transmit channel or switch from one frequency band to another frequency band. The switching time requirement must satisfy the GSM's system specifications. This switching time T_E is the time a PLL takes to settle the new output frequency to be within a specified accuracy E. In order to calculate T_E, the following equation is used [12]:

$$\theta_{err}(s) = \theta_{ref}(s)\cdot\frac{1}{1+G(s)H(s)} = \frac{\Delta\omega}{s^2}\cdot\frac{s}{s+\omega_p} = \frac{\Delta\omega}{s\cdot(s+\omega_p)} \tag{6.14}$$

Equation (6.14) can be obtained in time-domain as:

$$\theta_{err}(t) = \frac{\Delta\omega}{\omega_p}\cdot(1-e^{-\omega_p t}) = \theta_{err,ss}.(1-e^{-\omega_p t}) \tag{6.15}$$

By definition,

$$E = 1-\theta_{err}(t)/\theta_{err,ss} = e^{-\omega_p t} \tag{6.16}$$

For E = 1e-6, the switching time T_E is given by [13]:

$$T_E = -\frac{\ln E}{\omega_p} = \frac{2.2}{f_p} \tag{6.17}$$

The GSM specifications requires T_E <10 ms for Δf = 100 MHz, for E = 1e-6 [12].

Example

A PLL is designed for operation in the 2.40-2.48GHz ISM band with f_{ref} = 1MHz, N = 2400-2480, K_{pd} = 0.14 V/rad, and K_{vco} = 350MHz/V. Assume $F(s)$ = 1, calculate the settling time for the mid-band operating condition.

At the mid-band operating condition, f_{out} = 2.44GHz, N = 2440, hence ω_p = $K_{pd}K_{vco}/N$ = 20,082 rad/s and f_p = 3196Hz, and T_E is equal to 0.688ms.

6.3 Loop Bandwidth - Second Order PLL

The loop bandwidth is generally defined by the natural frequency, ω_n of the closed loop transfer function, which is different from the crossover frequency, ω_p defined by the unity magnitude of the open loop gain mentioned earlier. For the first-order, we have $\omega_n = \omega_p$. The distinction between these two values can be illustrated for the case of a second-order PLL, where the loop filter transfer function, $F(s)$ has a first-order.

$$F(s) = \frac{1}{1 + s/\omega_a} \tag{6.18}$$

where $\omega_a = 1/RC$ is the frequency of the pole or the – 3dB bandwidth of the simple R-C filter. The closed-loop transfer function is given by:

$$\frac{\theta_{out}(s)}{\theta_{ref}(s)} = \frac{G(s)}{1 + GH(s)} = \frac{\omega_a K_{pd} K_{vco}}{s^2 + \omega_a s + \omega_a K_{pd} K_{vco} / N} \tag{6.19}$$

Equation (6.19) can be re-written as:

$$\frac{\theta_{out}(s)}{\theta_{ref}(s)} = \frac{N\omega_n^2}{s^2 + 2\xi\omega_n s + \omega_n^2} \tag{6.20}$$

where the natural frequency is:

$$\omega_n = \sqrt{\frac{\omega_a K_{pd} K_{vco}}{N}} \tag{6.21}$$

and the damping factor is:

$$\xi = \frac{1}{2}\sqrt{\frac{\omega_a}{K_{pd} K_{vco} / N}} \tag{6.22}$$

Equations (6.21) and (6.22) show that the natural frequency ω_n is the geometric mean of the filter bandwidth ω_a and the open loop gain $K_{pd} K_{vco} / N$. Equation (6.20) has the response of a second order low pass filter which is flat between 0 and ω_n. The damping factor ξ is inversely proportional to $K_{pd} K_{vco} / N$ and influents the dynamic performance of the PLL. For $\xi = 1$, the system is critically damped, and becomes oscillatory for $\xi << 1$. For $\xi = \sqrt{2}/2$, the response is optimally flat, $\omega_a = 2 K_{pd} K_{vco} / N$, and $\omega_n = \omega_a / \sqrt{2}$.

The phase error transfer function can be derived from the closed loop transfer function as:

$$\frac{\theta_{err}(s)}{\theta_{ref}(s)} = 1 - \frac{\theta_{out}(s)/N}{\theta_{ref}(s)} = \frac{s^2 + 2\xi\omega_n s}{s^2 + 2\xi\omega_n s + \omega_n^2} \tag{6.23}$$

which tends to zero as $s \to 0$.

For an input frequency change of $\Delta\omega$, the steady-state phase error is:

$$\theta_{err,ss} = \lim_{s\to 0}[s \cdot \frac{\Delta\omega}{s^2} \cdot \frac{s^2 + 2\xi\omega_n s}{s^2 + 2\xi\omega_n s + \omega_n^2}] = \frac{2\xi\Delta\omega}{\omega_n} = \frac{\Delta\omega}{K_{pd} K_{vco} / N} \tag{6.24}$$

Similar to the case of the first-order PLL, the input change in frequency of $\Delta\omega$ is suppressed by the factor of $K_{pd} K_{vco} / N$ in the final phase error.

The crossover angular frequency ω_p is defined by the unity magnitude of the open loop gain. Since

$$OL(s) = \frac{K_{pd} \cdot K_{vco}}{(1 + s / \omega_a) N \cdot s} = \frac{\omega_n^2}{s^2 + \omega_a s} = \frac{\omega_n^2}{s^2 + 2\xi\omega_n s} \tag{6.25}$$

ω_p is given by: $|s^2 + 2\xi\omega_n s| = \omega_n^2$, or $\omega^4 + 4\xi^2\omega_n^2\omega^2 - \omega_n^4 = 0$. The only positive solution is: $\omega_p = \omega_n[\sqrt{4\xi^4 + 1} - 2\xi^2]^{1/2}$. For $\xi = \sqrt{2}/2$, we have: $\omega_p = \omega_n[\sqrt{2} - 1]^{1/2} = 0.64\omega_n$.

Example

For the same above example with f_{ref} = 1MHz, N = 2440, K_{pd} = 0.14 V/rad, and K_{vco} = 350MHz/V, if a second order PLL is designed for $\xi = \sqrt{2}/2$, $\omega_a = 2 K_{pd} K_{vco} / N$ = 40,164rad/s. Thus, the loop bandwidth is $\omega_n = \omega_a / \sqrt{2}$ = 28,400rad/s, or f_n = 4520Hz, and $\omega_p = \omega_n[\sqrt{2} - 1]^{1/2} = 0.64\omega_n$ = 18,278rad/s or f_p = 2,909Hz.

From Equations (6.21) and (6.22), we also have $\omega_n = \frac{2\xi K_{pd} K_{vco}}{N}$. If the loop is designed with $\omega_n \ll K_{pd} K_{vco} / N$, ξ will be small and the loop will have a poor transient response. In order to facilitate the independent selections of ω_n and ξ, a lead-lag filter is commonly used with the transfer function:

$$F(s)=\frac{1+s\tau_2}{1+s(\tau_1+\tau_2)} \tag{6.26}$$

The filter has a pole at $1/(\tau_1+\tau_2)$ and a zero at $1/\tau_2$. The closed-loop transfer function is given by:

$$\frac{\theta_{out}(s)}{\theta_{ref}(s)}=\frac{sN\omega_n(2\xi-N\omega_n/K_{pd}K_{vco})}{s^2+2\xi\omega_n s+\omega_n^2}=\frac{sN\omega_n^2\tau_2}{s^2+2\xi\omega_n s+\omega_n^2} \tag{6.27}$$

where

$$\omega_n=\sqrt{\frac{K_{pd}K_{vco}}{N(\tau_1+\tau_2)}} \quad \text{and } \xi=\frac{\omega_n}{2}(\tau_2+\frac{N}{K_{pd}K_{vco}}) \tag{6.28}$$

Most practical PLLs would have the loop gain $K_{pd}K_{vco}/N >> \omega_n$. Under this condition, the error transfer function is approximately equal to:

$$\frac{\theta_{err}(s)}{\theta_{ref}(s)}\approx\frac{s^2}{s^2+2\xi\omega_n s+\omega_n^2} \tag{6.29}$$

Apply the final value theorem for both the phase step and frequency step, θ_{err} tends to zero as $s\to 0$.

Example

For the same above example with f_{ref} = 1MHz, N = 2440, K_{pd} = 0.14 V/rad, and K_{vco} = 350MHz/V, we have: $K_{pd}K_{vco}/N$ = 20,082rad/s. If we select f_n = 2000Hz to give a transient time of τ_n = 0.5ms, we have ω_n = 12566rad/s. From equation (6.28), we have $\tau_1+\tau_2$ = 127.2 μs. For $\xi=\sqrt{2}/2$, we have τ_2 = 62.75μs, hence τ_1 = 64.45μs. The lead-lag filter of equation (6.26) is usually implemented by the circuit of Figure 6.4, where:

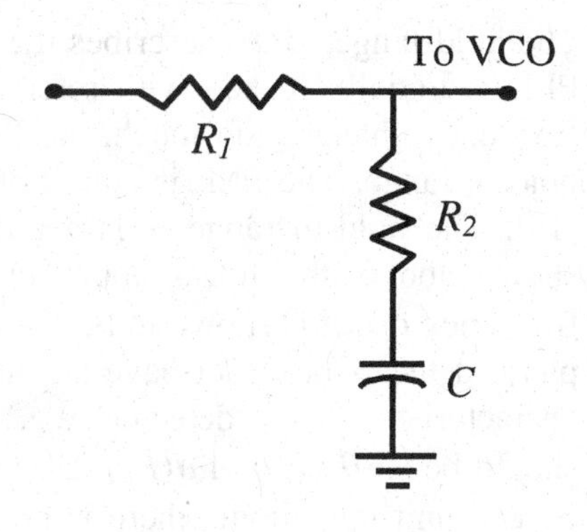

$\tau_1 = R_1C$, and $\tau_2 = R_2C$

Select $C = 1\text{nF}$, we have:

$R_1 = 64.45\text{k}\Omega$ and $R_2 = 62.75\text{k}\Omega$.

Figure 6.4 A lead-lag filter.

6.4 Acquisition

During startup, the PLL is initially in an unlocked condition, the process of achieving lock state is called acquisition. Since acquisition is inherently a non-linear process, its qualitative analysis is beyond the scope of this book. Some descriptive analysis will be given here, more information can be found in [1, 2]. For a lead lag filter, the parameters described below depend on the types of phase detectors, namely, multiplier, EXOR gate, FF (Flip-flop) and PFD (phase frequency detector) described in the next section.

The hold range, $\Delta\omega_H$, describes the PLL in a static or locked state. The PLL is initially locked with reference signal. If the reference signal's frequency changes too much, the PLL will lose lock at the edge of the hold-in range. The PLL is conditionally stable within the hold-in range [14]. The hold-in range is larger than both above defined ranges. As shown above, the linear approximation of the phase error due to a frequency offset is shown to be $\theta_{res,ss} = \Delta\omega/(K_{pd}K_{vco}/N)$. However, a real phase detector does not have an infinite linear range. For a sinusoidal-characteristic phase detector, e.g. the multiplier, the true expression should be $\sin\theta_{res,ss} = \Delta\omega/(K_{pd}K_{vco}/N)$ [6]. Since the sine function cannot exceed unit magnitude, there is no solution for $\Delta\omega > (K_{pd}K_{vco}/N)$. The hold-in range therefore equals $\Delta\omega_H = \pm K_{pd}K_{vco}/N$. Other types of phase detector, for example the charge pump phase-frequency detector, have a larger linear range and can therefore extend the hold-in range. $\Delta\omega_H$ is increased by a factor of $\pi/2$ for the EXOR PD, a factor of π for the FF PD, and 2π for the PFD as described in the next section. However, these definitions are only valid as long as the limit is set by the phase detector and not by other nonlinearities, such as the clipping in an operational amplifier (op-amp) or the VCO frequency tuning range.

If the initial VCO frequency is close enough to $N\times F_{ref}$, the PLL will lock with just a phase transient. The frequency range over which no cycles will be missed before the lock condition is obtained is called the lock range, $\Delta\omega_L$. If a reference frequency outside the lock range is applied, the pull-in process will be slower. The normal operation of the PLL is generally restricted to the lock range. Assuming a high loop gain, for the multiplier PD, we have $\Delta\omega_L \approx 2\xi\omega_n$. For other PDs, $\Delta\omega_L$ is increased by a factor of $\pi/2$ for the EXOR PD, π for the FF PD, and 2π for the PFD. The lock time is given by $T_L \approx 2\pi/\omega_n$.

The pull-in range, $\Delta\omega_{PI}$, describes the PLL in a dynamic state, and is the range within which a PLL will always become locked through the acquisition process [14]. If the reference frequency is outside the pull-in range, the PLL will not be able to lock onto the reference signal. The pull-in process is quite complex, assuming a high loop gain and a multiplier PD, we have $\Delta\omega_{PI} \approx \frac{4\sqrt{2}}{\pi}\sqrt{\frac{\xi\omega_n K_{pd} K_{vco}}{N}}$, and $T_{PI} \approx \frac{\pi^2 \Delta\omega^2}{16\xi\omega_n^3}$

is the pull-in time where $\Delta\omega$ is the initial offset at the PD input.

Example

For the previous example N = 2440, K_{pd} = 0.14 V/rad, and K_{vco} = 350 MHz/V, ω_n = 12566 rad/s, $\xi = \sqrt{2}/2$, we have: $\Delta\omega_H = K_{pd}K_{vco}/N$ = 20,082rad/s, $\Delta\omega_L \approx 2\xi\omega_n$ = 17,771rad/s, $T_L \approx 2\pi/\omega_n$ = 0.5ms, $\Delta\omega_{PI} \approx \frac{4\sqrt{2}}{\pi}\sqrt{\frac{\xi\omega_n K_{pd}K_{vco}}{N}}$ = 24,053rad/s. If Δf = 40kHz, $T_{PI} \approx \pi^2(\Delta\omega)^2/(16\xi\omega_n^3)$ = 27.8ms. Note that $T_{PI} >> T_L$.

An increment in the pull in range, and a reduction of the pull in time are achieved for the digital types of PDs. Assuming a high loop gain, we have $\Delta\omega_{PI} \approx \frac{\pi}{\sqrt{2}}\sqrt{\xi\omega_n K_{pd}K_{vco}/N}$, and $T_{PI} \approx \frac{4}{\pi^2}\frac{\Delta\omega^2}{\xi\omega_n^3}$ for the EXOR PD, while $\Delta\omega_{PI} \approx \pi\sqrt{2}\sqrt{\xi\omega_n K_{pd}K_{vco}/N}$ and $T_{PI} \approx \frac{1}{\pi^2}\frac{\Delta\omega^2}{\xi\omega_n^3}$ for the FF PD. For the PFD, the pull in rage becomes very wide, and in practice it is dictated by the frequency range of the VCO. The pull in time is dependent also on the supply voltage.

6.5 Phase Detector and Loop Filter

This section describes the phase detector, loop filter, and their various implementations. The transfer characteristics of the loop filter and the calculation of an optimum loop bandwidth will be presented. More information can be found in [1-4, 14-16].

6.5.1 Phase Detector

Phase detectors can be classified into three major categories: analog phase detector, digital phase detector, and phase-frequency detector. The analog phase detector or multiplier generates a DC component, which is dependent on the phase difference of input signals. The DC component is used for phase difference detection. The digital phase detector, such as the EXOR and the flip-flop phase detector, detects the phase difference of the input signals based on their zero crossing points. The phase-frequency detector is a sequential circuit, and it provides a frequency sensitive signal to improve the acquisition when the loop is out of lock.

6.5.1.1 Multiplier

From the trigonometric identity $\cos A \cos B = (1/2)\cos(A-B) + (1/2)\cos(A+B)$, the multiplier acts as a phase detector. If the input signals to the multiplier are $v_1 = A_1\cos(\omega_1 t + \theta_1)$ and $v_2 = A_2\cos(\omega_2 t + \theta_2)$, the output signal of a multiplier of gain A_d will be:

$$v_d = \frac{1}{2} A_d \cdot A_1 \cdot A_2 \cdot \{\cos[(\omega_1 - \omega_2)t + \theta_1 - \theta_2] + \cos[(\omega_1 + \omega_2)t + \theta_1 + \theta_2]\}$$

At the phase lock, both frequencies are the same and the DC component of the phase detector output equals $0.5 \times A_d \times A_1 \times A_2 \times \cos(\theta_1 - \theta_2)$, which indicates the phase difference between the two signals. The unwanted sum of the two frequencies at the output will be removed by the loop filter. One of the common implementation of the multiplier phase detector is the Gilbert multiplier [1].

6.5.1.2 EXOR gate

Figure 6.5 shows the EXOR gate phase detector. The operation of an EXOR gate phase detector is similar as an over-driven multiplier circuit, which has triangular phase detector characteristics.

Square wave inputs of 50% duty cycle are recommended for the EXOR phase detector. For other duty cycles, the detection range may be significantly reduced. In addition, it is possible to have the same output voltage for two different phase errors [17]. The output waveforms for inputs A and B are shown in Figure 6.6. The average value $\overline{C}$ of the output waveform is proportional to the phase difference between the input signals.

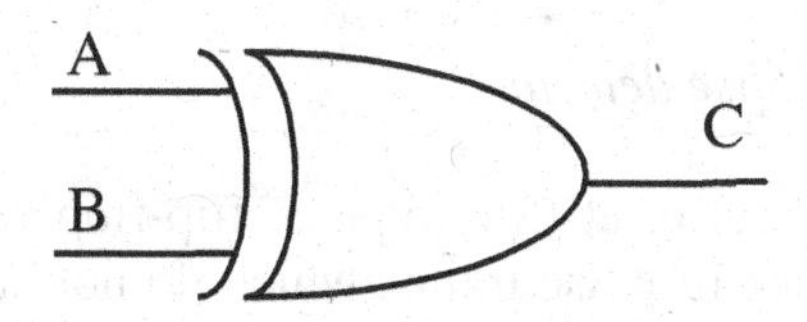

Figure 6.5 EXOR gate phase detector.

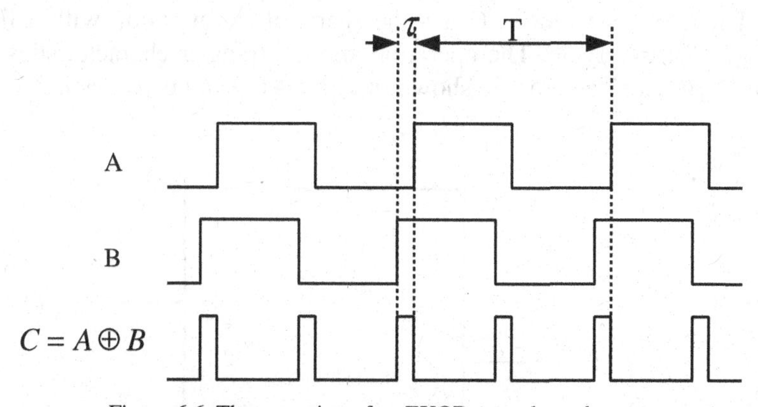

Figure 6.6 The operation of an EXOR gate phase detector.

The phase detector transfer characteristic is shown in Figure 6.7.

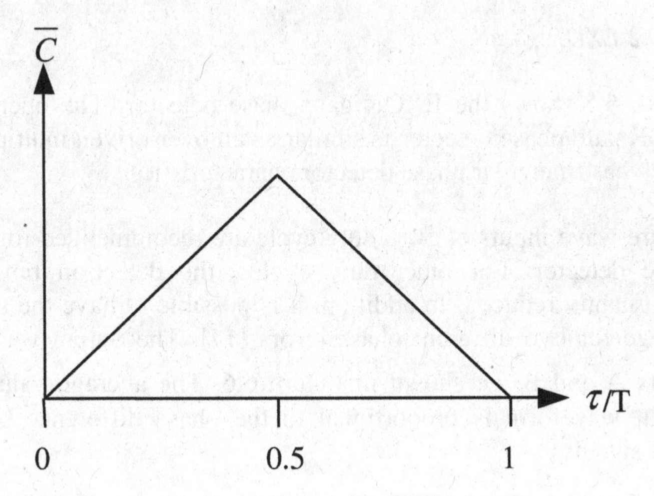

Figure 6.7 The transfer characteristics of an EXOR gate phase detector.

6.5.1.3 Flip-flop phase detector

An edge-sensitive set-reset (SR) type of flip-flop can be used to detect the phase difference of pulse trains, which do not have 50% duty cycle. The flip-flop phase detector is shown in Figure 6.8. Narrow pulses at input A set the output C, while narrow pulses at input B reset the output C. The average value of $\overline{C}$ has the shape of a saw-tooth, with a linear range of a full cycle. The operation and the transfer characteristics of a flip-flop phase detector are shown in Figures 6.9 and 6.10, respectively.

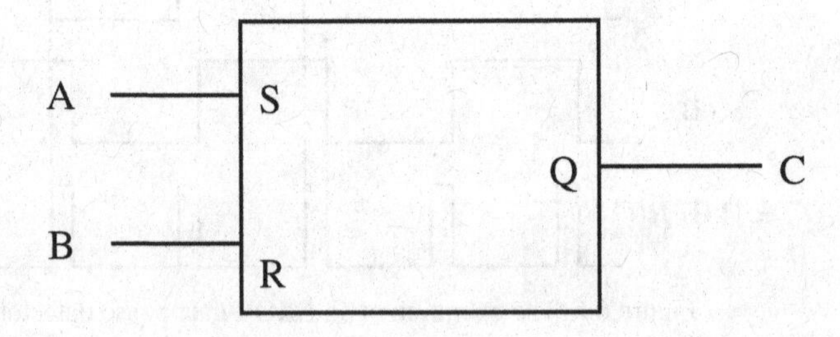

Figure 6.8 Flip-flop phase detector.

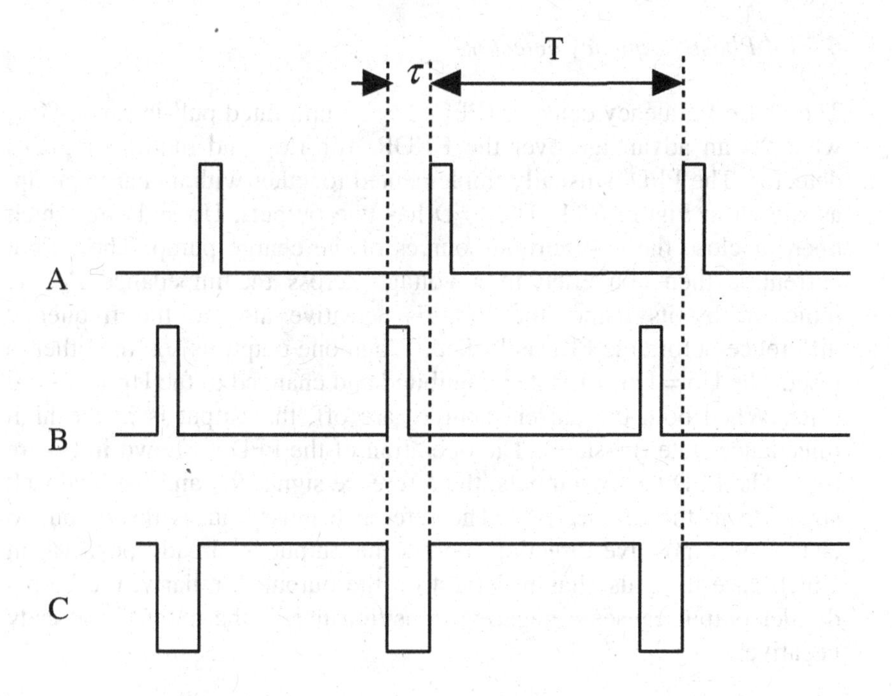

Figure 6.9 Operation of a flip-flop phase detector.

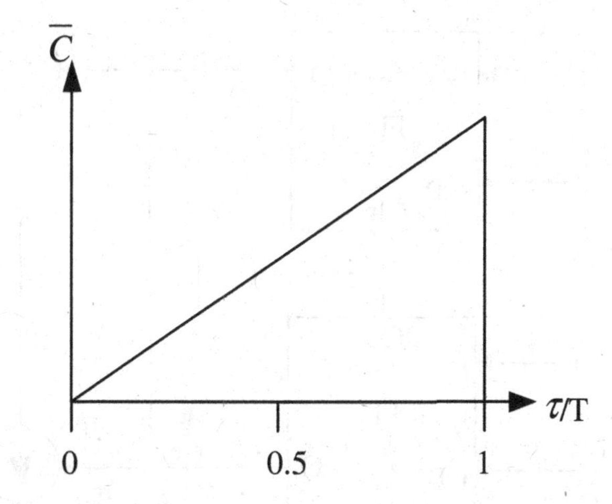

Figure 6.10 Transfer characteristics of a flip-flop phase detector.

6.5.1.4 Phase frequency detector

The phase frequency detector (PFD) has an unlimited pull-in range [16], which is an advantage over the EXOR, flip-flop and multiplier phase detector. The PFD is usually implemented together with a charge pump, as shown in Figure 6.11. The PFD has two outputs, Up and Dn, which open or close the two current sources of the charge pump. The output current is then converted to a voltage across the impedance Z_{lf}. As indicated by its name, the PFD is sensitive also to the frequency difference before the PLL is locked. When one output is set, the other is reset. The Up = Dn = 1 state is inhibited and changed to the Up = Dn = 0 state. When both the current sources are off, the output is in the high impedance state (tri-state). The operation of the PFD is shown in Figure 6.12. The PFD has two inputs, the reference signal R_{EF} and the feedback signal from the divider *Div*. The reference pulse causes the output to change to a positive direction, unless the output is already positive, in which case the pulse has no effect on the output. Similarly, the loop's divider output causes a negative transition unless the output is already negative.

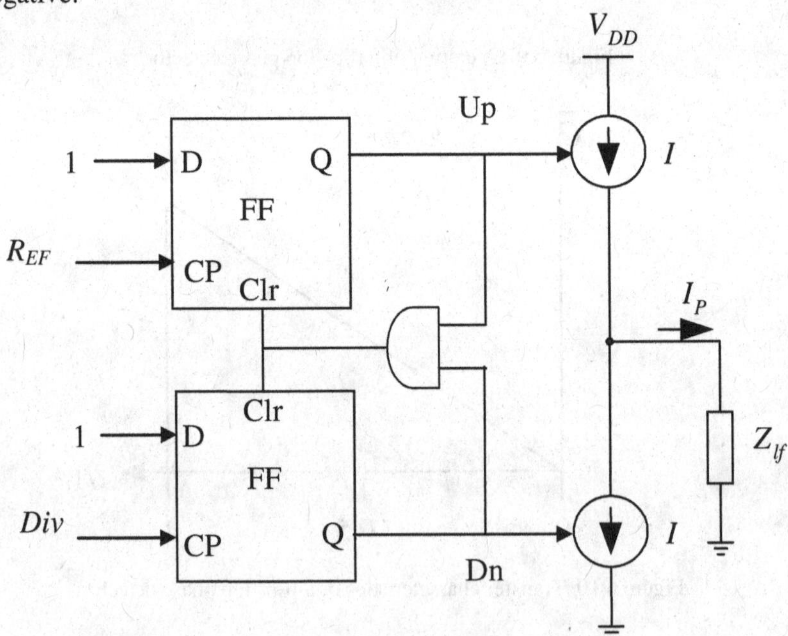

Figure 6.11 Phase frequency detector.

The transfer characteristics of the PFD are plotted in Figure 6.13 and have a linear phase range of 4π. The implementation of the PFD and the charge pump in the PLL results in an important characteristic. The PLL locks at a zero static phase difference of the two inputs; otherwise an indefinite accumulation of charge in Z_{lf} will occur.

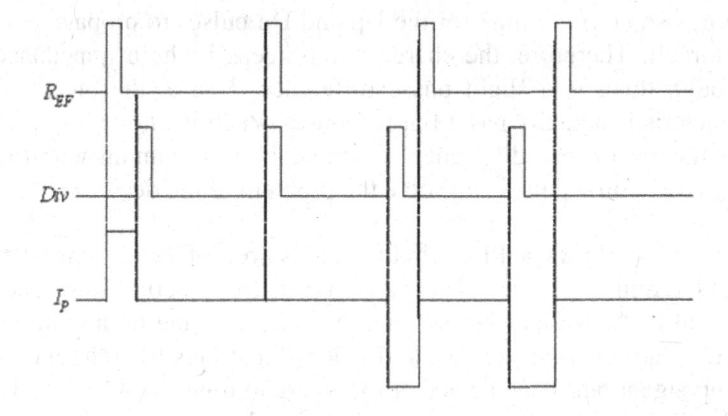

Figure 6.12 Operation of the phase frequency detector.

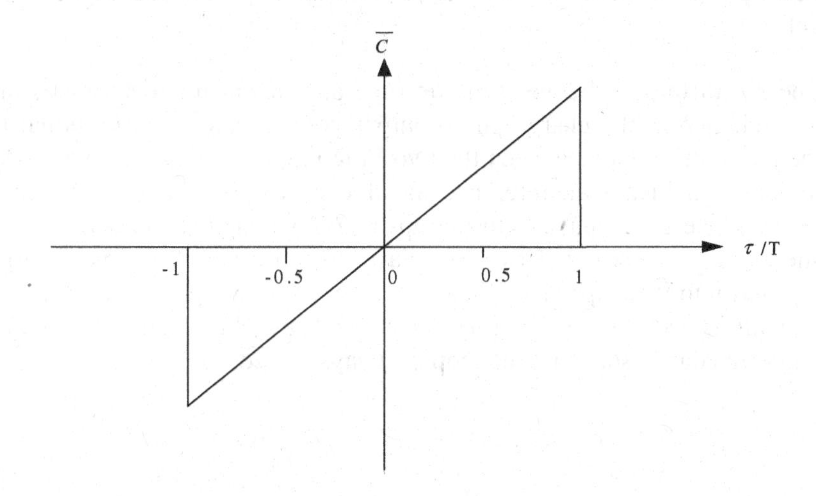

Figure 6.13 Transfer characteristics of the phase frequency detector.

The PFD in Figure 6.11 suffers from a "dead zone", which arises from the crossover distortion, where changes in gain occurring near the zero phase error [18]. If both the reference pulse and the divider pulse appear at the same time, none of the outputs becomes active and the charge-pump output is in high-impedance state. Even if the phase difference changes slightly, the phase detector will not respond immediately since it requires some finite time for the Up and Dn pulses to propagate through the circuit. Therefore, the charge pump keeps its high impedance state although there is a slight phase difference. Hence, the phase detector characteristic actually has a flat response, which is known as dead zone, near the zero phase difference. Giving a fixed minimum width to both the charge pump pulses can solve the problem of the dead zone.

Figure 6.14 shows a PFD circuit that is free of dead zone [19]. The output terminals Up and Dn are designed to be active low. The delay block after the 4-input NAND circuit determines the minimum width of the up- and down-pulses. If the divider output lags the reference signal, the up signal will become active for a certain time T_D, where T_D is equal to the time difference between the two signals, and the delay through the circuit, including the delay caused by the extra delay stage. Similarly, the down-pulse will also become active for a short period due to the extra delay.

The net difference between the up-time and down-time determines the total change in the charge pump output voltage, and is proportional to the phase difference between the two PFD inputs. For the case when the divider signal leads the reference signal, even without a phase difference between the two inputs, both the up- and down-signals are active for a short period determined by the delay stage. However, the net charge injected into the impedance Z_{lf} is zero, since the two pulses are of equal magnitude and opposite polarity. In this case, the charge pump is not in a high-impedance state and the loop is always closed.

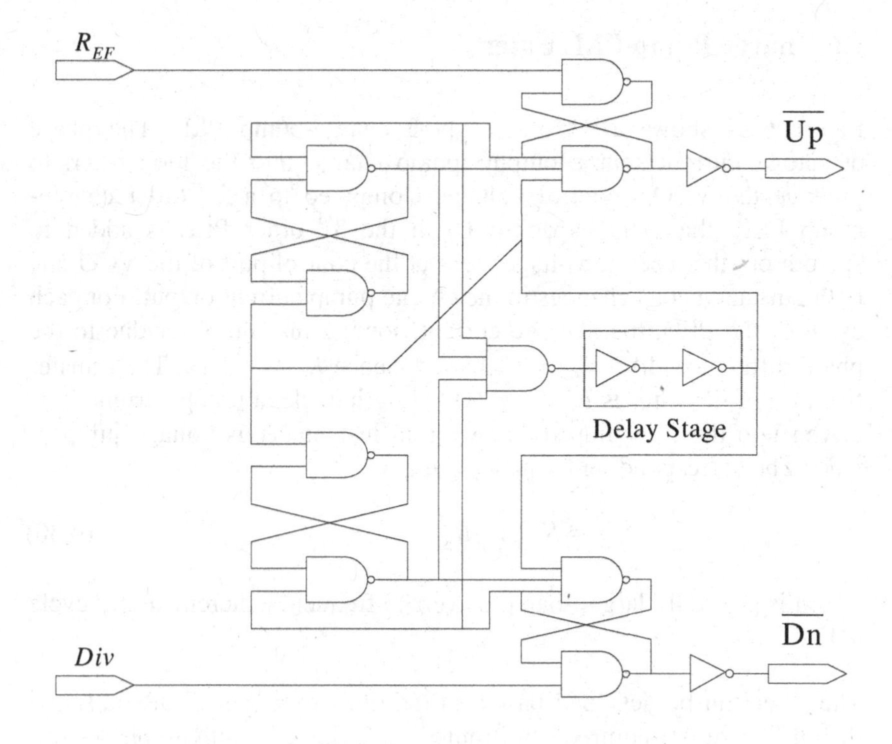

Figure 6.14 Phase frequency detector without dead zone.

6.5.2 Loop Filter

There are two types of loop filters, active and passive. An active filter uses op-amps to generate a tuning voltage higher than that generated by a passive filter. The op-amp itself provides the DC amplification necessary to develop a high control voltage required by the VCO in wide band applications. Passive filter has the advantages of reduced noise and lower circuit complexity. They are formed by only R (resistor), C (capacitor) elements, and often used as the charge pump loads to generate the control voltage proportional to the phase error. The charge pump passive loop filter is used widely for wireless applications, and is referred to as the current source loop filter [20]. This is in contrast to the voltage source loop filter for an active loop filter.

6.6 Charge Pump PLL Filter

Figure 6.15 shows a 3rd order type-2 charge-pump PLL. The phase detector's current source outputs pump charge into the loop filter, to produce the VCO's control voltage. Compared to a 2nd order charge-pump PLL, the extra capacitor C_1 in the 3rd order PLL is added to smooth out the discrete voltage steps at the control port of the VCO due to the instantaneous changes in the charge pump current output. For each cycle of the PFD, the average current flowing into the filter due to the phase difference $\Delta\phi_i$ is $I_{avg} = I_{cp}\Delta\phi_i/2\pi$, hence $K_{pd} = I_{cp}/2\pi$. The transfer function of the filter is $F(s) = Z(s)$. At each cycle, a pump current I_{cp} is driven into the filter impedance with an instantaneous voltage jump of $I_{cp}R_2$. The corresponding frequency jump is

$$|\Delta\omega| = K_{vco} I_{cp} R_2 \tag{6.30}$$

which is generally larger than the average frequency increment per cycle [21].

The filter can be designed based on the open loop gain bandwidth and the phase margin required. Positioning the point of minimum phase shift at the unity gain frequency of the open loop response as shown in Figure 6.16 ensures the loop stability. The phase relationship between the pole and zero also allows the determination of the loop filter component values. The phase margin θ_p is defined as the difference between 180° and the phase of the open loop transfer function at the unity-gain frequency f_p. The phase margin is chosen between 30° and 70° for most applications [22]. The larger the phase margin, the more stable the loop. However, the transient response is slower and requires a longer switching time. A loop with a low phase margin may still be stable but could exhibit oscillator problems associated with an undamped loop, such as longer switching time and increased noise. A phase margin of 45° is a good compromise between desired stability and the other generally undesired effects. The impedance of the filter in Figure 6.15 is

$$Z(s) = \frac{s \cdot C_2 \cdot R_2 + 1}{s^2 \cdot C_1 \cdot C_2 \cdot R_2 + s \cdot C_1 + s \cdot C_2} = \frac{1 + sT_2}{s(1 + sT_1)(C_1 + C_2)} \tag{6.31}$$

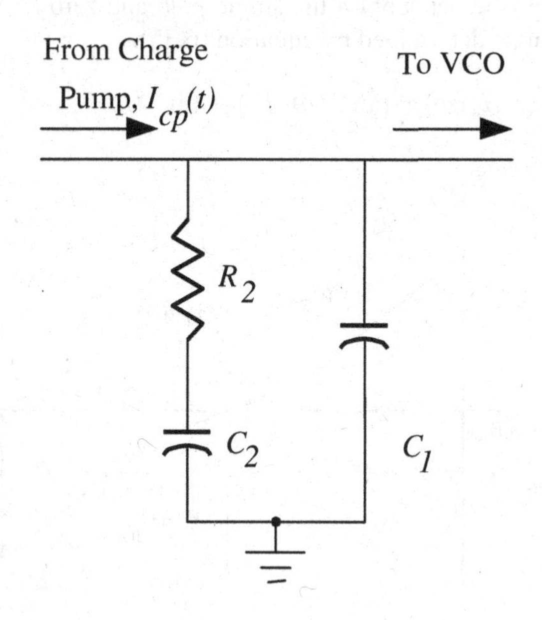

Figure 6.15 A 3rd order, type-2 charge pump PLL filter.

where the time constants that determine the pole and zero frequencies of the filter transfer function are defined by the following equations

$$T_1 = R_2 \cdot \frac{C_1 \cdot C_2}{C_1 + C_2} \tag{6.32}$$

$$T_2 = R_2 \cdot C_2 \tag{6.33}$$

Thus, from equation (31), the 3rd order PLL open loop gain can be written as

$$OL(s)|_{s=j\omega} = \frac{K_{pd} \cdot K_{vco} \cdot (1 + j\omega \cdot T_2)}{\omega^2 \cdot C_1 \cdot N \cdot (1 + j\omega \cdot T_1)} \cdot \frac{T_1}{T_2} \tag{6.34}$$

There are three poles in equation (6.34), where two of the poles are contributed by the filter, while the third pole is contributed by the integrator of the VCO. From equation (6.34), it can be seen that the phase term is dependent on the single pole and zero such that the phase margin can be determined by equation (6.35).

$$\theta_p(\omega) = \tan^{-1}(\omega \cdot T_2) - \tan^{-1}(\omega \cdot T_1) \tag{6.35}$$

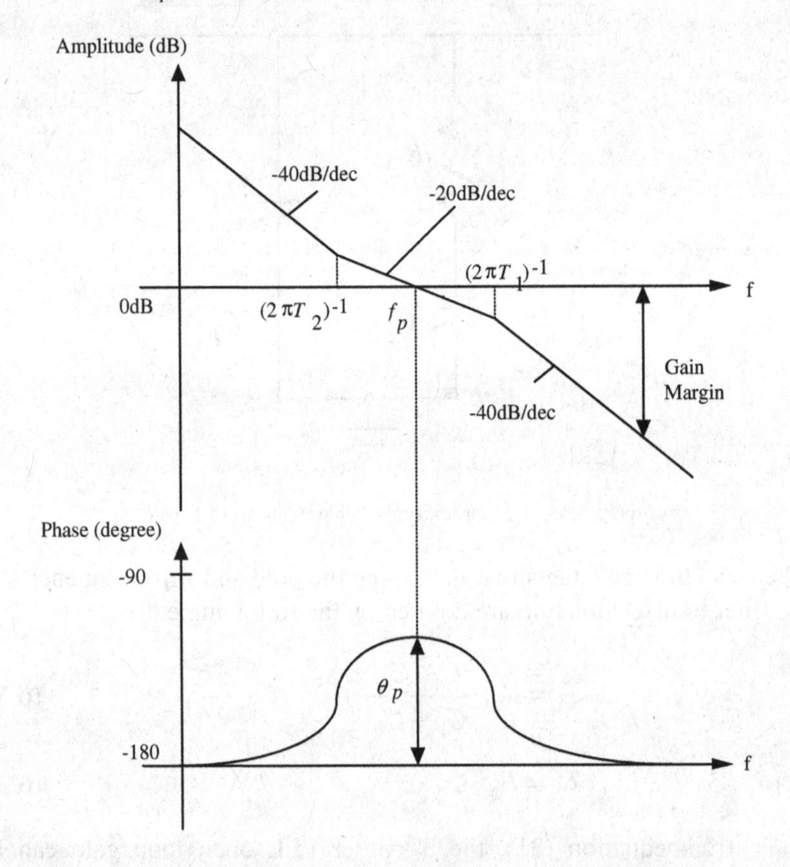

Figure 6.16 Bode plot of the open loop response for a 3rd order, type-2 charge pump PLL filter.

By equating the derivative of the phase margin to zero, Eqn. (6.35) becomes

$$\frac{d\theta_p}{d\omega} = \frac{T_2}{1+(\omega \cdot T_2)^2} - \frac{T_1}{1+(\omega \cdot T_1)^2} = 0 \tag{6.36}$$

Thus, the loop bandwidth ω_p can be given by

$$\omega_p = (T_1 \cdot T_2)^{-1} \tag{6.37}$$

Sometimes, the 3rd order structure does not provide sufficient rejection to the reference spur. The reference spur is caused by the current switching noise in the dividers and that in the charge pump at the reference rate F_{ref}. In wireless communications, the phase detector operation frequency is generally a multiple of the RF channel spacing. These spurious sidebands can cause noise in adjacent channels. This is usually the case in the TDMA digital cellular standards, such as GSM or IS-54. A narrow loop filter has advantage of better attenuation of the reference spur, but the requirement of the sub-milliseconds switching between channels makes a relatively wide loop filter mandatory.

One solution is to use an additional low pass pole for more attenuation of the unwanted spur. The use of a passive loop filter eliminates noise contribution from the op-amp in an active filter. This is critical due to the strict phase error and integrated phase noise requirements. For example, the integrated phase noise requirement for the SONET's OC-192 specification is 1ps root-mean-square value (*rms*). The recommended filter configuration is shown in Figure 6.17.

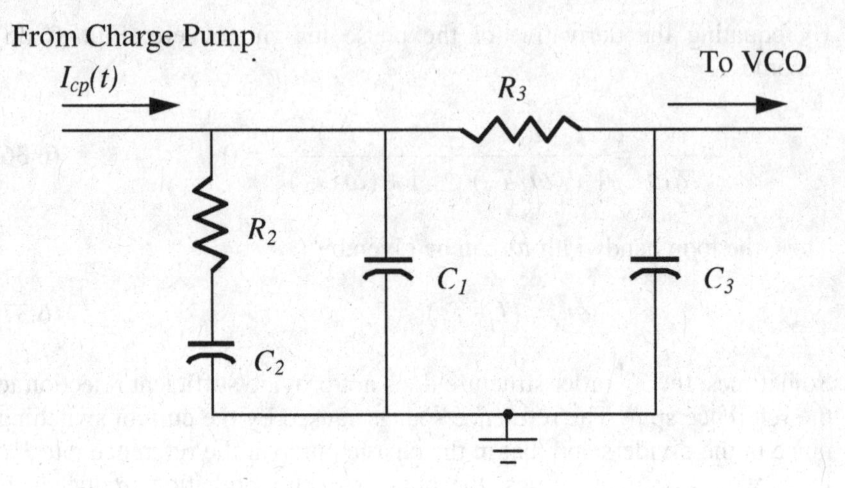

Figure 6.17 A 4th order, type-2 charge pump PLL filter.

The additional pole must be lower than the reference frequency to significantly attenuate the reference spur, but must be at least 5 times higher than the loop bandwidth to maintain the loop stability [20]. The additional filter time constant can be defined by equation (6.38). The bode plot of the open loop response for the 4th order charge pump PLL is shown in Figure 6.18.

$$T_3 = (R_3 \cdot C_3)^{-1} \tag{6.38}$$

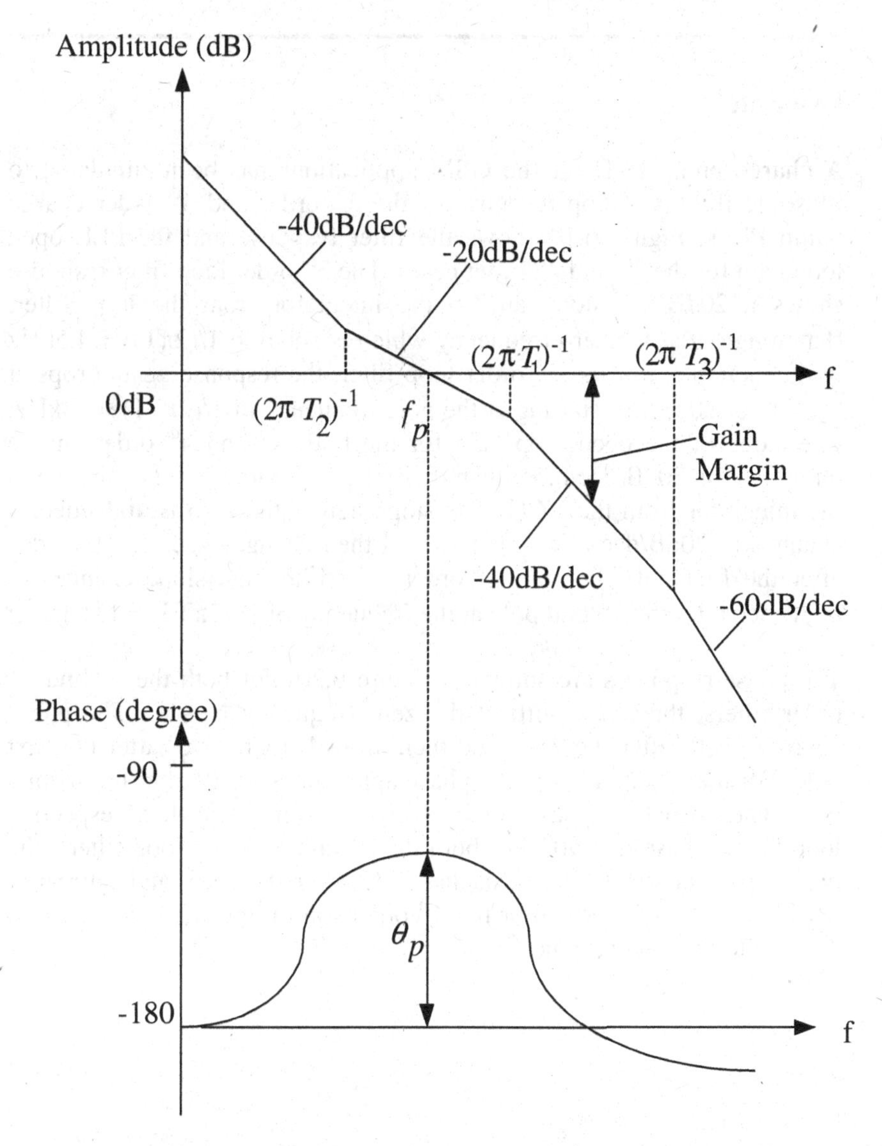

Figure 6.18 Bode plot of the open loop response for a 4th order, type-2 charge pump PLL.

Example

A charge pump PLL for the GSM applications has been simulated to illustrate the open loop response of the 3rd order and 4th order charge pump PLLs. Figure 6.19 shows the filter response and the PLL open loop gain for the 3rd and 4th order cases. The 3rd order loop filter response shows a 20dB/dec slope due to the integrator from the loop filter, flattening out at the zero frequency which is equal to $1/(2\pi T_2) = 4.5$kHz in the example. For the 4th order loop filter, the response again drops at the rate of 20dB/dec starting at the pole frequency of $1/(2\pi T1) = 54$kHz. The slope of the open loop gain for the both 3rd and 4th order charge pump PLL is 20dB/dec more than their respective loop filters gain due to the integrator from the VCO. The slope before the zero is 40dB/dec, it changes to 20dB/dec after the zero and then changes back to 40dB/dec after the first pole. For the 4th order loop filter, the slope changes to 60dB/dec after the second pole at the frequency of $1/\ (2\pi T3) = 1$MHz.

The phase responses are shown in Figure 6.20. For both the 3rd and 4th order filters, the phase shift at the zero frequency is –90°. The phase approaches 0° after the zero, and then drops back to –90° after the first pole. For the 4th order filter, the phase approaches –180° after the second pole. The open loop phase for both cases differs from their respective loop filters phase by –90°. For both the 3rd and 4th order loop filters, the phase starts at –180°. It approaches –90° after the zero, and returns to –180° after the first pole. For the 4th order loop filter, the phase goes to –270° after the second pole.

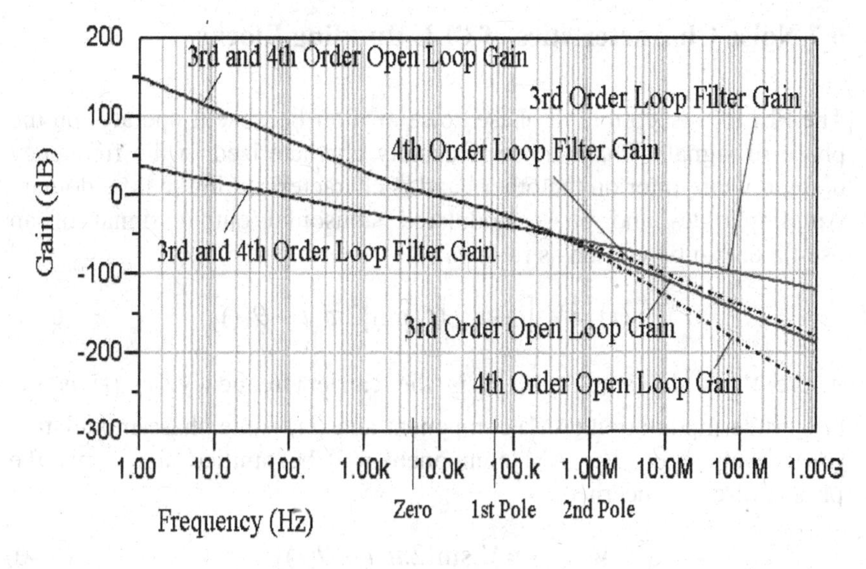

Figure 6.19 Loop filter gain and open loop gain for 3rd and 4th order charge pump PLLs.

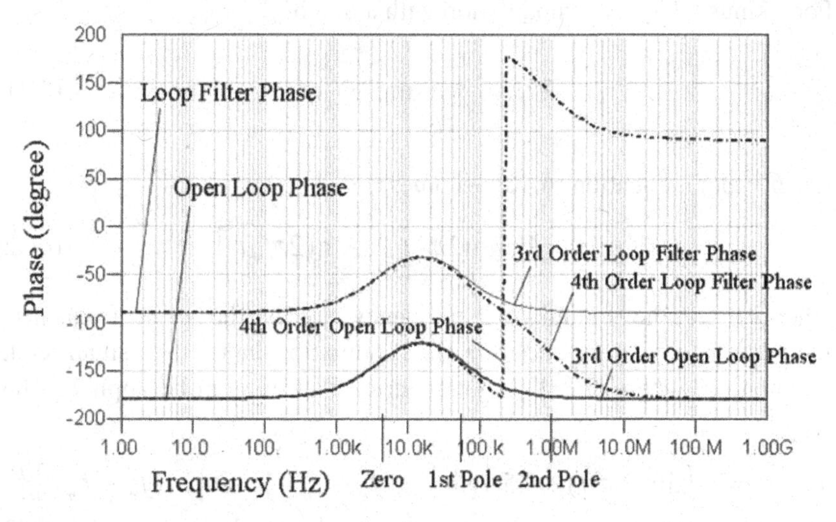

Figure 6.20 Loop filter phase and open loop phase for 3rd and 4th order charge pump PLLs

6.7 Noise Characteristics of PLL Building Blocks

The PLL is susceptible to phase noise or jitter because it operates on the phase of signals. Phase noise is usually characterized in the frequency domain while jitter on the other hand is characterized in the time domain. An amplitude- and phase-modulated sinusoidal output signal of an oscillator can be written as follows

$$V_{out}(t) = V_o\left[1 + v_{am}(t)\right]\sin\left[2\pi f_c t + \theta(t)\right] \tag{6.39}$$

where V_o is the amplitude, f_c is the carrier frequency, $v_{am}(t)$ is the amplitude-modulation (AM) component and $\theta(t)$ is the phase-modulation (PM) component. The AM component will be omitted since only the phase noise is concerned.

$$V_{out}(t) = V_o\sin\left[2\pi f_c t + \theta(t)\right] \tag{6.40}$$

For a sinusoidal-angle modulation with a rate of f_m

$$\theta(t) = \frac{\Delta f}{f_m}\cdot\sin(2\pi f_c t) \tag{6.41}$$

Let $\beta = \Delta f/f_m$, equation (6.40) becomes

$$V_{out}(t) = V_o\sin\left[2\pi f_c t + \beta\sin(2\pi f_m t)\right] \tag{6.42}$$

where f_m is the modulation frequency, Δf is the peak frequency-modulation deviation, and β is the modulation index. For small-angle modulation, where $\beta < 1$, trigonometric identities can be applied. This leads to

$$V_{out}(t) = V_o\left\{\sin(2\pi f_c t) + 0.5\beta\left\{\sin\left[2\pi(f_c + f_m)t\right] - \sin\left[2\pi(f_c - f_m)t\right]\right\}\right\} \tag{6.43}$$

From this expression, it can be concluded that a small-angle deviation gives rise to sidebands on each side of the carrier with an amplitude of $\beta/2$. Therefore, phase noise can be regarded as an infinite number of single FM sidebands [23].

Phase noise indicates the error or random deviation of the frequency of the oscillator output signal. In the ideal case, without phase noise, the output spectrum of an oscillator working at the frequency of f_c is a single line as shown in Figure 6.21(a). Phase noise is exhibited as a skirt around the oscillating frequency in the power spectrum. Phase noise is defined as the noise power in a unit bandwidth at an offset of Δf from the center frequency f_c divided by the carrier power as shown in Figure 6.21(b).

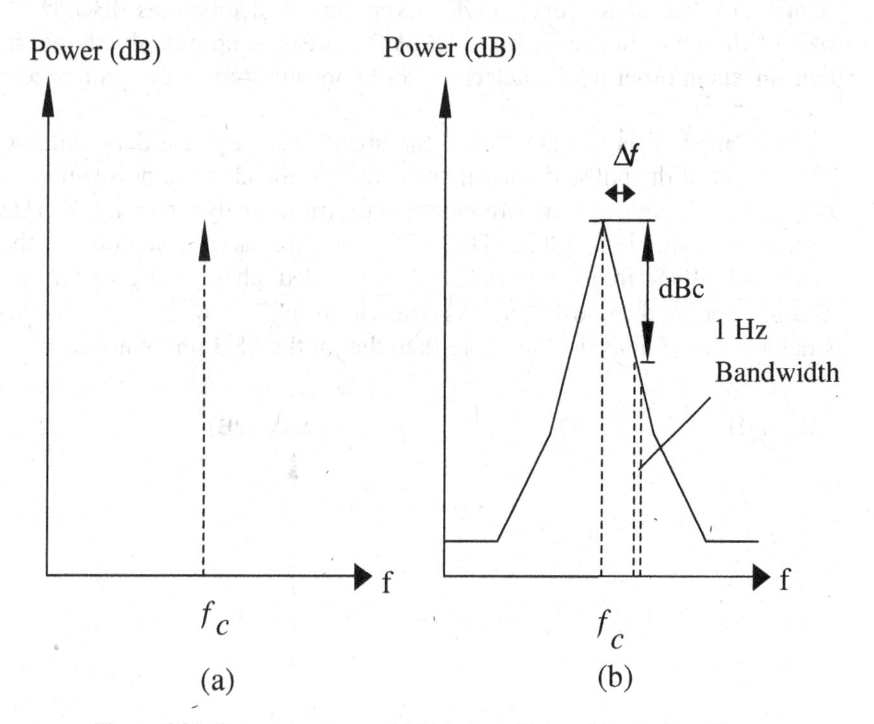

Figure 6.21 Output spectrum of (a) ideal oscillator; (b) actual oscillator.

Phase noise can be measured in different ways. One way is to use a spectrum analyzer where the total power of the signal would firstly be measured. Since the noise is small, this is essentially equal to the carrier power [23]. Then the phase noise power would be measured with the receiver tuned to a particular offset Δf from the carrier. The ratio of these two measurement results, expressed in decibels, is the normalized power spectral density (PSD) in one sideband at a frequency offset Δf referred to the carrier. This is known as the single-sideband (SSB) phase noise relative to the carrier level, $L(\Delta f)$, expressed in decibels as dBc/Hz. A plot of the phase noise as function of the offset is commonly shown in data sheets in order to characterize oscillators and frequency synthesizers.

Another method is to demodulate the signal with a phase demodulator. The output of the phase demodulator is the baseband phase noise and can be analyzed with a low frequency spectrum analyzer, with a 1Hz resolution-bandwidth filter. The resulting plot as a function of the baseband offset frequency is the double-sided phase noise spectrum, $S(\Delta f)$, expressed in dBc/Hz. As shown in Figure 6.22, the double-sideband phase noise is 3dB more than that of the SSB phase noise.

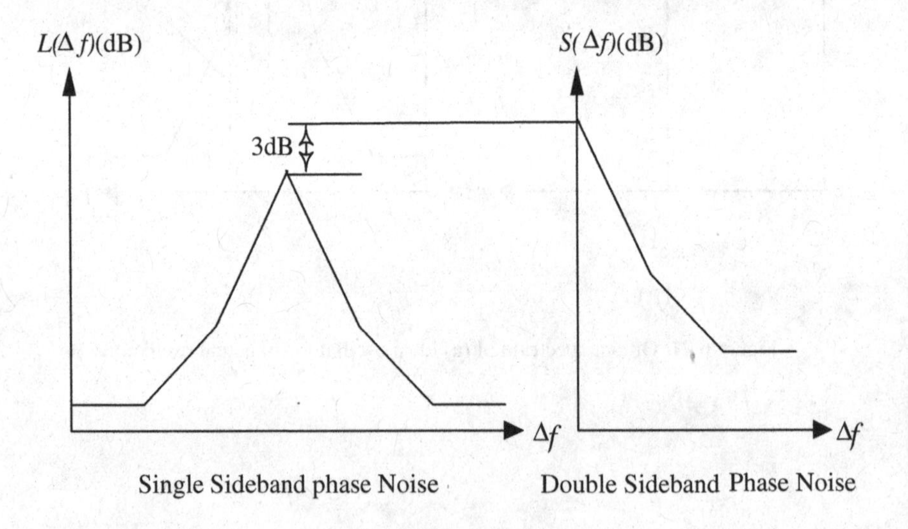

Figure 6.22 Single sideband and double sideband phase noise.

In the case where the noise of the PLL circuit is lower than that of the analyzer, there are alternative phase noise measurement systems. Phase noise can be measured by comparing the PLL signals in quadrature to a signal source with a noise level lower than that of the PLL signals and it is calculated at the output of the mixer using an FFT analyzer [24]. A new method to measure phase noise is the delay line method [24], which implements a frequency discriminator by mixing the signal with its delayed replica using a coaxial cable.

If the input signal or the building blocks of a PLL exhibit noise, the output signal will also suffer from noise. All loop components, which include the VCO, LF, phase detector and frequency divider, may contribute to phase noise [25]. The objectives are to understand how the spectrum of a given noise source is shaped as it propagates to the output and its effects on the total phase noise. Four important noise sources will be examined: (1) the VCO, (2) the reference signal, (3) the frequency divider, and (4) the LF. Note that the noise contribution of the phase detector is not being considered here. The reason is that at a low operating frequency, phase detector can be designed to have negligible effects on the overall phase noise of the PLL [26].

Figure 6.23 can also be used to describe the relationship among individual phase noise sources, where noise generated by the VCO, noise included in the reference signal, noise generated by the frequency divider and the loop filter noise are represented by θ_{vco}, θ_{ref}, θ_{div} and θ_{lf}, respectively. The typical phase noise plot of the VCO noise, reference signal noise and frequency divider noise for a frequency synthesizer in the GSM application is illustrated in Figure 6.23 [23]. The loop filter noise will be discussed later in detail.

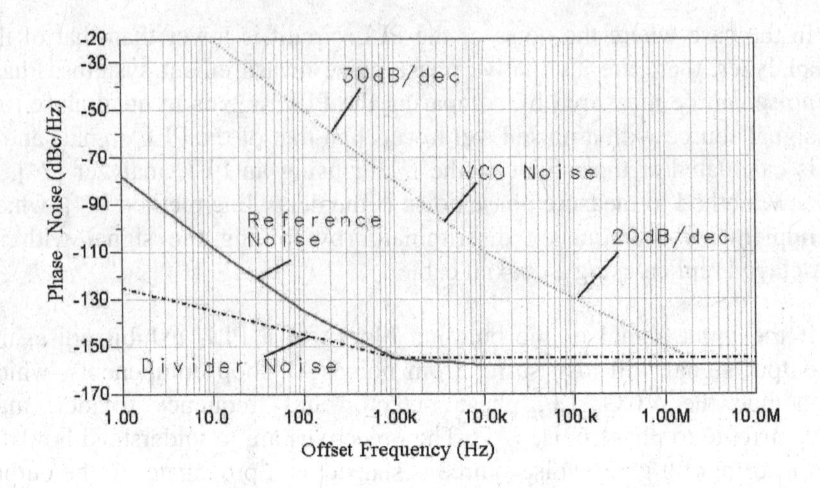

Figure 6.23 Phase noise plot of the noise sources in a PLL.

6.7.1 Phase Noise of VCO

The closed-loop transfer function from a VCO signal is

$$\frac{\theta_{out}(s)}{\theta_{vco}(s)} = \frac{1}{1+OL(s)} = \frac{N \cdot s}{N \cdot s + K_{pd} \cdot F(s) \cdot K_{vco}} \tag{6.44}$$

For simplicity, assuming that the loop filter has a constant transfer function, $F(s) = K_{lf}$, equation (6.44) becomes:

$$\frac{\theta_{out}(s)}{\theta_{vco}(s)} = \frac{1}{1+K_F/(Ns)} = \frac{N \cdot s}{N \cdot s + K_{pd} \cdot K_{lf} \cdot K_{vco}} = \frac{s}{s+\omega_c} \tag{6.45}$$

where $\omega_c = K_F/N$ is defined as the crossover frequency (or loop bandwidth), i.e., the frequency at which the open loop gain is equal to one.

Equation (6.45) shows that the noise transfer function from the VCO to the output has a high pass characteristic. Noise at high frequencies passes without being attenuated, because the feedback action of the loop is too slow to suppress these noise components. Note that although it is assumed that the loop filter has a constant transfer function, the analysis is applicable for higher order loop filter. A 4th order charge pump PLL frequency synthesizer for the GSM application has been simulated to illustrate the noise properties of the PLL. The Leeson-Cutler phase noise model [27-29], is used for modeling the VCO noise in the example. The spectral density of a VCO, $S_{vco}(\Delta\omega)$, is found to be

$$S_{vco}(\Delta\omega)=\frac{2FkT}{P_s}\cdot(1+\frac{\omega_{c3}}{\Delta\omega})[1+(\frac{\omega_0}{2Q_L\Delta\omega})^2] \tag{6.46}$$

where F is an empirical parameter often called the device excess factor, k is the Boltzmann's constant, T is the absolute temperature, P_s is the average power dissipated in the resistive part of the tank, ω_0 is the oscillation frequency, Q_L is the effective quality factor of the tank, $\Delta\omega$ is the offset from the carrier and ω_{c3} is the frequency of the corner between the $1/f^3$ and $1/f^2$ regions. Let $A = FkT/P_s$, equation (6.46) can be rearranged as:

$$S_{vco}(\Delta\omega)=(\frac{A\omega_{c3}}{4Q_L^2\Delta\omega})(\frac{\omega_0}{\Delta\omega})^2+(\frac{A}{4Q_L^2})(\frac{\omega_0}{\Delta\omega})^2+A(\frac{\omega_{c3}}{\Delta\omega})+A \tag{6.47}$$

Let $k_3 = A\omega_{c3}\omega_0^2/(4Q_L^2)$, $k_2 = A\omega_0^2/(4Q_L^2)$, $k_1 = A\omega_{c3}$, and $k_0 = A$, equation (6.47) can be simplified to be:

$$S_{vco}(\Delta\omega)=\frac{k_3}{\Delta\omega^3}+\frac{k_2}{\Delta\omega^2}+\frac{k_1}{\Delta\omega}+k_0 \tag{6.48}$$

One of the advantages of equation (6.48) is its close resemblance to the actual phase noise characteristics of the oscillator, which will be discussed in detail in the next chapter. Another advantage is that the four coefficients of equation (6.48) can be manually adjusted to yield the correct numerical value of VCO phase noise at all offset frequencies.

The coefficients k_3, k_2, k_1 and k_0 of the VCO noise in Figure 6.23 are experimentally determined using asymptotic lines with a slope of -30dB/dec, -20dB/dec, -10dB/dec and 0dB/dec. The values were obtained to be $10^{0.7}$, 10^{-3}, $10^{-14.5}$ and $10^{-15.5}$, respectively. For example, using equation (6.48), the calculated phase noise at 1kHz offset is

$$10\log \times S_{vco}(10^3) = -82.209 \text{ dBc/Hz} \tag{6.49}$$

which agrees with value of VCO noise at 1kHz offset in Figure 6.23.

The closed loop transfer function of the VCO noise is shown in Figure 6.24. By multiplying the square of the closed loop transfer function of the VCO given in equation (6.45) with equation (6.48), the contribution of the VCO noise to the total output noise of a PLL can be obtained as

$$S_{T\_VCO}(\Delta\omega) = S_{VCO}(\Delta\omega) * \left|\frac{\theta_{out}(s)}{\theta_{vco}(s)}\right|^2 \tag{6.50}$$

Figure 6.25 shows that while the VCO is very noisy close to the carrier, the noise is significantly attenuated within the loop bandwidth, which is 3.3kHz for this example. The magnitude of the VCO closed loop transfer function determines the amount of noise attenuation. On a close inspection of Figure 6.24, the loop has a gain of 40dB/dec from DC to the zero at 1.7kHz. From the zero to the gain crossover frequency at 3.3kHz, the loop can reject the VCO noise by 20dB/dec. The VCO has a 30dB/dec slope from DC to 5kHz, where it changes to a 20dB/dec slope. Close to the gain crossover frequency, the VCO noise is increased by 30dB/dec but the loop can only reject noise by 20dB/dec. Therefore, there exists a net increase of 10dB/dec in the total output noise. At the zero, the loop can reject the noise by 40dB/dec. Thus, taking into account the 30dB/dec slope of the VCO noise, there is a net decrease of 10dB/dec in the total output noise.

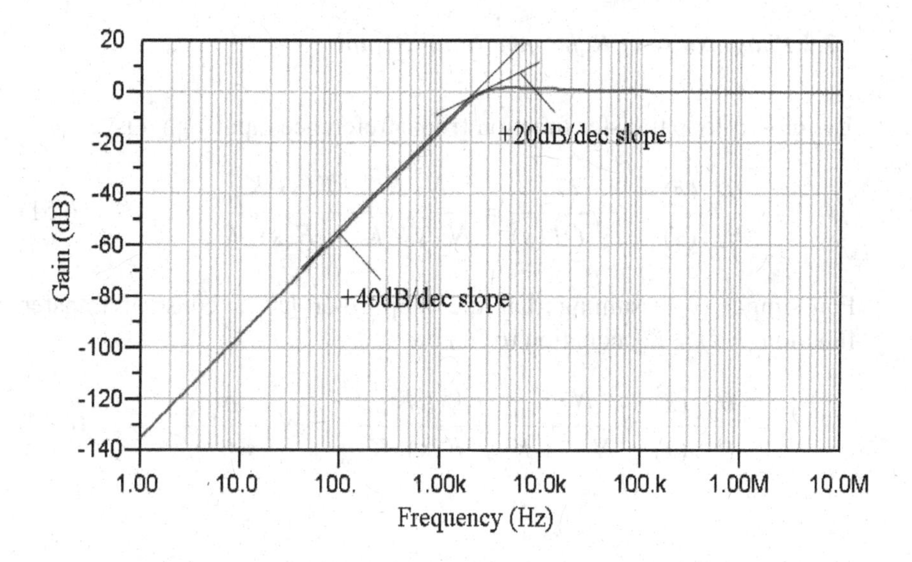

Figure 6.24 Closed loop transfer function of the VCO noise.

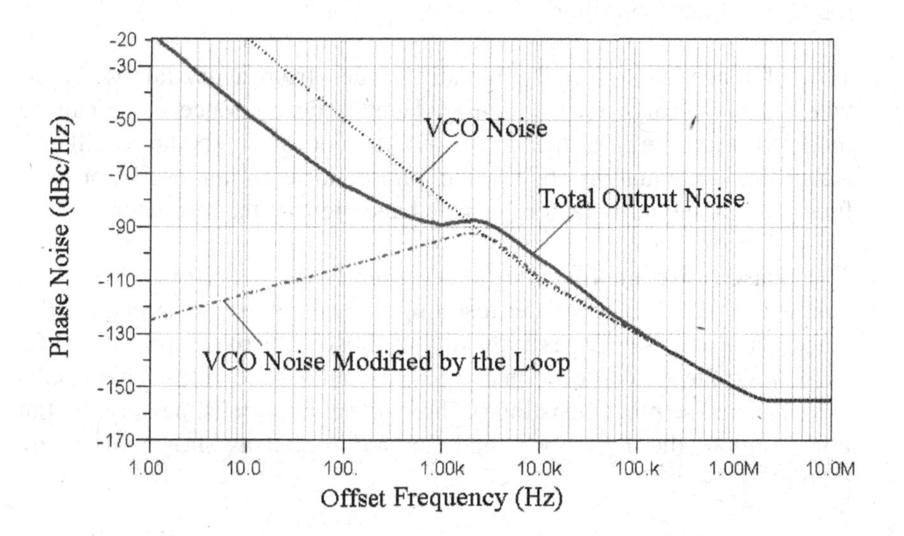

Figure 6.25 Effect of the PLL on VCO noise.

6.7.2 Phase Noise of Reference Input Signal

The closed-loop transfer function from a reference input signal is:

$$\frac{\theta_{out}(s)}{\theta_{ref}(s)} = \frac{G}{1+OL(s)} = \frac{N\cdot K_{pd}\cdot F(s)\cdot K_{vco}}{N\cdot s + K_{pd}\cdot F(s)\cdot K_{vco}} \tag{6.51}$$

For simplicity, assuming that the loop filter has a constant transfer function, $F(s) = K_{lf}$, equation (6.51) becomes

$$\frac{\theta_{out}(s)}{\theta_{ref}(s)} = \frac{N\cdot K_{pd}\cdot F(s)\cdot K_{vco}}{N\cdot s + K_{pd}\cdot F(s)\cdot K_{vco}} = N\cdot\frac{\omega_c}{s+\omega_c} \tag{6.52}$$

Equation (6.52) shows that the transfer function from the reference input signal to the output signal has a low pass characteristic superimposed on a multiplication with a factor *N*. Note that although it is assumed that the loop filter has a constant transfer function, the analysis is applicable for higher order loop filter.

The reference input signal generally comes from a crystal oscillator, which has typically a very high *Q*. Hence, the reference noise can be modeled according to the phase noise plot of a crystal oscillator. According to equation (6.46), the phase noise of an oscillator is a function of the quality factor *Q* or the bandwidth of the resonator.

The phase noise improves with a higher *Q*, or a narrower resonator bandwidth. A lower *1/f* corner also improves the noise performance. If a resonator has a low *Q*, as typically in most inductive-capacitive (LC) oscillators, the *1/f* corner is inside the resonator bandwidth. This results in a phase noise plot consisting of three different regions, namely the flat noise region, the $1/f^2$ region and the $1/f^3$ region as shown in Figure 6.26(a).

For a high Q resonator, the resonator bandwidth is smaller than the *1/f* corner. Therefore, after the flat region, the *1/f* corner is reached first and the phase noise increases at a rate of 10dB/dec, which is then followed by the $1/f^3$ region as shown in Figure 6.26(b). However, the explanation is rather simplified since a more accurate model will include four regions [23]. Similar to the VCO noise, the reference noise can be modeled by

$$S_{ref}(\Delta\omega) = \frac{k_3}{\Delta\omega^3} + \frac{k_2}{\Delta\omega^2} + \frac{k_1}{\Delta\omega} + k_0 \tag{6.53}$$

where the coefficients k_3, k_2, k_1 and k_0 of the reference noise in Figure 6.23 are experimentally determined using asymptotic lines with a slope of -30dB/dec, -20dB/dec, -10dB/dec and 0dB/dec and the values were obtained to be $10^{-7.8}$, $10^{-9.8}$, $10^{-12.7}$ and $10^{-15.8}$, respectively.

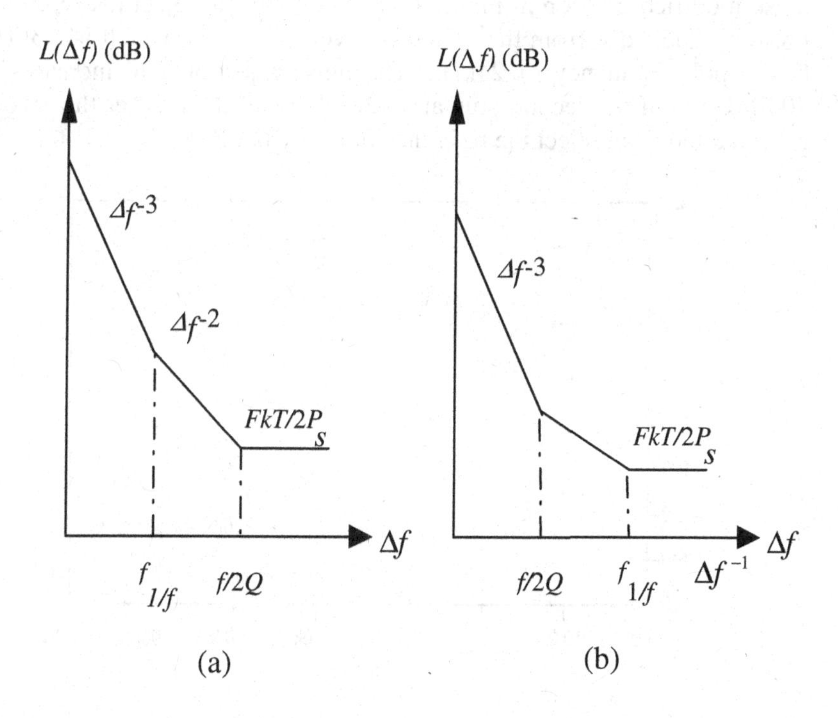

Figure 6.26 Phase noise plots (a) low Q; (b) high Q.

The closed loop transfer function of the reference noise is shown in Figure 6.27. By multiplying the square of the closed loop transfer function of the reference noise given in equation (6.52) with equation (6.53), the contribution of the reference noise to the total output noise of a PLL can be obtained as

$$S_{T\_REF}(\Delta\omega) = S_{ref}(\Delta\omega) * \left|\frac{\theta_{out}(s)}{\theta_{ref}(s)}\right|^2 \tag{6.54}$$

Figures 6.27 and 6.28 show that the magnitude of the closed loop transfer function of the reference noise determines the amount of noise rejection. The multiplied reference noise is the result of the reference noise multiplied by the division ratio N of the divider, as shown in Equation (6.52). The multiplied reference noise is then modified by the closed loop transfer function of the reference noise, denoted as reference noise modified by loop in Figure 6.28. The loop can reject the reference noise by 20dB/dec from the gain crossover frequency, which is 3.3kHz, to the pole frequency of 24kHz. The noise rejection rate increases to 40dB/dec until the second pole at 0.42MHz is reached. After the second pole, the loop can reject the reference noise by 60dB/dec.

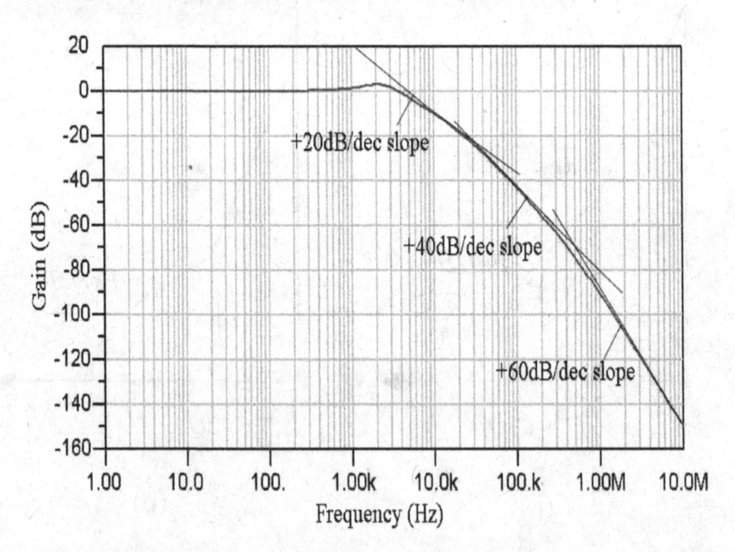

Figure 6.27 Closed loop transfer function of the reference noise.

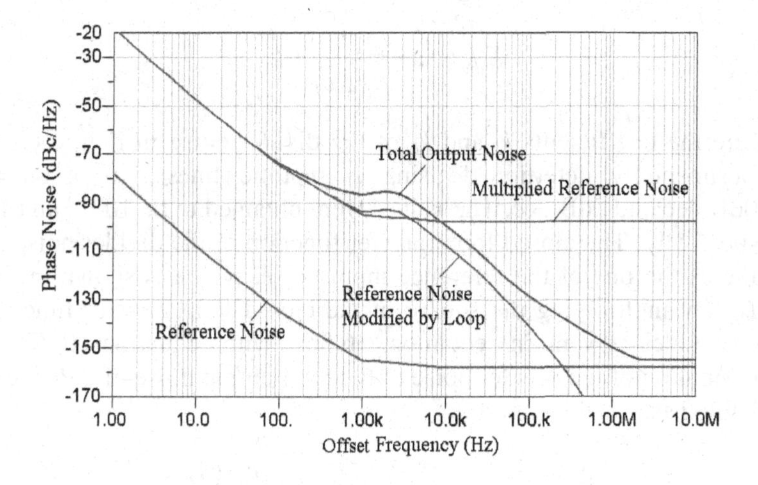

Figure 6.28 Effect of the PLL on the reference noise.

6.7.3 Phase Noise of Frequency Divider

The closed-loop transfer function from a frequency divider signal is

$$\frac{\theta_{out}(s)}{\theta_{div}(s)} = -\frac{G}{1+OL(s)} = -\frac{N \cdot K_{pd} \cdot F(s) \cdot K_{vco}}{N \cdot s + K_{pd} \cdot F(s) \cdot K_{vco}} \tag{6.55}$$

For simplicity, assuming that the loop filter has a constant transfer function, $F(s) = K_{lf}$, equation (6.55) becomes

$$\frac{\theta_{out}(s)}{\theta_{div}(s)} = -\frac{N \cdot K_{pd} \cdot F(s) \cdot K_{vco}}{N \cdot s + K_{pd} \cdot F(s) \cdot K_{vco}} = -N \cdot \frac{\omega_c}{s+\omega_c} \tag{6.56}$$

Equation (6.52) and equation (6.56) are essentially the same except for the difference in polarity. So the PLL will have the same effect on the frequency divider noise and reference input noise. The divider can be modeled by equation (6.57) [25].

$$S_{div}(\Delta\omega) = \frac{k_1}{\Delta\omega} + k_0 \tag{6.57}$$

where the coefficients k_1 and k_0 of the divider noise in Figure 6.23 are experimentally determined using asymptotic lines with a slope of -10dB/dec and 0dB/dec. The values were obtained to be $10^{-12.5}$ and $10^{-15.5}$, respectively. The closed loop transfer function of the divider noise is the same as the one of the reference input noise, which is shown in Figure 6.28. By multiplying the square of the closed loop transfer function of the divider given in equation (6.56) with equation (6.57), the contribution of the divider noise to the total output noise of a PLL can be obtained as

$$S_{T\_DIV}(\Delta\omega) = S_{div}(\Delta\omega) * \left| \frac{\theta_{out}(s)}{\theta_{div}(s)} \right|^2 \tag{6.58}$$

The effect of the PLL on divider noise is shown in Figure 6.29. The multiplied divider noise is the result of the divider noise multiplied by the divider division ratio N as shown in equation (6.56). The multiplied divider noise is then modified by the divider closed-loop transfer function, which is denoted as the divider noise modified by loop in Figure 6.29. From equation (6.58), it is expected that the divider noise will degrade by a factor of $10\log N^2$ within the loop bandwidth. Thus the multiplied divider noise with a divider's division ratio of 1000 is 60dB more than the divider noise as shown in Figure 6.29. Similarly, the multiplied reference noise is 60dB more than the reference noise as shown in Figure 6.28. Since the reference noise and the divider noise are non-correlated noise sources, it is possible to combine the effects of the reference noise and divider noise by power summation as shown in Figure 6.30.

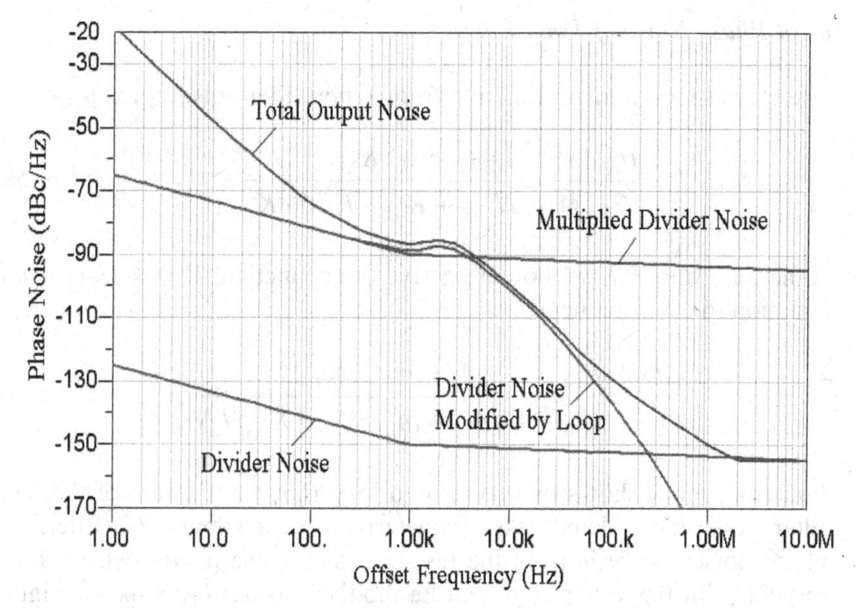

Figure 6.29 Effect of the PLL on the divider noise.

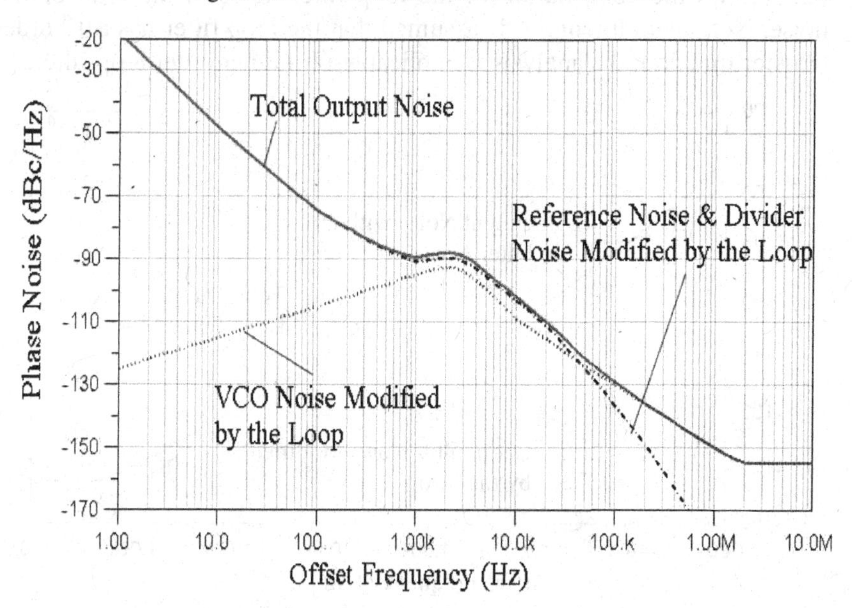

Figure 6.30 Effect of the PLL on the divider noise and the reference noise.

6.7.4 Phase Noise of Loop Filter

The closed-loop transfer function from a loop filter noise signal is

$$\frac{\theta_{out}(s)}{\theta_{lf}(s)} = \frac{N \cdot K_{vco}}{N \cdot s + K_{pd} \cdot F(s) \cdot K_{vco}} \tag{6.59}$$

If the loop filter has a 1st order passive filter function, $F(s) = 1/(1+s/\omega_p)$, equation (6.59) becomes

$$\frac{\theta_{out}(s)}{\theta_{lf}(s)} = -\frac{(\omega_p + s) \cdot K_{vco}}{s^2 + s\omega_p + \omega_p \cdot K_{pd} \cdot K_{vco} / N} \tag{6.60}$$

Equation (6.60) shows that the closed loop transfer function of the loop filter noise has a band-pass characteristic. In a simple RC filter, the major noise source is from the resistors. The noise power density for a resistor with the value of R, can be modeled using $S_{res} = 4kTR$. Figure 6.31 shows the contribution of the loop filter noise to the total output noise. Note that although it is assumed that the loop filter has a 1st order transfer function, the analysis is applicable for higher order loop filter.

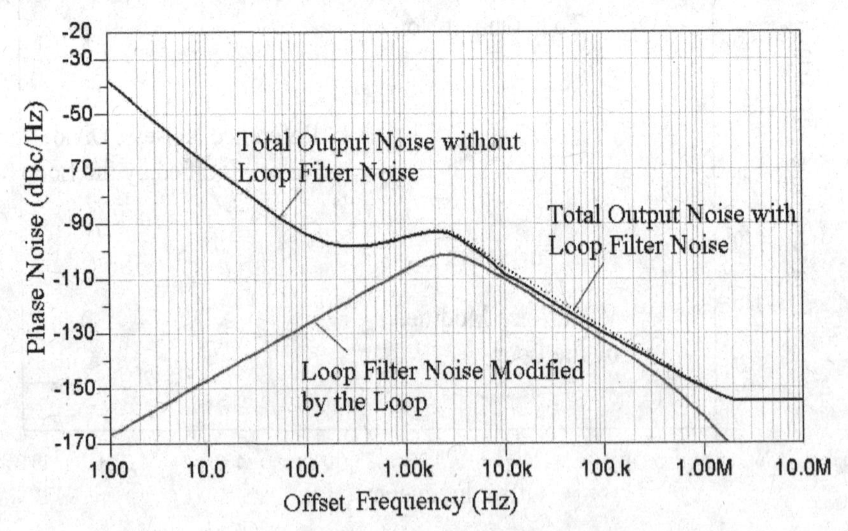

Figure 6.31 Contribution of the loop filter noise to the total output noise.

In Figure 6.31, the band-pass characteristic of closed loop transfer function of the loop filter noise is shown. It is evident that the loop filter noise is not negligible as it affects the total output noise of the PLL. The value of the resistor must be minimized in order to reduce the loop filter noise. However, a capacitor with larger value will be required to obtain the same cut-off frequency with the use of a resistor with smaller value. Hence, there is a tradeoff between the capacitor and resistor in the design of the RC filter.

6.7.5 Optimum Loop Bandwidth

After analyzing the noise properties of the PLL's blocks and the loop filter characteristic, the optimum loop bandwidth can now be obtained. The PLL's noise sources can be separated into three categories as shown in Figure 6.32, noise sources outside the loop, noise sources inside the loop and noise source contributed by the loop filter.

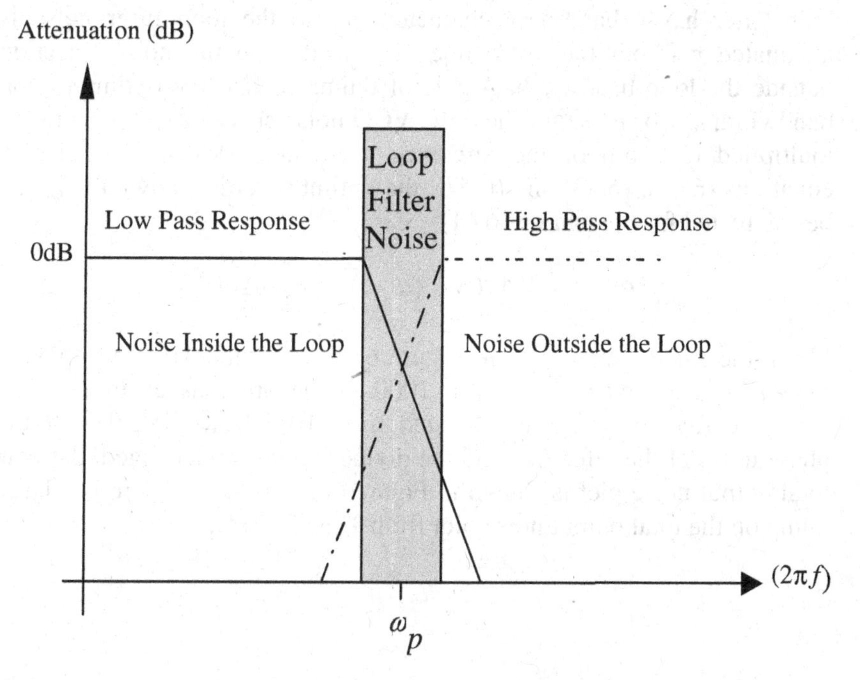

Figure 6.32 Phase noise contributions in a PLL.

The VCO noise is said to be the noise source outside the loop because the VCO noise is not modified by the loop outside the loop bandwidth. Inside the loop bandwidth, the loop attempts to reject the noise. In order to achieve the most rejection of VCO noise, the loop bandwidth should be as wide as possible and the zero frequency should be placed as close to the unity gain frequency ω_p as possible. However, moving the zero closer to the gain crossover frequency will decrease the phase margin and increase the peaking effect.

Both the reference noise and the divider noise are the noise sources inside the loop because these noise sources are not modified by the loop inside the loop bandwidth. Outside the loop bandwidth, the loop attempts to reject the noise. The phase detector noise can also be included in this category. Making the loop bandwidth narrower will increase the noise rejection inside the loop bandwidth. Moving the pole closer to the gain crossover frequency will increase the noise rejection within the loop but decrease the phase margin and increase peaking effect. Note that the loop filter has a band-pass characteristic, so the loop filter noise is attenuated by both the noise rejection inside and the noise rejection outside the loop bandwidth. A rule of thumb to find the optimum loop bandwidth f_{opt} for noise is where the VCO noise curve intersects with the multiplied (N) sum of the reference noise and divider noise. Using equations (6.48), (6.53) and (6.57), the optimum loop bandwidth f_{opt} can be calculated from equation (6.61)

$$S_{vco}(\omega_{opt}) = N^2 \cdot (S_{ref}(\omega_{op}t) + S_{div}(\Delta\omega_{opt})) \tag{6.61}$$

Using the same example as in Figure 6.29, and supposing the division ratio N is changed to 100 (N = 1000 in the previous example). The reference frequency is thus changed from 10MHz to 100MHz. If the phase noise of the reference and the divider remains unchanged, the new total output noise plot is shown in Figure 6.33. Note that there is a large bump on the total output noise plot from 1kHz to 8kHz.

In this region, the contribution of the multiplied reference noise and divider noise is about 13dB lower than the total output noise. The total output noise at this region is mostly from the VCO noise. This shows that the original loop bandwidth of 3.3kHz is not optimized. By applying equation (6.61), the optimum loop bandwidth, f_{opt} can be derived to be 8.5kHz. The new noise plot using the f_{opt} of 8.5kHz is shown in Figure 6.34.

It is shown that the total output noise has been improved by about 10dB in the region from 1kHz to 8kHz, which is due to the wider loop bandwidth. The loop is now able to reject more VCO noise in this region. Note that noise in the region from 8kHz to 20kHz has increased by about 1dB. This is because less reference noise and divider noise is being attenuated in this region due to the wider loop bandwidth.

6.8 Summary

In this chapter, a brief overview of the PLL was presented. The total output noise is critical to the performance of the PLL and is determined by the noise sources presented in PLL blocks and by their closed loop transfer functions. The order and type of the loop as well as the loop bandwidth determine the transient performance of the PLL.

Various types of phase detectors along with their advantages and disadvantages have been discussed. An analysis of the 3rd order and the 4th order charge pump PLL was presented, and the optimum loop bandwidth was shown. The design of VCO will be presented in the following chapter.

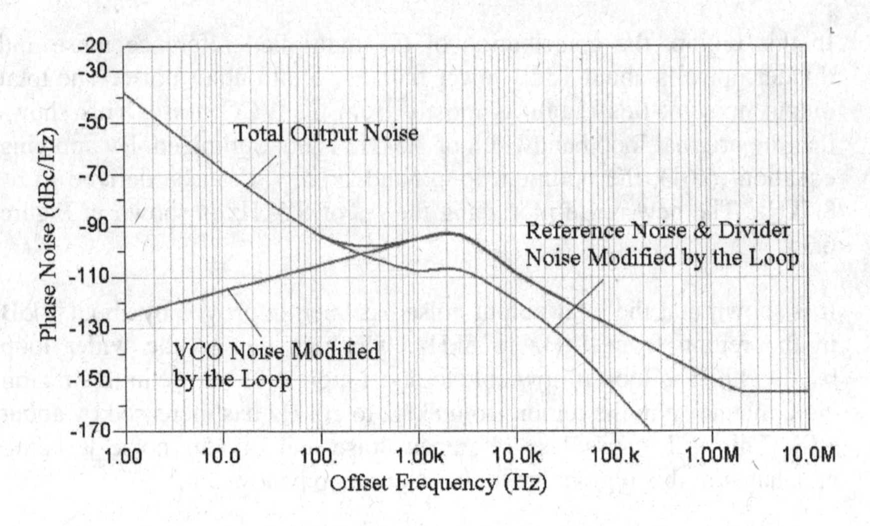

Figure 6.33 Phase noise contribution in a PLL with N=100 and loop bandwidth of 3.3kHz.

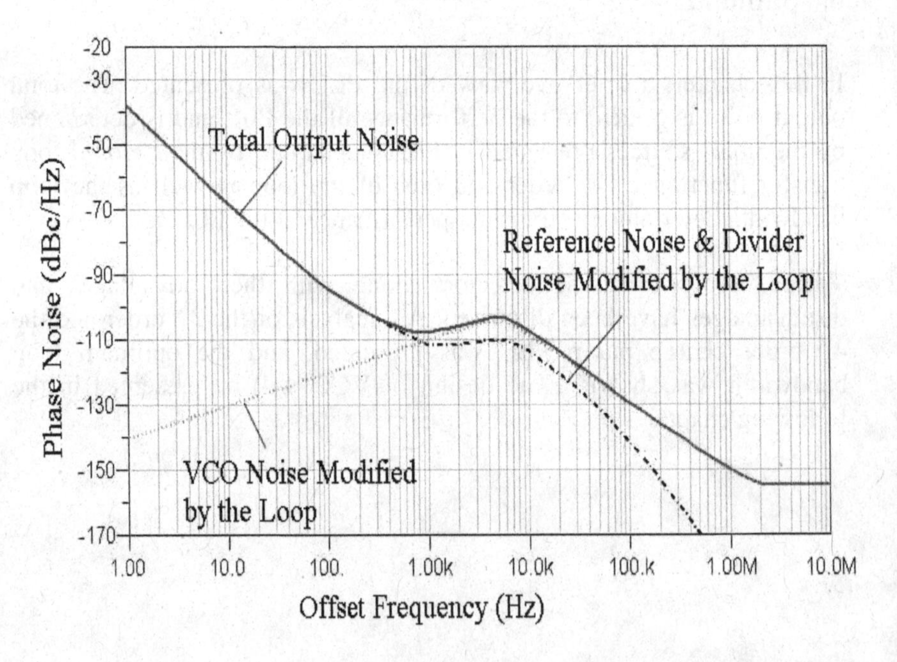

Fig. 6.34 Phase noise contribution in a PLL with N=100 and loop bandwidth of 8.5kHz.

References

[1] F. M. Gardner, "Phaselock Techniques", 2nd *Edition, New York Wiley & Sons*, 1979.

[2] W. F. Egan, "Frequency Synthesis by Phase Lock", *New York Wiley & Sons*, 1981.

[3] V. Manassewitsch, "Frequency Synthesizers – Theory and Design", *Wiley-Interscience*, pp. 283-284, 1987.

[4] D. H. Wolaver, "Phase-Locked Loop Circuit Design", *New Jersey Prentice Hall*, 1991.

[5] D. G. Wilson, W. Rhee, and B. S. Song, "Low Power RF Receiver Front Ends and Frequency Synthesizers for Wireless", *IEEE ISCAS*, Tutorial Workshop, 1996.

[6] Craninckx and M. S. J. Steyaert, "A Fully Integrated CMOS DCS-1800 Frequency Synthesizer", *IEEE Journal of Solid-State Circuits*, vol. 33, no. 12, pp. 2054-2065, Dec. 1998.

[7] Y. Koo, H. Huh, Y. Cho, J. Lee, J. Park, K. Lee, D-K. Jeong, and W. Kim, "A Fully Integrated CMOS Frequency Synthesizer With Charge-Averaging Charge Pump and Dual-Path Loop Filter for PCS- and Cellular-CDMA Wireless Systems", *IEEE Journal of Solid-State Circuits*, vol. 37, no. 5, pp. 536-542, May 2002.

[8] M. Banu and A. Dunlop, "A 660-Mb/s CMOS Clock Recover Circuits with Instantaneous Locking for NRZ Data and Burst-Mode Transmission", *ISSCC Digest of Technical Papers, San Francisco*, pp. 102-103, Feb. 1993.

[9] B. Razavi and J. Sung, "A 6-GHz 60-mW BiCMOS Phase Locked Loop with 2-V Supply", *ISSCC Digest of Technical Papers, San Fransisco*, pp. 114-115, Feb. 1994.

[10] A. Pottbacker and U. Langmann, "An 8-GHz Silicon Bipolar Clock Recovery and Data-Generator IC", *IEEE Journal of Solid-State Circuits*, vol. 29, no. 12, pp. 1572-1578, Dec. 1994.

[11] M. Wurzer, J. Bock, H. Knapp, W. Zirwas, F. Schumann, and A. Felder, "A 40-Gb/s Integrated Clock and Data Recover Circuit in 50-GHz f_T Silicon Bipolar Technology", *IEEE Journal of Solid-State Circuits*, vol. 34, no. 9, pp. 1320-1324, Sep. 1999.

[12] J. Craninckx and M. Steyaert, "Wireless CMOS Frequency Synthesizer Design", 1st *Edition, Boston Kluwer Academic Publishers*, 1998.

[13] P. Hugues, "UMA1021M Low Voltage Frequency Synthesizer", *Philips Semiconductors Application Note AN96083*, 1996.

[14] R. E. Best, "Phase-Locked Loops: Theory and Applications", *New York CRC Press*, 1997.

[15] F. M. Gardner, "Charge-Pump Phase Locked Loops," *IEEE Trans. Comm.*, vol. 28, pp. 1849-1859, Nov. 1980.

[16] M. Souyuer and R. G. Meyer, "Frequency Limitation of a Conventional Phase-Frequency Dectector", *IEEE Journal Of Solid-State Circuits*, vol. 25, no. 4, pp. 1019-1022, Aug. 1990.

[17] U. L. Rohde, "Microwave and Wireless Synthesizers", *New York Wiley & Sons*, 1997.

[18] A. Hill and J. Surber, "The PLL Dead Zone and How to Avoid It", *RF Design*, pp. 131-134, Mar. 1992.

[19] D. Mijuskovic, M. J. Bayer, T. F. Chomicz, N. K. Garg, F. James, P. W. McEntarfer, and J. A. Porter, "Cell Based Fully Integrated CMOS Frequency Synthesizers", *IEEE Journal of Solid-State Circuits*, vol. 29, no. 3, pp. 271-279, Mar. 1994.

[20] W. O. Keese, "An Analysis and Performance Evaluation of a Passive Filter Design Technique for Charge Pump PLLs", *National Semiconductor Application Note AN012473*, 1996.

[21] H. Meyr and G. Ascheid, "Synchronization in Digital Communication, Vol. 1: Phase-, Frequency Locked Loops and Amplitude Control", *New York Wiley & Sons*, 1990.

[22] P. Hugues, "UMA1021M Low Voltage Frequency Synthesizer", *Philips Semiconductors Application Note AN96083*, 1996.

[23] E. Drucker, "Model PLL Dynamics and Phase-Noise Performance", *Microwave & RF*, pp. 73-117, Feb. 2000.

[24] C. Barrett, "Fractional/Integer-N PLL Basics", *Texas Instruments Technical Brief SWRA029*, Aug. 1999.

[25] V. F. Kroupa, "Noise Properties of PLL Systems", *IEEE Trans. Comm.*, vol. 30, no. 10, pp. 2244-2252, Oct. 1982.

[26] A. Hajimiri, "Noise in Phase-Locked Loops", *2001 Southwest Symposium on Mixed-Signal Design*, pp. 1-6, 2001.

[27] C E. J. Baghdady, R. N. Lincoln, and B. D. Nelin, "Short-Term Frequency Stability: Characterization, Theory, and Measurement", *Proceedings of IEEE*, vol. 53, pp. 704-722, Jul. 1965.

[28] L. S. Cutler and C. L. Searle, "Some Aspects of the Theory and Measurement of Frequency Fluctuations in Frequency Standards", *Proceedings of IEEE*, vol. 54, pp. 136-154, Feb. 1966.

[29] D. B. Leeson, "A Simple Model of Feedback Oscillator Noise Spectrum", *Proceedings of IEEE*, vol. 54, pp. 136-154, Feb. 1966.

One chip, many applications.

Kiat Seng YEO

CHAPTER 7

RF CMOS Prescalers

The most apparent feature that differentiates a phase-locked loop (PLL) frequency synthesizer from other phase-locked loops lies in the frequency divider. Much attention and efforts have been given to this circuit block to minimize its power consumption and size and to improve its speed. Figure 7.1 shows a PLL frequency synthesizer with a frequency divider in the feedback loop.

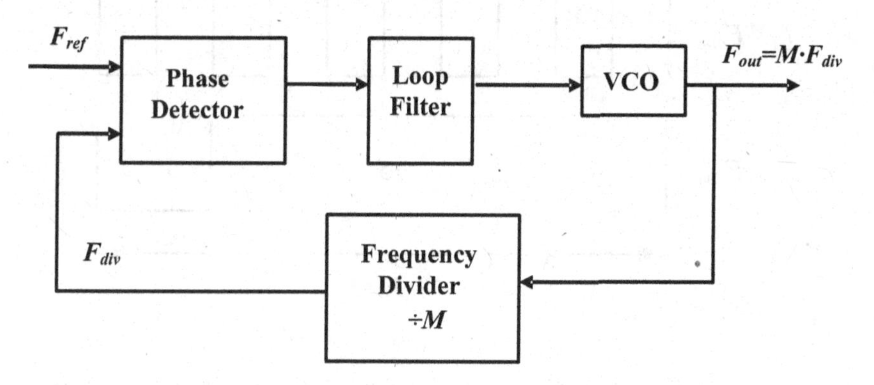

Figure 7.1 A PLL frequency synthesizer with a divide-by-M frequency divider.

Due to the limitation of the operating frequency of the phase detector (PD), the high output frequency of the voltage-controlled oscillator (VCO) at frequency F_{out} has to be divided down to a lower frequency F_{div}. A frequency divider with a division value of M is used to obtain

$$F_{div} = F_{out} / M \tag{7.1}$$

In order to obtain a variable, controllable division ratio (modulus), a counter divider can be used as shown in Figure 7.2(a). A certain preset number M can be loaded into the counter, which counts the input pulses. A reset signal will be generated when M pulses are reached. Then the counter is reset to zero and the counting process restarts. The division ratio then equals to M. Figure 7.2(b) shows the waveforms of F_{div} and F_{out} for $M = 5$. Frequency F_{out} is divided by 5 to become F_{div}.

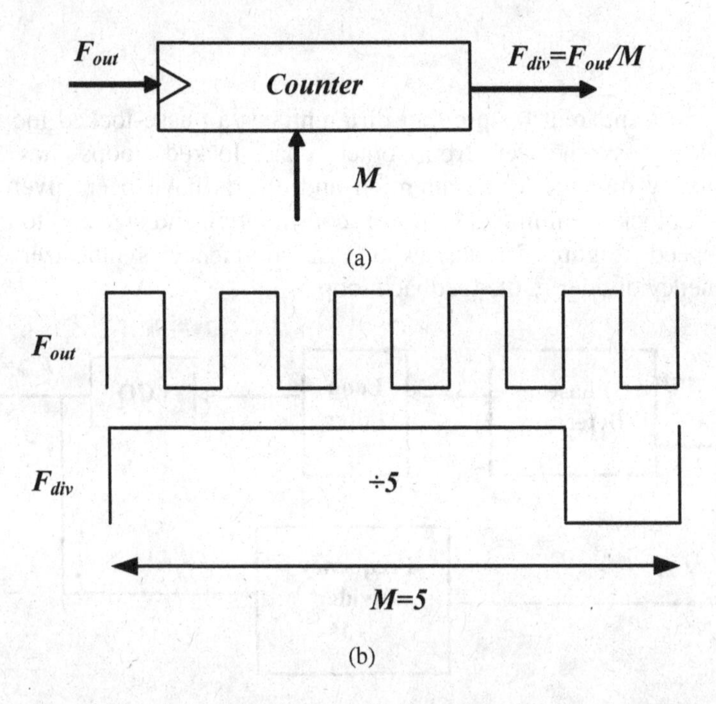

Figure 7.2 (a) A counter divider; (b) The waveforms of F_{div} and F_{out}.

For the example waveforms shown in Figure 7.2(b), the counter counts the rising edge of the input pulses. During the counting process, the output waveform of F_{div} remains logic high. When 5 input rising edges are counted, the counter will reset. The output waveform will become logic low until the next rising edge.

The limitation of the counter divider is its low operating frequency. When the input frequency of the divider is too high to permit a proper operation of the programmable divider or counter, a prescaler can be used.

The design of prescaler is the most challenging part of the high speed frequency divider design because it operates at the highest input frequency. The basic concepts for the prescaler will be discussed in Section 7.1. This will be followed by a brief discussion of various implementations of a DFF, which is the most important basic building block of the prescaler, in Section 7.2. Then, the detailed descriptions of a high-speed prescaler as a design example will be presented in Section 7.3. The summary will be given in Section 7.4.

7.1 Prescaler

A prescaler divides the input frequency by fixed ratios, and can therefore operate at higher frequencies because it does not suffer from the delays involved in counting and resetting. Adding a few high-speed prescaler stages will lower the speed requirement for the subsequent counter stages. In general, a prescaler consists of several divide-by-2 units cascaded in series to achieve a division ratio of 2^k, where k is the number of divide-by-2 units.

Example

Show the circuit and waveforms of a divide-by-4 prescaler followed by a counter divider with a division ratio of 5.

Solution

By inserting a prescaler formed by two divide-by-2 units cascaded in series before the counter divider will lower the input frequency to the counter by $2^2 = 4$ times. A divide-by-2 unit can be formed by a flip-flop (FF). Figure 7.3 shows a prescaler consisting of two DFFs performing a divide-by-4 operation, which is then connected to a counter divider.

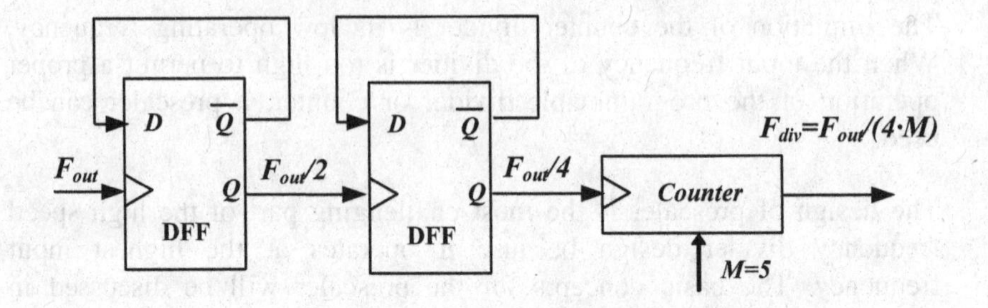

Figure 7.3 A divide-by-4 prescaler followed by a counter divider.

The waveforms of $F_{out}/2$ and $F_{out}/4$ are shown in Figure 7.4(a), and the output waveform of the counter divider F_{div} is shown in Figure 7.4(b).

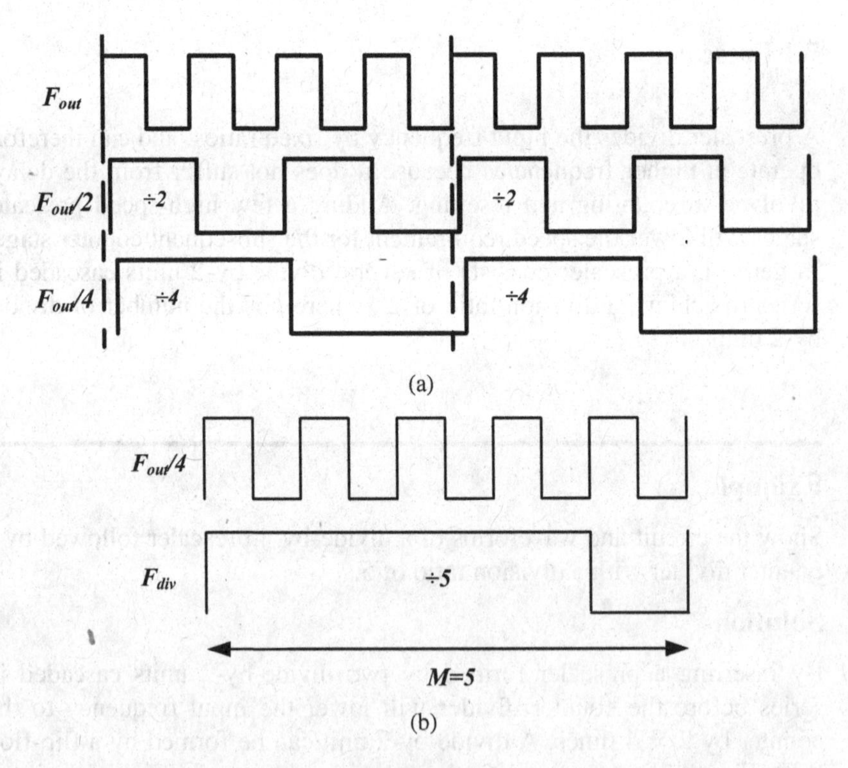

Figure 7.4 (a) Waveforms of $F_{out}/2$ and $F_{out}/4$; (b) Waveform of F_{div}.

It can be seen from Figure 7.4(a) that there is a delay between F_{out}, $F_{out}/2$ and $F_{out}/4$ due to the asynchronous nature of the prescaler in this configuration. The frequency of $F_{out}/2$ and $F_{out}/4$ are 2 and 4 times less than that of F_{out} respectively. Figure 7.4(b) showed that the counter output F_{div} is 5 times less than that of $F_{out}/4$. This means the frequency of F_{div} is 20 times less than that of F_{out}.

A few important insights can be deduced from the above example. Due to the prescaler

$$F_{div} = F_{out} /(2^k \cdot M) \tag{7.2}$$

where k is the number of divide-by-2 units used to form the prescaler and M is the division ratio of the counter divider. In the above example, $k = 2$ and $M = 5$, thus $F_{div} = F_{out}/20$.

For a PLL frequency synthesizer, the two input frequencies to the phase detector, namely F_{div} and the reference frequency F_{ref}, are equal during phase lock. If this PLL frequency synthesizer switches from one channel to another channel by changing the value M to the next value M+1, F_{ref} must be a factor $2k$ times lower than the channel spacing which is $2kF_{ref}$.

A smaller F_{ref} implies a lower loop bandwidth, which is often undesirable in terms of the PLL transient response [1]. In order to resolve the above-mentioned resolution problem, a Dual-Modulus Prescaler (DMP) can be used.

7.1.1 Dual-Modulus Prescaler

A dual-modulus prescaler extends a fixed-ratio prescaler with some extra logic circuits, which enable a selectable division by N or by (N + 1). Due to the extra functionality, the speed of the circuit is slowed down. Notably, there are a few proposals for specialized circuits, for example, NOR/flip-flop combination circuits [2], which have made much improvement in the performance of the DMP. Nevertheless, these types of DMPs still cannot match the speed of the asynchronous prescaler.

Figure 7.5 shows the circuit of a synchronous DMP, which enables the selectable division by 4 or 5.

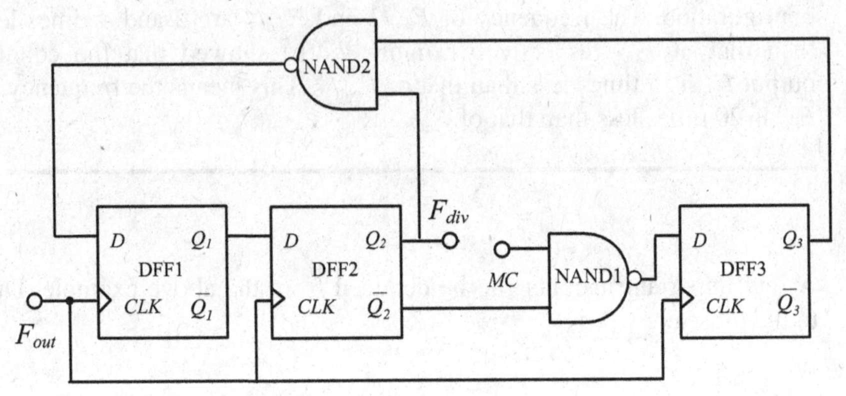

Figure 7.5 A synchronous divide-by-4/5 DMP.

This synchronous circuit consists of two NAND gates and three DFFs. The division ratio of this circuit can be either 4 or 5 depending on the signal *MC*. Figure 7.6 shows circuit and waveforms of the divide-by-4/5 prescaler with divide-by-4 operation when $MC = 0$. Note that the output Q_2 of DFF2 is F_{div}.

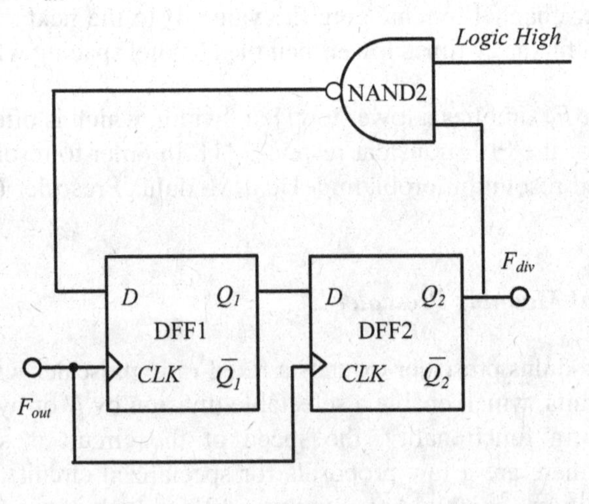

(a)

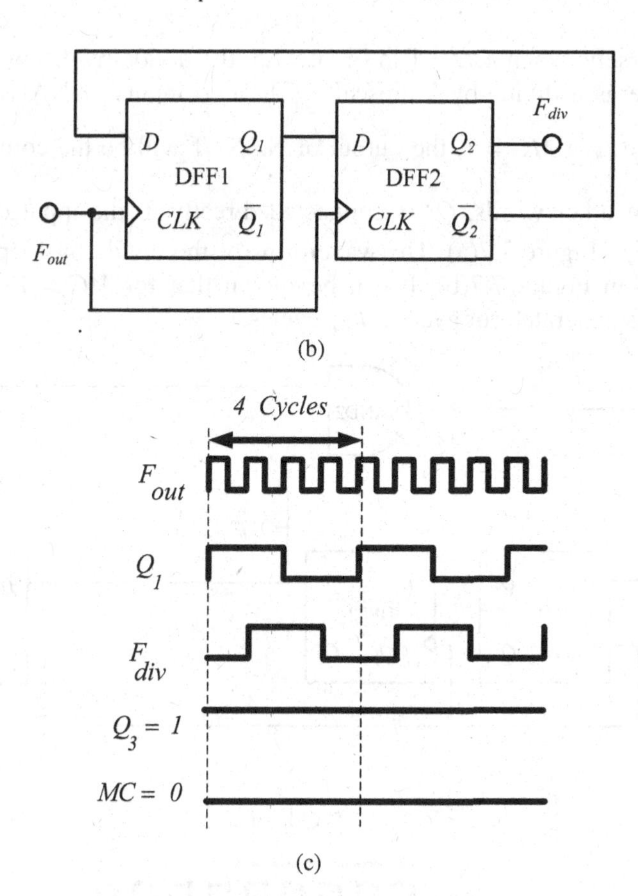

Figure 7.6 The divide-by-4/5 prescaler when *MC*=0
(a) circuit; (b) equivalent circuit; (c) waveforms.

When $MC = 0$, the output of NAND1 is at logic high, and the output Q_3 of DFF3 is also at logic high. This output is connected to one of the inputs of NAND2. The other input of NAND2 is connected to the output Q_2 of DFF2. Hence, the output of NAND2 will be the complement of Q_2 since the other input is at logic high. In other words, Figure 7.6(a) can be replaced by an equivalent circuit where NAND2 and the connection of Q_2 of DFF2 are removed, while $\overline{Q_2}$ from DFF2 is connected back to D of DFF1 as shown in Figure 7.6(b). This forms a basic synchronous divide-by-4 circuit. Note that DFF3 in Figure 7.5 does not take part in the divide-by-4 operation.

Conversely, when $MC = 1$ in Figure 7.5, the divide-by-4/5 prescaler will operate as a divide-by-5 prescaler. The two inputs of NAND1 are MC and $\overline{Q}_2$. For $MC = 1$, the output of NAND1 will be the complement of $\overline{Q}_2$. In other words, Q_2 is connected directly to the input of DFF3 as shown in Figure 7.7(a). The waveforms of the divide-by-5 operation are shown in Figure 7.7(b). It can be shown that for $MC = 1$, one output cycle is generated for every 5 F_{out} cycles.

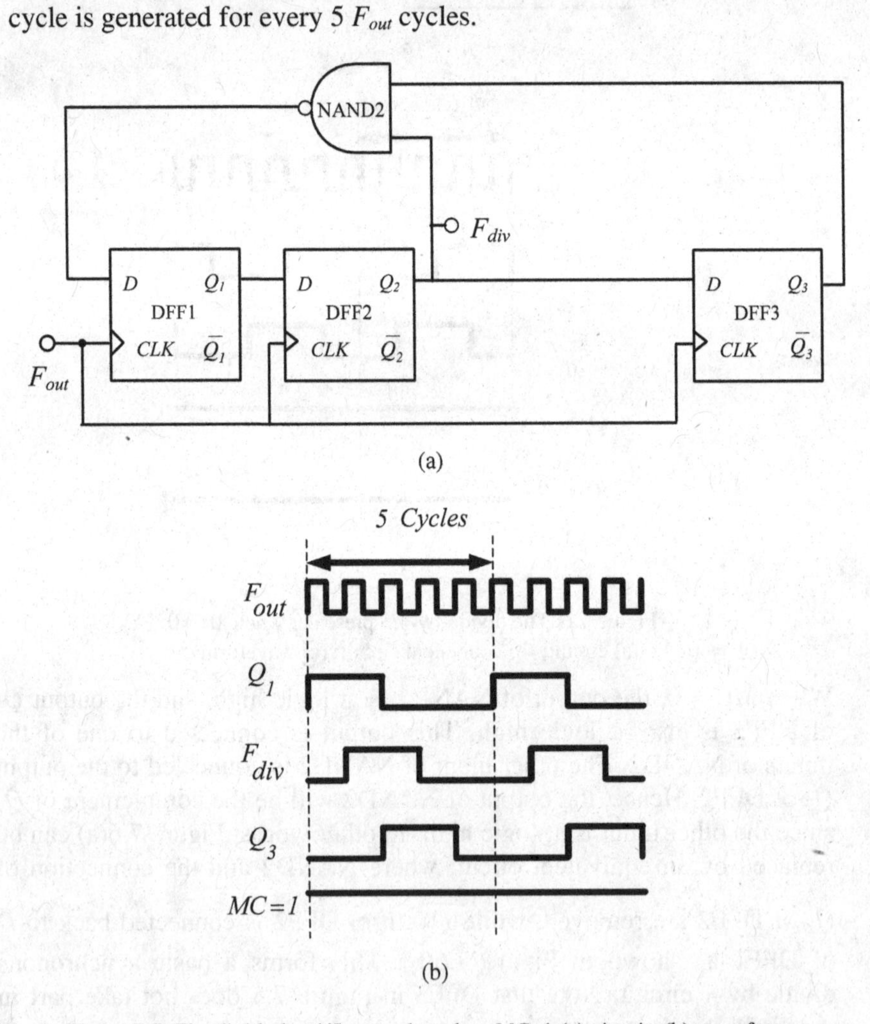

Figure 7.7 The divide-by-4/5 prescaler when MC=1 (a) circuit; (b) waveforms.

In order to increase the division modulus, several asynchronous divide-by-2 units that operate at lower frequencies can be combined to the divide-by-4/5 (or 2/3) unit that operates at the highest frequency. Figure 7.8 shows a conventional divide-by-32/33 prescaler. This divide-by-32/33 prescaler consists of one synchronous divide-by-4/5 prescaler unit, three asynchronous divide-by-2 units and one AND gate.

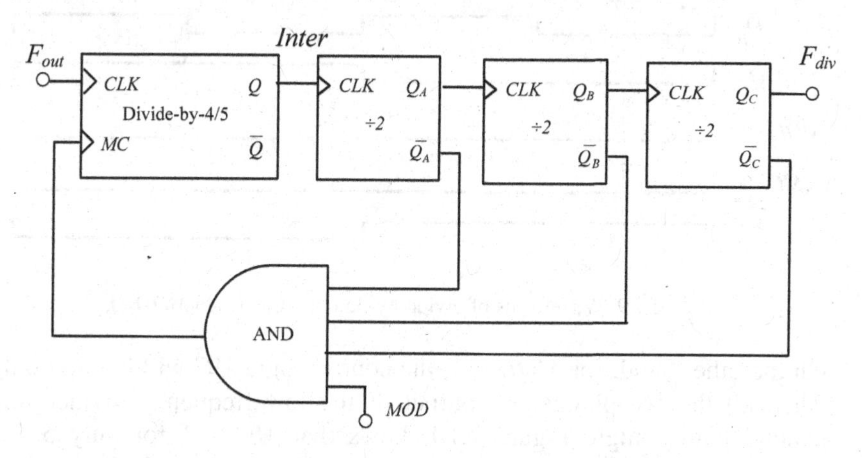

Figure 7.8 A conventional divide-by-32/33 prescaler using divide-by-4/5 prescaler.

The input F_{out} is divided by either 4 or 5 depending on the logic value of *MC* to get the output at node *Inter* of the synchronous divide-by-4/5 prescaler unit. The purpose of the three low frequency divide-by-2 circuits (÷2) is to further divide the signal *Inter* by 8, so that a divide-by-32/33 can be achieved at output F_{div} with only one divide-by-4/5 unit. Hence, output F_{div} is equal to input F_{out} divided by either 32 or 33 depending on *MOD*, which determines the value of *MC*. From Figure 7.8, it can be observed that the equation of *MC* is given by

$$MC = MOD * \overline{Q}_A * \overline{Q}_B * \overline{Q}_C \tag{7.3}$$

From equation (7.3), *MC* will be low when *MOD* = 0 and the divide-by-4/5 prescaler will operate as a divide-by-4 prescaler as shown in Figure 7.6. The three divide-by-2 units will further divide the output signal *Inter* of the prescaler by 8 to achieve a divide-by-32 operation at F_{div}. Figure 7.9 shows the waveforms of the divide-by-32 operation.

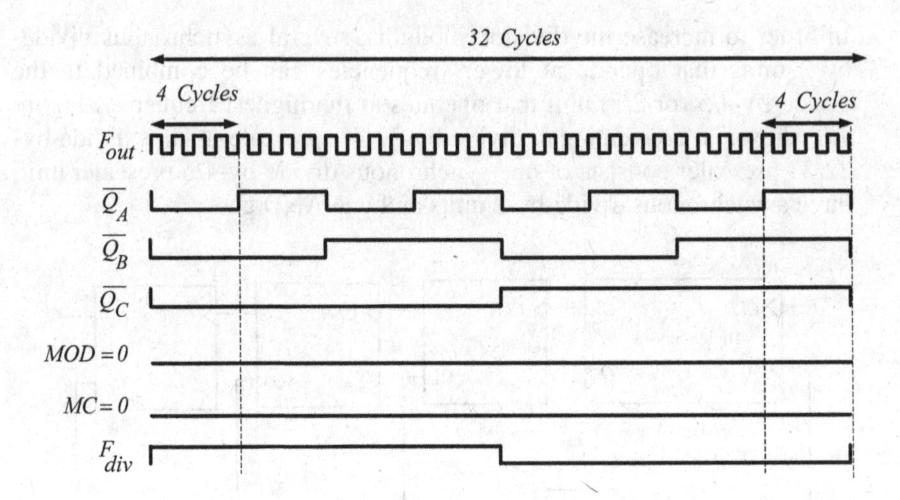

Figure 7.9 Waveforms of divide-by-32 operation when *MOD*=0.

On the other hand, for *MOD* = 1, the control logic *MC* will be high only when all the complementary output of the low frequency dividers are simultaneously high. Figure 7.10 shows that *MC* = 1 for only 5 F_{out} cycles and low for the rest of 28 cycles. Note that once *MC* is high, the divide-by-4/5 unit will perform a divide-by-5 operation.

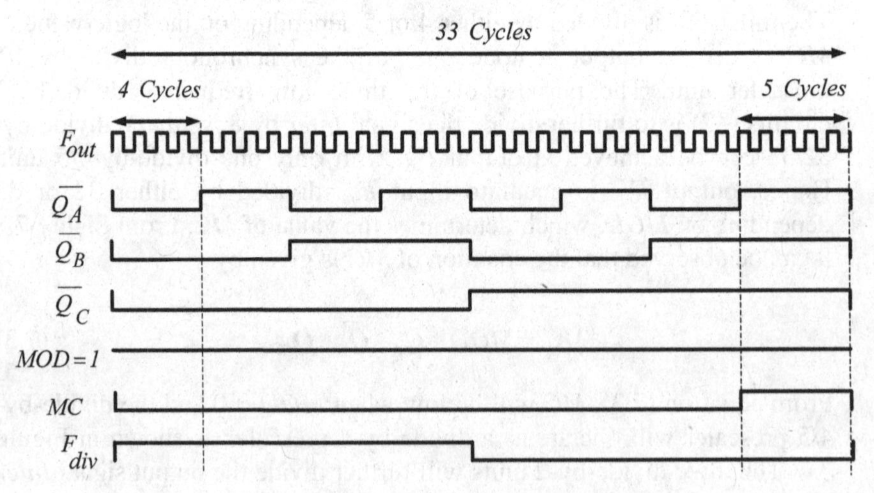

Figure 7.10 Waveforms of divide-by-33 operation when *MOD*=1.

Example

Design a divide-by-16/17 prescaler based on a divide-by-4/5 unit.

Solution

It is rather intuitive to modify the divide-by-32/33 prescaler into a prescaler with a different division ratio. For example, one low frequency divide-by-2 circuit can be removed and slight modifications on AND gate can be done as shown in Fig. 7.11. A divide-by-16/17 prescaler can be obtained with one synchronous divide-by-4/5 prescaler unit and two low frequency divide-by-2 units. Conversely, to achieve a divide-by-64/65, one additional low frequency divide-by-2 unit should be added to the divide-by-32/33 prescaler.

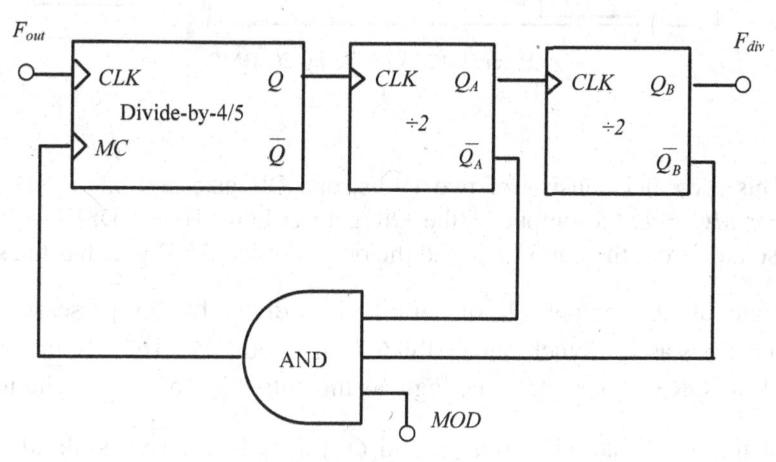

Figure 7.11 A divide-by-16/17 prescaler.

In the divide-by-4/5 unit as shown in Figure 7.5, the three DFFs are operating at a high input frequency, thus these DFFs will require high power consumption. In order to reduce power consumption, the number of DFFs working at high input frequency must be reduced. In other words, power consumption can be reduced with the DMP divide-by-4/5 to be replaced by a DMP divide-by-2/3. Figure 7.12 shows a DMP divide-by-2/3 unit.

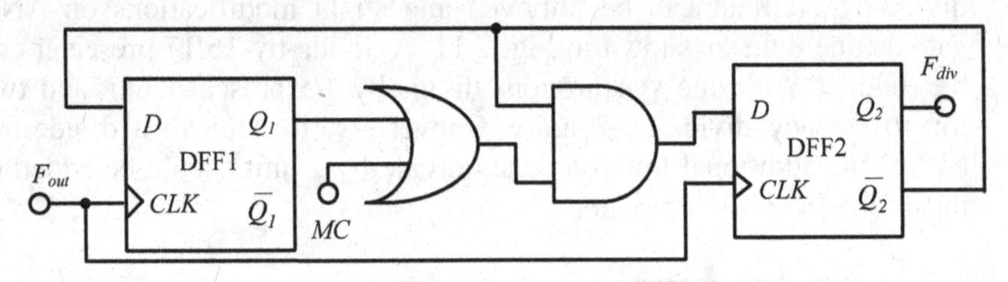

Figure 7.12 A divide-by-2/3 DMP.

This prescaler consists of two DFFs, one OR gate and one AND gate. For $MC = 1$, the output of the OR gate is high. Hence DFF1 is being isolated from the output F_{div} and the output of the AND gate has the same logic as the output $\overline{Q_2}$ of DFF2. The divide-by-2/3 prescaler now functions as an asynchronous divide-by-2 circuit. For $MC = 0$, the output of the OR gate has the same logic as the output Q_1 of DFF1. The inputs of the AND gate are from Q_1 and $\overline{Q_2}$. This forms a basic divide-by-3 circuit.

Figure 7.13 shows a divide-by-32/33 using a divide-by-2/3 circuit. In this implementation, only two DFFs in the divide-by-2/3 circuit are working at high input frequency thus reducing the power consumption.

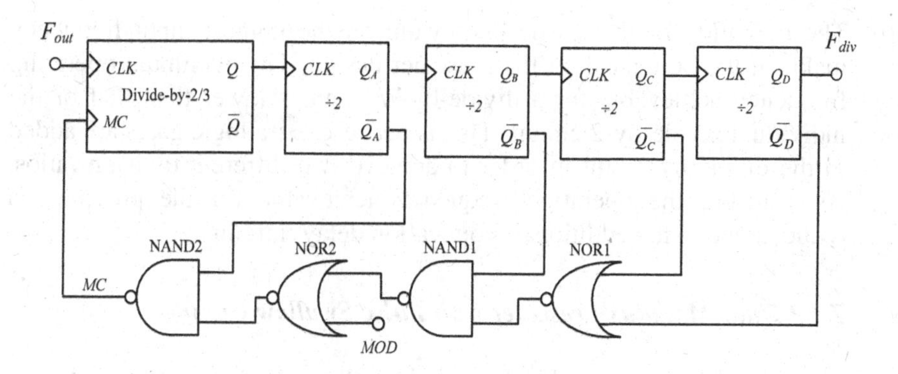

Figure 7.13 Another implementation of a divide-by-32/33 prescaler using divide-by-2/3 prescaler.

Instead of using one AND gate with many inputs, two NOR gates and two NAND gates can be used to generate the control logic *MC* as shown in Figure 7.13. This is a better option as it requires lower voltage headroom which is suitable for low voltage high speed operation. The equation for the control logic *MC* that controls the division ratio of the divide-by-2/3 unit is given by

$$MC = MOD + \overline{Q}_C + \overline{Q}_D + Q_A + Q_B \quad (7.4)$$

For *MOD* = 1, *MC* will always be high and the divide-by-2/3 unit will perform the divide-by-2 operation. The output of the divide-by-2/3 unit will be divided further by 16 through the four asynchronous divide-by-2 units. Hence, for *MOD* = 1, the circuit in Figure 7.13 will operate as a divide-by-32 prescaler.

On the other hand, for *MOD* = 0, *MC* will be low for 3 F_{out} cycles and high for the remaining 30 cycles. Note that *MC* will be low only if $\overline{Q}_C$, $\overline{Q}_D$, Q_A and Q_B are all low. Hence, for *MOD* = 0, the circuit in Figure 7.13 will operate as a divide-by-33 circuit.

The operation of the divide-by-2/3 unit at the highest input frequency makes it the bottleneck of the prescaler design. The maximum operating frequency achievable for a divide-by-2/3 unit is lower than that of the individual divide-by-2 circuit. This is because extra logic gates are added to the divide-by-2 unit in order to achieve two different division ratios. As a result, the operating frequency achievable for the prescaler is reduced due to the additional propagation delay introduced.

7.1.2 Dual-Modulus Prescaler with Pulse Swallow Counter

Note that all those DMPs described in the previous section can only divide by N and $N+1$. In order to handle all integer division ratios M, a DMP with pulse swallow counter can be used as shown in Figure 7.14. It incorporates a DMP, a Programmable Counter (P-counter) and a Swallow Counter (S-counter). Note that $F_{out} = F_{div} \cdot M$, where M is an integer.

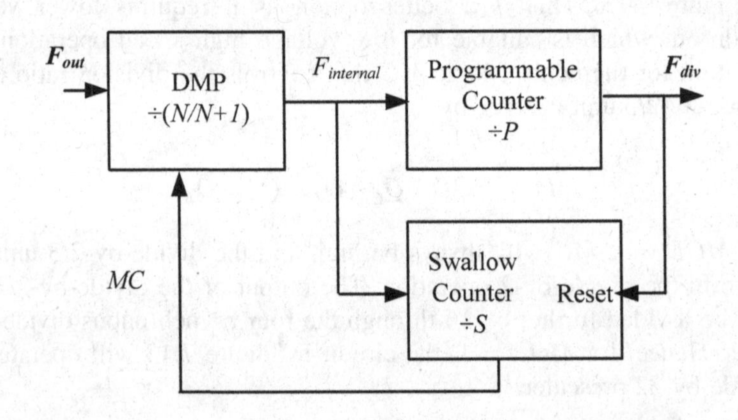

Figure 7.14 A dual-modulus prescaler with swallow counter.

When the DMP divides the input frequency by (N + 1), the S-counter counts the DMP output pulses till a number S is reached. It then changes the DMP modulus control, which starts dividing by N.

The DMP output pulses are also counted in the P-counter, which resets itself and the S-counter when P pulses have been counted. Both counters restart counting and the DMP divides by $(N + 1)$ again. Therefore, during one output period, the DMP has divided the input frequency by $(N + 1)$ for S times and divided the input frequency by N for $(P – S)$ times, which yields an overall division ratio of M

$$M = P.[(N + 1).\frac{S}{P} + N.\frac{P - S}{P}] = P.N + S \tag{7.5}$$

For a proper reset of the P-counter, P must always be larger than the largest value of S. Note that if the value of P increases by 1, the division ratio M will increase by N. Thus, to achieve continuous division ratio, S must be a variable from 0 to $(N – 1)$. Since $S \leq P$, it can be obtained that $N\text{-}1 \leq P$.

Example

Show the waveforms of the dual-modulus prescaler with swallow counter in Figure 7.14 to achieve a divide-by-8 operation. The DMP has a division ratio of 2/3.

Solution

From equation (7.5), with $M = 8$, $N = 2$, it can be obtained that $8 = P \cdot 2 + S$. As $P \geq S$, P and S are set to be 3 and 2 respectively. Its waveforms are shown in Figure 7.15. $F_{interval}$ is taken at the output of the DMP.

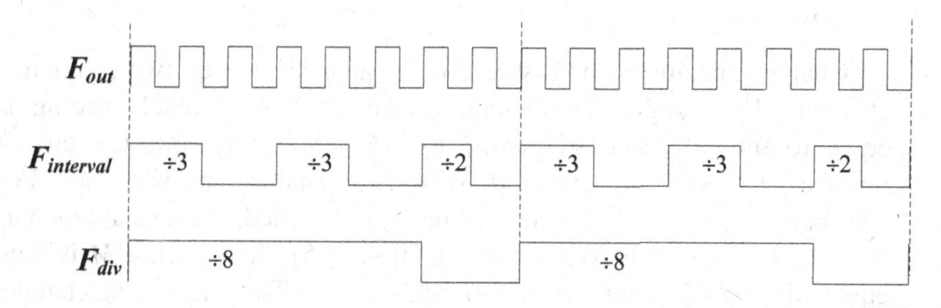

Figure 7.15 Waveforms of a dual-modulus prescaler with swallow counter.

7.1.3 Integer-N Architecture through Dual-Modulus Prescaler with Pulse Swallow Counter

An architecture that generates an output frequency F_{out} to be the integer multiple of the reference frequency F_{ref} is called the integer-N architecture. In other words, $F_{out} = F_{ref} \cdot M$, where M is an integer. Recall that for a frequency synthesizer, F_{div} is equal to F_{ref} during phase-lock. The variable division ratio M of the integer-N architecture can be implemented through the dual-modulus prescaler with swallow counter.

Example

(a) Calculate the largest channel spacing, F_{ch}, for an integer-N frequency synthesizer with a minimum frequency of 2.438GHz using a prescaler with N/N+1 = 128/129.

(b) If the P-counter of a pulse swallow counter has 11 bits, what is the maximum division ratio M achievable for an integer-N divider? Given that $N = 32$. If F_{ch} is 30kHz, what is the highest possible frequency?

(c) If given that the output frequency F_{out}, F_{ch} and N are 2.4GHz, 200kHz and 64 respectively, calculate the value of P and S.

Solution

(a) Channel spacing is the distance in frequency between two frequency channels. For integer-N architecture, the smallest channel spacing is equal to the reference frequency F_{ref}. As $F_{out} = F_{ref} \cdot M$, the largest possible F_{ref} can be obtained with the smallest possible M. For continuous division, $N>S$ and $P \geq S$ must be fulfilled. As the largest S is N-1, the smallest P is N-1. From equation (7.5), the smallest M is then equal to $(N\text{-}1)N$, where P is N-1 and S = 0. The maximum channel spacing $F_{ref} = \cdot F_{out} / M = F_{out} /[(N-1)N] = 2.438GHz /[127 \cdot 128] = 150kHz$.

(b) The largest value of P is 2^{11}-1 = 2047, while the largest value of S is N-1. Hence, the maximum division ratio M achievable is $PN + S$ = 2047(32) + 31 = 65535. The corresponding frequency is MF_{ch} = 65535 x 30kHz = 1.96605GHz.

(c) The division ratio $M = F_{out} / F_{ref} = 2.4GHz / 200kHz = 12000$. M/N = 12000/64 = 187.5, which is the maximum value of P assuming that $S = 0$. Let P to be 187, $S = M - PN = 12000 - 187 \text{x} 64 = 32$.

7.2 DFFs for Prescaler

In the design of a high speed prescaler, the operating frequency and power consumption are the key considerations that are mostly determined by the design of DFFs. In general, a DFF can be implemented through MOS Current Mode Logic (MCML) or through CMOS dynamic circuits. Two widely popular CMOS dynamic circuits are the True-Single-Phase-Clock (TSPC) DFF and its variant the Extended True-Single-Phase-Clock (E-TSPC) DFF. In this section, an analysis of the operating frequency and the power consumption of the MCML and CMOS dynamic circuits will be presented.

7.2.1 MCML

MCML circuits are widely used in mixed signal circuit design. Figure 7.16 shows an MCML inverter. Its operation is based on the differential input pair *M1* and *M2*. If V_{gs} of *M1* is larger than V_{gs} of *M2*, the current passing through *M1* will be larger than that of *M2*. Therefore, the node *OUT* will be pulled down and $\overline{OUT}$ is pulled up through the load resistor. The voltage difference between these two nodes is defined as the output voltage swing, which is given by $I_{bias}R_L$, where I_{bias} and R_L are the biasing current and load resistance respectively. At this state, the output voltages of the node *OUT* and the node $\overline{OUT}$ are $V_{dd} - I_{bias}R_L$ and V_{dd} respectively.

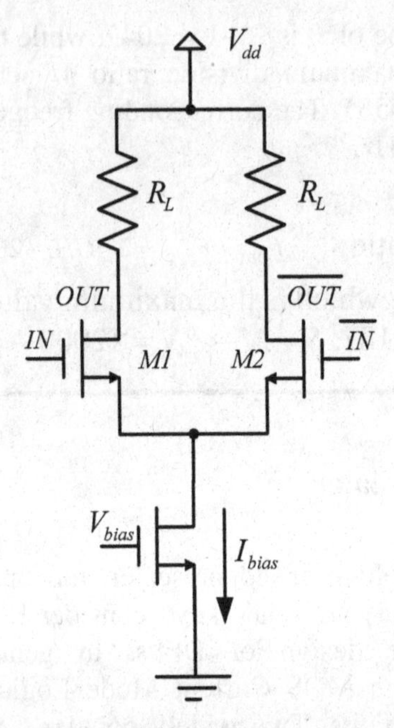

Figure 7.16 An MCML inverter.

The noise performance of the MCML circuit is better than that of the single-ended logic circuit due to the differential topology. Moreover, the operating speed of the MCML circuit is improved due to the small input and output voltage swings. However, the MCML circuit has a fixed biasing current, which results in higher power consumption compared to a CMOS dynamic logic that only consumes power during charging and discharging.

The speed of a DFF is closely related to the propagation delay. Propagation delay is defined as the delay between 50% transition points of the input and output waveforms. The propagation delay is given as [3]

$$t_P = \left[-\frac{d[A(s)/ds]}{A(s)} \right]_{S=0} \tag{7.6}$$

where $A(s)$ is the transfer function of the DFF in consideration.

In digital circuit design, the propagation delay is generally analyzed based on the time constant. The pull-up and pull-down network in the MCML circuit can also be expressed as a simple first order RC response. The output voltage swing and propagation delay of the MCML circuit is given by [4]

$$\Delta V = R_L I_{bias} \tag{7.7}$$

$$t_p = C_L R_L = \frac{C_L \Delta V}{I_{bias}} \tag{7.8}$$

where C_L is the load capacitance. The power consumption of the MCML circuit is given by

$$Power_{MCML} = V_{dd} I_{bias} \tag{7.9}$$

There is a trade-off between the power consumption and the propagation delay. For example, smaller load capacitance and resistance are required to achieve shorter propagation delay. In order to drive a small load capacitance and resistance, a large current is needed, which leads to higher power consumption.

Due to its small propagation delay compared to other logic families, the MCML is suitable for high speed applications [5]. In order to optimize the power consumption of an MCML circuit for a given operating frequency, considerations must be given to numerous parameters such as the voltage swing, resistors' size, transistors' dimension and biasing current, etc [6].

For an MCML circuit, lower power consumption can be achieved by applying a low supply voltage. However, reduction of the supply voltage will reduce the output impedance of the current source and may shift the NMOS out of the saturation region and reduce the mid-swing gain [7]. As shown in equation (7.8), the output voltage swing of an MCML circuit should be as small as possible for high speed operation. However, the output voltage cannot be reduced indefinitely as it is needed to drive the next stage. Its lower voltage swing limit is determined by the gain and the switching current requirements. In the case where the output swing is small, the following stage is required to have a high gain. This means large transistor size is necessary, which will increase the load capacitance of the output and increase the propagation delay [8].

7.2.2 CMOS Dynamic Circuit

In this section, propagation delay and power consumption of the CMOS dynamic circuit will be analyzed. The propagation delay of a CMOS circuit is given by [7]

$$t_{p_{cmos}} = \frac{t_{pHL} + t_{pLH}}{2} \tag{7.10}$$

where t_{pLH} and t_{pHL} are the propagation delay of the "low to high" and "high to low" transitions respectively.

Figure 7.17 shows a CMOS inverter, which forms the basic circuit of the CMOS dynamic circuit [7].

When "high to low" transition occurs, NMOS of the inverter on left hand side turns on. The output voltage swing is $V_o = V_{dd}/2$ and the biasing current is $I_{bias} = \frac{1}{2}\mu_n C_{ox} \frac{W}{L}(V_{dd} - V_{tn})^2$.

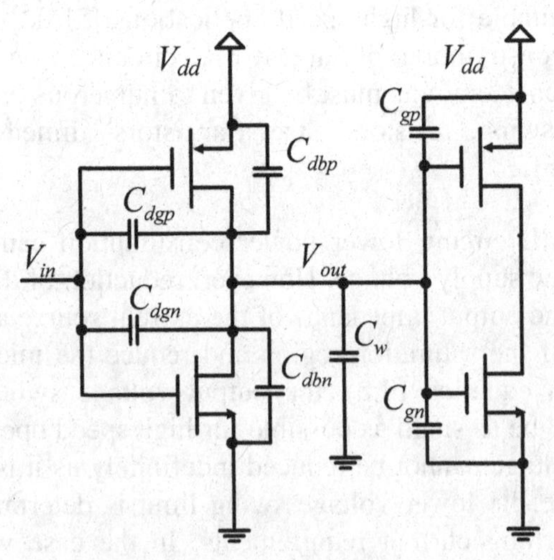

Figure 7.17 CMOS inverter.

When "low to high" transition occurs, PMOS of the inverter on left hand side turns on. The output voltage swing is $V_o = V_{dd}/2$ and biasing current is $I_{bias} = \frac{1}{2}|\mu_p|C_{ox}\frac{W}{L}(V_{dd} - |V_{tp}|)^2$.

From equation (7.8), a simple first order propagation delay can be obtained for t_{pHL} and t_{pLH}

$$t_{pHL} = \frac{C_L V_{dd}}{\mu_n C_{ox}\frac{W}{L}(V_{dd} - V_{tn})^2} \tag{7.11}$$

$$t_{pLH} = \frac{C_L V_{dd}}{|\mu_p|C_{ox}\frac{W}{L}(V_{dd} - |V_{tp}|)^2} \tag{7.12}$$

where C_L is the total load capacitance, which is given by

$$C_L = (C_{dgp} + C_{dgn} + C_{dbp} + C_{dbn}) + (C_{gp} + C_{gn}) + C_w \tag{7.13}$$

From the above equations, two methods can be implemented in order to reduce the propagation delay, which determines the operating frequency. The first method is to reduce the load capacitance. The gate and diffusion capacitances which are linearly proportional to the transistor's gate area can be reduced by choosing a smaller aspect ratio (*W*/*L*) and minimum channel length. However, reduction of *W*/*L* increases the propagation delay since the charging/discharging currents are reduced. The second method is to increase the supply voltage, which will increase the power consumption.

There are three major sources of the power consumption in digital CMOS circuits that are summarized in the following equation [9]

$$P_{avg} = P_{switching} + P_{short-circuit} + P_{leakage} = \alpha_{0\to1} C_L V_{dd}^2 f_{clk} + I_{sc}V_{dd} + I_{leakage}V_{dd} \tag{7.14}$$

The first term represents the switching power consumption, where C_L is the load capacitance, f_{clk} is the clock frequency and $\alpha_{0\to1}$ is the node transition activity factor, which is the average number of times that the node makes a power consuming transition in one clock period. The node transition activity factor $\alpha_{0\to1}$ is determined by many factors such as the logic function, the logic style, and the signal statistics, etc. [9]. The second term is due to the direct-path short circuit current I_{sc}, which arises when both NMOS and PMOS transistors are simultaneously active, conducting a current directly from the supply to ground. Last term is due to the leakage current $I_{leakage}$, which can arise from the substrate injection and subthreshold effects and is primarily determined by the fabrication technology used.

In a well-designed CMOS digital circuit, the switching power consumption is the dominant component of the total power consumption. The supply voltage is the dominant parameter of the switching power consumption. If the output swing is not rail to rail, the switching power is given by [10]

$$P_{switching} = \alpha_{0\to1} C_L V_{dd} V_{swing} f_{clk} \tag{7.15}$$

In the CMOS circuit, due to the non-ideality of the input signal, there will be a transition period during which the PMOS and NMOS transistors are conducting current simultaneously. This causes a short circuit (direct path) from supply voltage to ground. The short circuit power consumption is usually ignorable in the CMOS circuit [7]. However, in some high speed digital circuits, e.g. the E-TSPC logic [11], the PMOS and NMOS may not be configured complementarily. The direct path exists over a significant portion of the circuit operation besides the rising and falling edges. As a result, the short circuit power will be comparable or even larger than the switching power.

Example

Show how the load capacitance can be determined.

Solution

The capacitive load of a circuit affects its propagation delay and power consumption. The load of a CMOS circuit is usually given by the gate capacitance of the transistors in the following stage. The equations to determine the gate capacitance in different operating regions are available in many analog design textbooks [12] and will not be repeated here. However, to calculate the value of the gate capacitance from the equations is rather tedious as it requires the knowledge of a number of technology parameters that are often not explicitly available. Fortunately, through simulation, such values can be easily obtained. The gate capacitance is given by

$$C_G = \frac{\mathrm{Im}(Y_{11})}{2 \cdot \pi \cdot f} \tag{7.16}$$

where $\mathrm{Im}(Y_{11})$ is the imaginary value of the Y_{11} parameter and f is the frequency of interest. The value of Y_{11} can be obtained from S-parameter simulation.

Figure 7.18 shows the S-parameter simulation setup in order to determine the value of the gate capacitance of an NMOS at different operating regions. In this setup, the drain (D) voltage is set at 1.8V and the gate (G) voltage is varied from -2V to 2V. Resistor R_1 is a large resistor to isolate the RF signal at the gate from the DC bias.

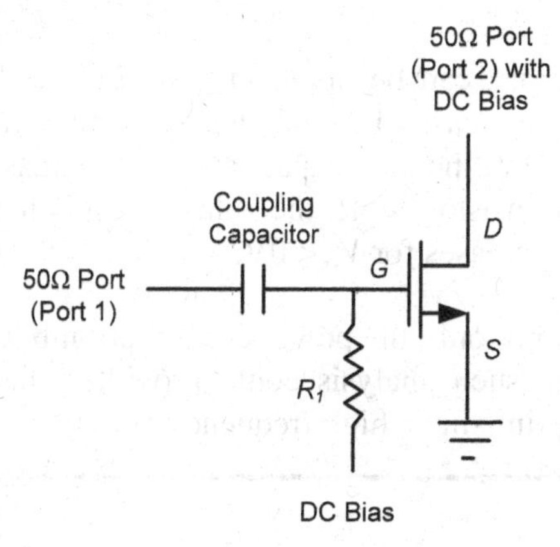

Figure 7.18 S-parameter simulation setup to find gate capacitance C_G.

Figure 7.19 shows the gate capacitance C_G versus gate voltage V_G plot for a 0.18μm CMOS technology transistor with a dimension of 400μm/0.18μm.

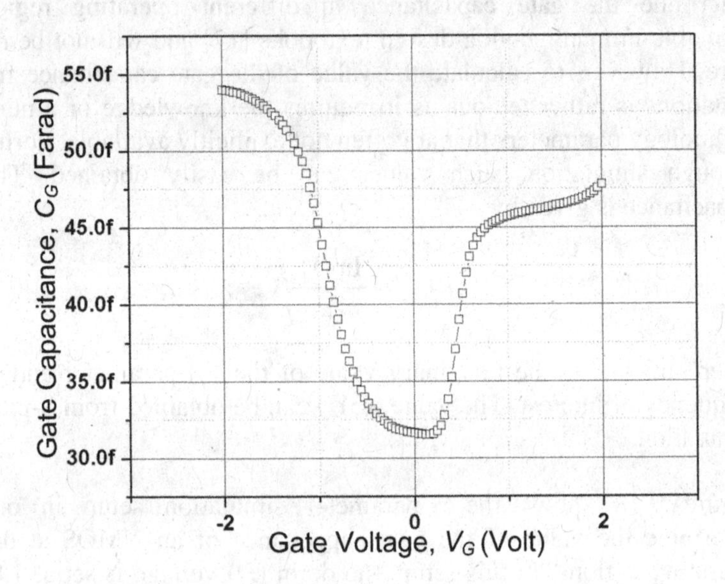

Figure 7.19 The plot of gate capacitance C_G versus gate voltage V_G.

Several observations can be obtained from Figure 7.19. Firstly, as the gate voltage V_G increases beyond the transistor threshold voltage V_T, which is about 0.4V, the gate capacitance C_G increases. Secondly, at V_G = 0V when the transistor is off, the value of C_G is not equal to 0 and C_G increases as V_G decreases for $V_G < 0V$.

Since C_G affects not only the power consumption but also the speed of a particular circuit, such analysis could provide valuable information in designing and optimizing a high frequency circuit.

7.3 Design and Optimization of CMOS Dynamic Circuit (CDC) Based Prescaler

As discussed in the previous section, the MCML circuit, which has high power consumption, is commonly used to achieve a high operating frequency. On the other hand, a CDC based circuit like the TSPC dynamic circuit, which only consumes power during the switching, has a lower operating frequency [13]. In this section, the TSPC and E-TSPC in [13] to be analyzed will be defined as TSPC and Type 1 E-TSPC respectively. Another type of E-TSPC circuit is proposed to increase the operating frequency in some circuit designs [11]. However, this causes additional power consumption. In this section, the E-TSPC in [11] will be defined as Type 2 E-TSPC.

7.3.1 E-TSPC Based Divide-by-2 Unit

So far, the impacts of the modified topologies over the operating frequency and power consumption have not been fully investigated. In this section, the power consumption and operating frequency of the E-TSPC circuits will be evaluated. Two major sources of the power consumption of the E-TSPC divide-by-2 unit, namely the short circuit power and the switching power, will be calculated and simulated. Based on the analyses, a divide-by-2/3 unit will be described in Section 7.3.3. This design example can achieve low power consumption by reducing the switching activities and the short circuit current of the DFFs in the E-TSPC circuit. In addition, a dual-modulus prescaler designed and implemented with the divide-by-2/3 circuit example will be discussed.

Example

Compare the performance of TSPC DFF and E-TSPC DFF in terms of propagation delay.

Solution

The toggled TSPC DFF [14] is the most popular divide-by-2 unit in the low power high speed frequency divider design, while Type 2 E-TSPC DFF is proposed to increase the operating frequency. Figure 7.20(a) and (b) show the topologies of the TSPC DFF and the E-TSPC DFF respectively.

When performing a divide-by-2 function, the output *S3* is fed back to *D*. The operation of both divide-by-2 units is shown in Figure 7.21.

The high-to-low transition's propagation delay of the TSPC unit as shown in Figure 7.20(a) is given by [7]

$$t_{pHL} = 0.69\tau = 0.69 R_{on} C_L = 0.69 R_{onNM4} C_L \tag{7.17}$$

where

$$C_L = C_{dbNM4} + 2C_{gdNM4} + C_{dbPM5} + 2C_{gdPM5} + C_{gNM2} + C_{gPM1} \tag{7.18}$$

and R_{onNM4} is the equivalent resistance of NM4 during the high-to-low transition. To account for the miller effect, the gate-to-drain capacitance (C_{gd}) is multiplied by a factor of 2 for all internal nodes and output node. The factor of 2 is obtained by assuming equal output and input swing.

For the E-TSPC shown in Figure 7.20(b), this delay is given by

$$t_{pHL} = 0.69\tau = 0.69 R_{on} C_L = 0.69 R_{onNM3} C_L \tag{7.19}$$

where

$$C_L = C_{dbNM3} + 2C_{gdNM3} + C_{dbPM3} + 2C_{gdPM3} + C_{gPM1} \tag{7.20}$$

and R_{onNM3} is the equivalent resistance of *NM3* during the high-to-low transition.

It can be observed that the load capacitance of the E-TSPC unit has been reduced compared to that of the TSPC unit. Consequently the propagation delay of the E-TSPC unit will be smaller.

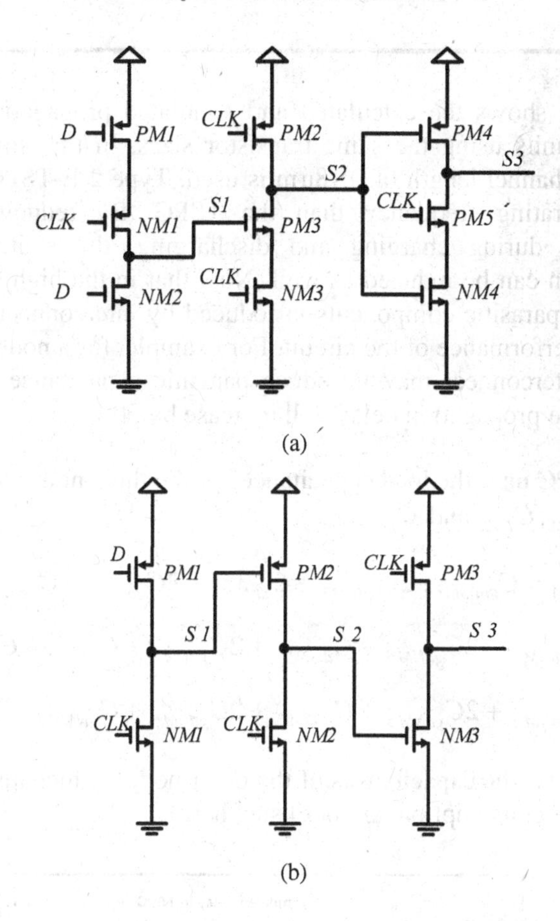

Figure 7.20 Dynamic DFFs (a) TSPC; (b) E-TSPC.

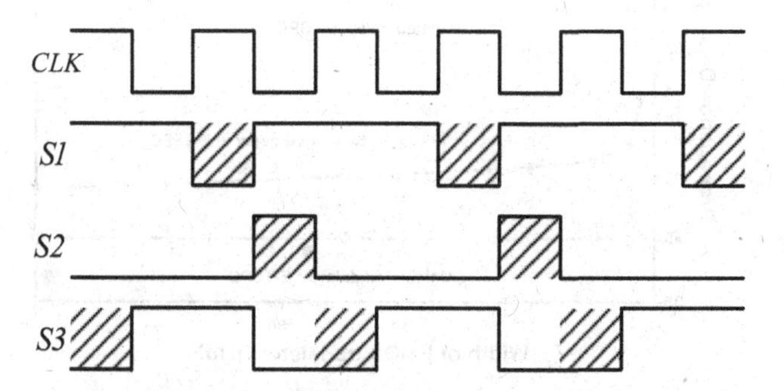

Figure 7.21 Operation of divide-by-2 function.

Figure 7.22 shows the calculated and simulated propagation delay for these two units using the same transistor sizes. In this simulation, the minimum channel length of 0.18μm is used. Type 2 E-TSPC achieves a higher operating frequency than the TSPC. By reducing the load capacitance during charging and discharging, the switching power consumption can be reduced as well. Note that in the high speed circuit design, the parasitic components introduced by interconnect will greatly affect the performance of the circuit. For example, for a node capacitance of 15fF, interconnect may introduce parasitic capacitance of 4.5fF. In this case, the propagation delay will increase by 30%.

For the TSPC unit, the load capacitances of the three nodes *S1*, *S2* and *S3*, namely C_{LS1}, C_{LS2} and C_{LS3} are

$$C_{LS1} = C_{dbPM1} + 2C_{gdPM1} + C_{dbNM1} + 2C_{gdNM1} + C_{gPM2} \tag{7.21}$$

$$C_{LS2} = C_{dbNM3} + 2C_{gdNM3} + C_{dbPM3} + 2C_{gdPM3} + C_{gPM4} + C_{gNM5} \tag{7.22}$$

$$C_{LS3} = C_{dbNM5} + 2C_{gdNM5} + C_{dbPM5} + 2C_{gdPM5} + C_{gNM2} + C_{gPM1} \tag{7.23}$$

For simplicity, the capacitances of the other nodes, which also contribute to the power consumption, are not listed here.

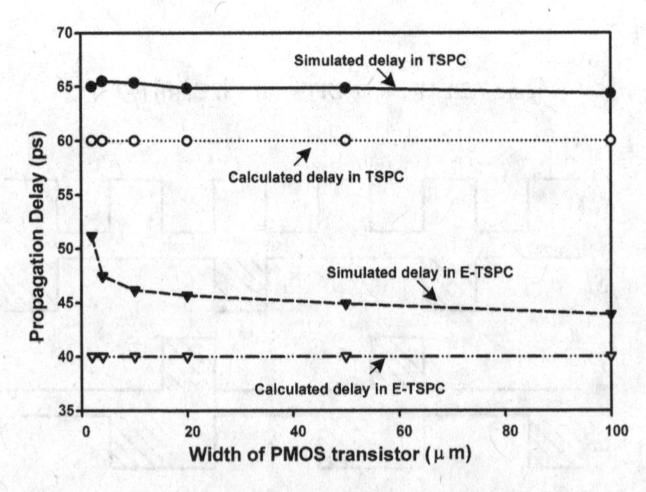

Figure 7.22 Propagation delays of the TSPC and the E-TSPC divide-by-2 units.

For the E-TSPC unit, the load capacitances of each stage have been reduced as follows

$$C_{LS1} = C_{dbPM1} + 2C_{gdPM1} + C_{dbNM1} + 2C_{gdNM1} + C_{gPM2} \tag{7.24}$$

$$C_{LS2} = C_{dbPM2} + 2C_{gdPM2} + C_{dbNM2} + 2C_{gdNM2} + C_{gNM3} \tag{7.25}$$

$$C_{LS3} = C_{dbPM3} + 2C_{gdPM3} + C_{dbNM3} + 2C_{gdNM3} + C_{gPM1} \tag{7.26}$$

With the lower load capacitances, the E-TSPC unit has a lower switching power compared to that of the TSPC unit.

However, in the E-TSPC unit, there is a period during which a direct path from the supply voltage to ground is established in the divide-by-2 operation. During this period, the PMOS and NMOS transistors are turned on simultaneously. The shaded areas in Figure 7.21 mark the transition periods during which the short circuit takes place.

The behavior of the short circuit in a single stage of the E-TSPC DFF is analyzed in Figure 7.23 and Figure 7.24. The short circuit current depends on the aspect ratio *W/L* of the PMOS and the NMOS. For the case where the length of the PMOS and the NMOS are the same, the short circuit current depends on the ratio of W_p / W_n. Figure 7.23 shows the schematic of the single stage of the E-TSPC DFF and Figure 7.24 shows the simulation results of the short circuit current versus the ratio of W_p / W_n, where W_p and W_n are the widths of the PMOS and the NMOS transistors respectively.

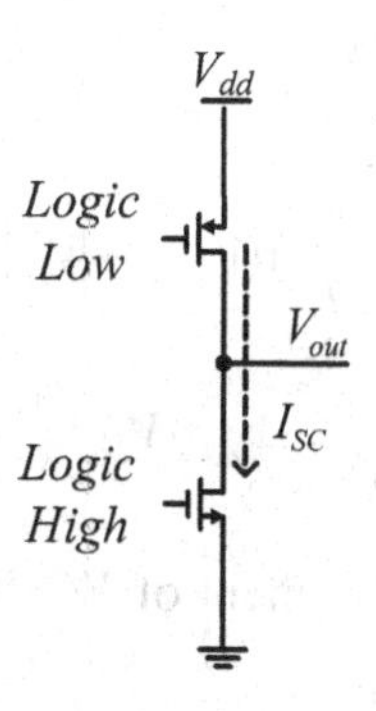

Figure 7.23 Schematic of short circuit in the E-TSPC logic.

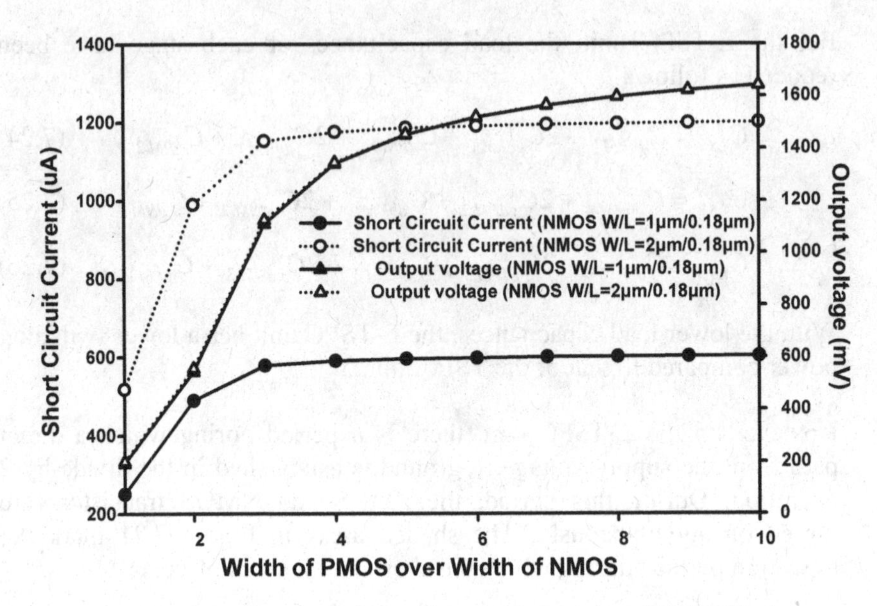

Figure 7.24 Simulation results for short circuit power in the E-TSPC logic.

If $W_p / W_n \leq \mu_n / \mu_p$, the PMOS transistor operation region changes from saturation region to triode region with the increment of W_p / W_n. When $W_p / W_n \geq \mu_n / \mu_p$, the NMOS shifts from the triode region to the saturation region with the increment of W_p / W_n. When both the NMOS and the PMOS are in the triode region, the short circuit current is given by

$$\begin{aligned} I_{short} &= \mu_n C_{ox} \frac{W_n}{L} [(V_{GSn} - V_{tn}) V_{DSn} - \frac{V^2{}_{DSn}}{2})] \\ &= \mu_p C_{ox} \frac{W_p}{L} [(V_{SGp} - |V_{tp}|) V_{SDp} - \frac{V^2{}_{SDp}}{2})] \end{aligned} \tag{7.27}$$

where $V_{DSn} = V_{out}$ and $V_{SDp} = V_{dd} - V_{out}$.

V_{out} increases with the increment of W_p / W_n before the NMOS transistor reaches the saturation region.

As marked in Figure 7.21, there is a quarter of the period during which a direct path is established from supply voltage to ground for all the three stages in the E-TSPC unit. From the calculations of the short circuit current, the short circuit power in the E-TSPC unit can be determined by the product of the supply voltage and the short circuit current.

From the above analyses, the two sources of power consumption in the E-TSPC unit exhibit different characteristics. The short circuit current and the power of each stage are determined by the sizes of the MOS transistors as shown in equation (7.27). While the switching power is linearly proportional to the input frequency for a fixed size of the MOS transistors as shown in equation (7.15).

The two types of the power consumption can be determined using the process parameters [7]. One stage of the E-TSPC unit as shown in Figure 7.23 is examined. For simplicity, W_p / W_n is 2 [15] and the channel length for all the transistors is 0.18μm. The input signals of Figure 7.23 are logically low for the PMOS and logically high for the NMOS. As a result, the PMOS and NMOS transistors are simultaneously turned on, and the short circuit power consumption in one stage of the E-TSPC DFF can be determined. In order to determine the switching power of the E-TSPC DFF, the same inverter but with a square wave input signal is simulated. To evaluate the difference in switching power of the E-TSPC and the TSPC due to different capacitive loads, the load of one PMOS for the E-TSPC unit and the load of one PMOS plus one NMOS for the TSPC unit are used in the calculations and simulations.

Figure 7.25 shows the results of the two power consumption sources in one E-TSPC unit and one TSPC unit. The E-TSPC unit has lower switching power consumption. However, its short circuit power consumption is much larger than the switching power consumption. Within the operating frequency range, the E-TSPC unit has larger total power consumption than that of the TSPC unit as the TSPC unit does not have any short circuit power consumption.

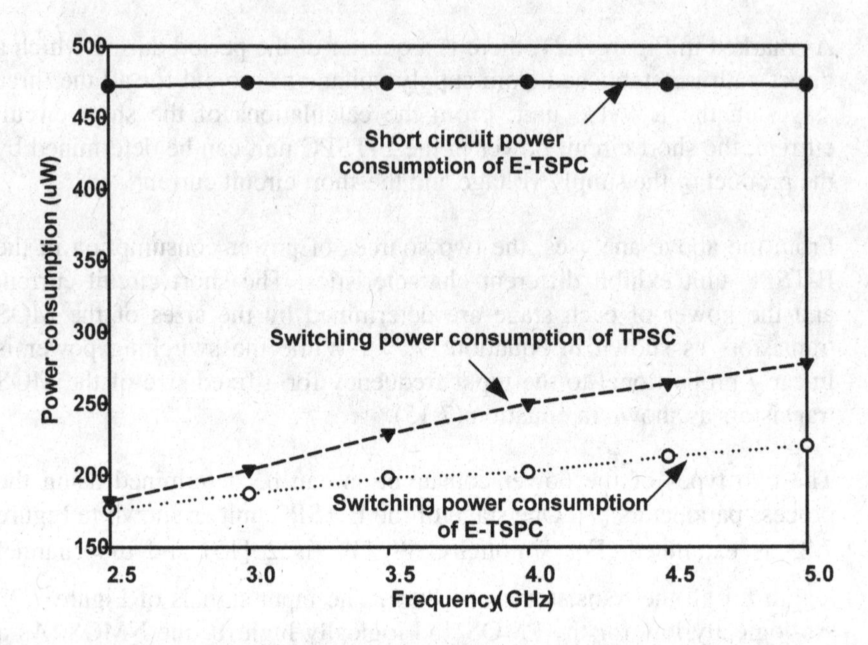

Figure 7.25 Power consumption in one E-TSPC unit and one TSPC unit.

7.3.2 E-TSPC Based Divide-by-2/3 Unit

The E-TSPC divide-by-2 unit has the merit of high operating frequency compared to the traditional TSPC divide-by-2 unit. An E-TSPC based divide-by-4/5 unit is used to form a high speed prescaler in Type 2 design. To reduce the number of components working at full-speed, the divide-by-2/3 is used in Type 1 design. Since the divide-by-2/3 unit consists of two toggled DFFs and additional logic gates, one way to effectively reduce the delay and power consumption is to integrate the logic gates into the divide-by-2/3 unit [16]. In Type 1 design, a gate-integrated dual-modulus prescaler based on the dynamic circuit has been proposed to achieve the high operating frequency at low power consumption. This design uses two DFFs while Type 2 design divide-by-4/5 unit uses three DFFs. Type 1 divide-by-2/3 unit is shown in Figure 7.26.

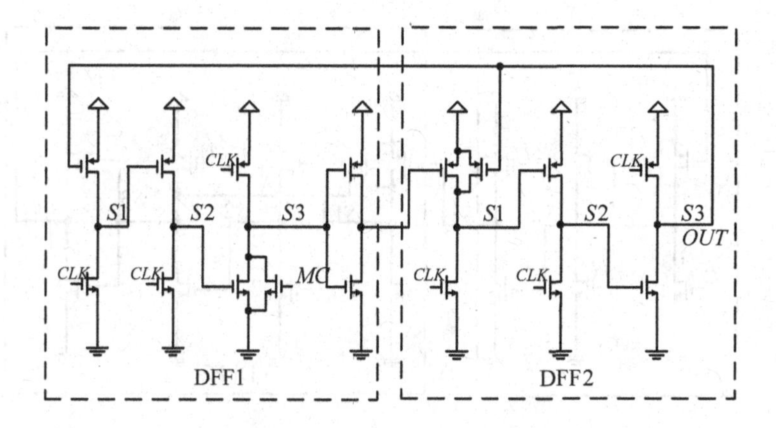

Figure 7.26 Type 1 divide-by-2/3 unit.

When the modulus control signal *MC* is logically low, it performs a divide-by-3 function. When *MC* is logically high, the output of DFF1 will be always high and disabled to achieve the divide-by-2 function. However, the nodes *S1* and *S2* of DFF1 still have switching activities since the output of DFF2 is fed back to DFF1. So both the DFFs switch at half the input frequency even though DFF1 does not participate in the divide-by-2 function. Such a topology introduces unnecessary power consumption. Moreover, during a quarter of the output period, the short circuit power consumption still exists in DFF1 as discussed in the previous section.

7.3.3 Design Example

The design of a low power divide-by-2/3 unit is a complex process. During the divide-by-2 operation, it is not necessary for both DFFs to operate at full-speed since only one toggled DFF is needed to perform the divide-by-2 function. If only one DFF is active during the divide-by-2 operation, a reduction in power consumption will be achieved. The output of the DFF1 is manually pulled down by the *MC* controlled NMOS, but DFF1 still works at full speed in Type 1 design. To reduce the unnecessary power consumption, a divide-by-2/3 unit example is shown in Figure 7.27.

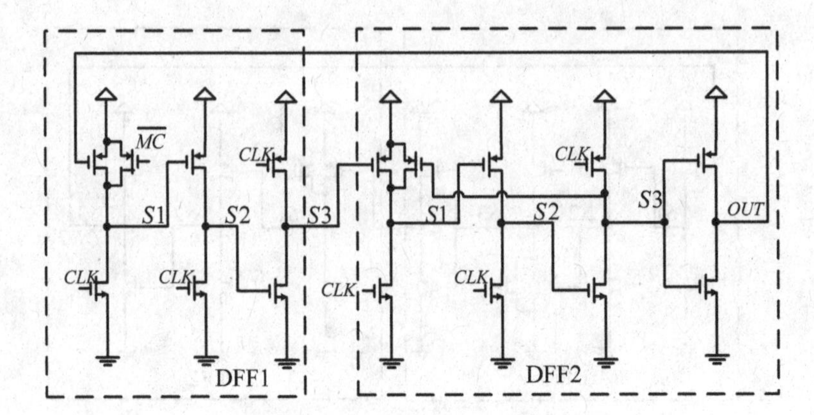

Figure 7.27 The divide-by-2/3 unit as design example.

In contrast to Type 1 design, in this topology, two AND gates are used instead of one OR gate and one AND gate [16] to achieve a symmetrical architecture. By changing the *MC* controlled NMOS at the output of the DFF1 in Type 1 design to an $\overline{MC}$ controlled PMOS as shown in Figure 7.27, DFF1 is blocked at the input. As a result, node *S1* in DFF1 will be logic "1" when *MC* = 1. DFF1 has only one short circuit path, which is in the first stage, while the following stages have no switching activities or short circuits. DFF2 still functions as a toggled divide-by-2 unit. Hence, the divide-by-2/3 unit example has a significant power consumption reduction in the divide-by-2 operation. For the divide-by-3 operation, due to the complementary logic type, the short circuit power consumption in DFF1 is also reduced.

7.3.4 Simulation and Silicon Verifications

Thus far the performance of Type 1 E-TSPC design is the best in the literature. To confirm the reliability of the divide-by-2/3 unit example, a performance comparison of the design example with Type 1 E-TSPC unit was carried out. Both the circuits were designed using 0.18μm CMOS technology and the simulations were performed using the Cadence SpectreRF simulator. The PMOS and NMOS transistors of these two units were of the same size and were based on basic low frequency model (non-RFMOS model).

Figure 7.28 shows the simulation results of the power consumption versus the operating frequency of the two divide-by-2/3 units. An example frequency of 3.5GHz will be taken as the operating frequency in the following comparison.

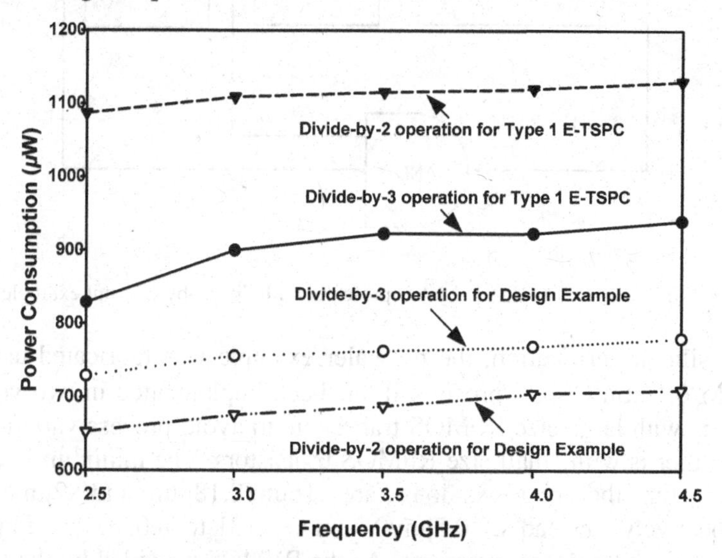

Figure 7.28 Power consumption versus operating frequency for the divide-by-2/3 unit example and Type 1 divide-by-2/3 unit.

In the divide-by-3 operation, the unit example has about 17% power consumption reduction compared to that of Type 1 E-TSPC unit. In the divide-by-2 operation, the unit example consumed about 40% less power consumption than that of Type 1 E-TSPC unit due to the reduced switching activities and reduced short circuit current in DFF1 as shown in Figure 7.27. If the divide-by-2 and divide-by-3 operations are of equal probabilities in the dual-modulus prescaler, the unit example can achieve about 30% reduction in the total power consumption.

To further verify the advantages of this divide-by-2/3 unit example, a divide-by-8/9 dual-modulus prescaler using the same architecture [13] but with the divide-by-2/3 unit example was implemented. In this divide-by-8/9 prescaler as shown in Figure 7.29, the divide-by-2/3 unit example is followed by two stages of the toggled TSPC divide-by-2 units.

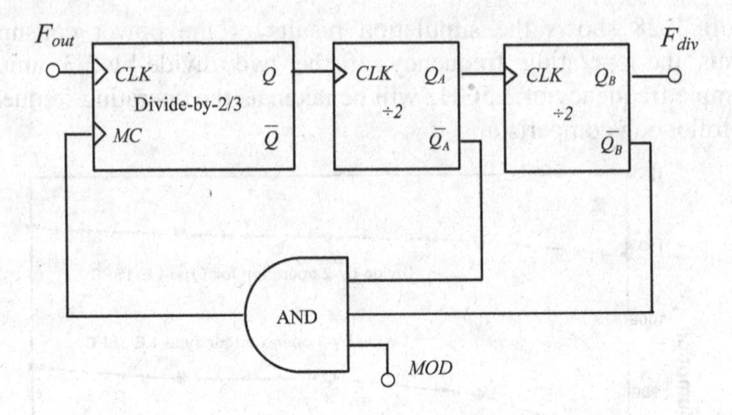

Figure 7.29 The divide-by-8/9 prescaler with divide-by-2/3 unit example.

For silicon verification, the prescaler example was fabricated using the CSM 0.18μm CMOS process. It has been implemented in two versions: one is with large size RFMOS transistors to avoid process variation, and the other is with small size RFMOS transistors. The minimum transistor sizes for the two versions are 16μm/0.18μm and 2μm/0.18μm respectively, instead of 0.5μm/0.2μm in [13] to reduce the impact of parasitic and process variations. As the RFMOS provided by the foundry was not scalable at the time of fabrication, a ratio of 2/1 was used for PMOS/NMOS dimension as in [15]. The operating speed of the design example can be further increased if the proper transistor sizing is carried out [13]. Figure 7.30 shows the die photograph of the divide-by-8/9 prescaler example (the larger transistors version).

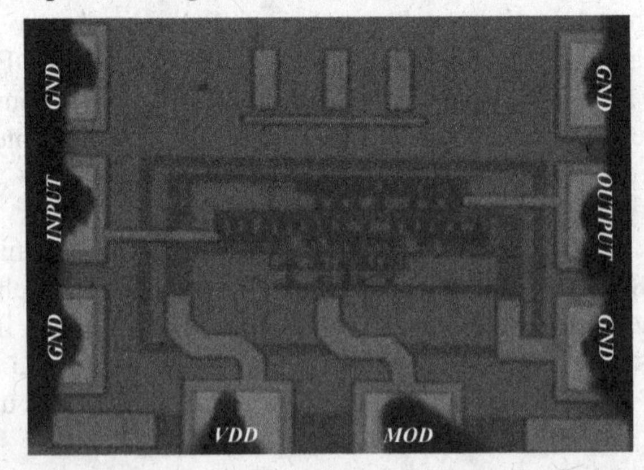

Figure 7.30 Die photograph of the divide-by-8/9 prescaler example.

On-wafer tests were carried out using an RF probe station. The input signal for the measurement was from the Antristu 68347C 10MHz-20GHz signal generator, while the output signals were captured by Lecroy Wavemaster 8600A 6G oscilloscope. The power consumption of the measured chip was 28mW with the supply voltage of 1.8V for a 4.2GHz input frequency in the first version because of the large size of MOS transistors. In the prescaler with smaller transistor sizes, the power consumption was reduced to 3.3mW for an input frequency of 4GHz at 1.8V supply voltage as shown in Figure 7.31.

Based on the concepts described in Section 7.2, calculations of the power consumption were performed. Due to the simplified model in the calculations, the power consumption is lower than the simulated results even though the parasitic capacitors have been added. On the other hand, there are some discrepancies between the post-layout simulation results and the measurement result due to the inaccurate models used and the process variations. The post-layout simulation of the prescaler [13] designed with Type 1 E-TSPC is presented as well for fair comparison. The two prescaler circuits are same except the divide-by-2/3 unit and all the transistor sizes are also same.

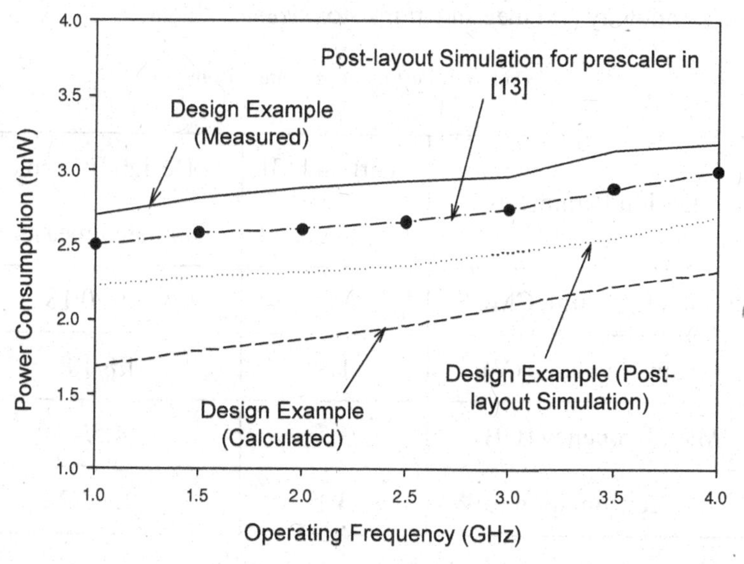

Figure 7.31 Power consumption of the prescalers versus frequency.

Figure 7.32 shows the output waveform of the prescaler example with an input frequency of 4GHz.

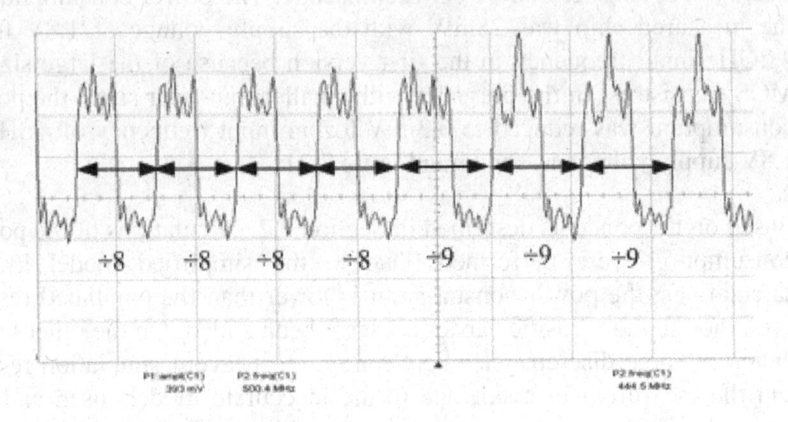

Figure 7.32 Output waveform of the divide-by-8/9 prescaler example with large transistors size.

Table 1 compares the performance of the prescaler example with that in [13]. For fair comparison, the work in [13] was re-simulated using the same technology and the same transistor sizes.

Table 7.1 Performance Comparisons.

Design Parameters	Work in [13] Re-simulated	Design Example (Simulated/measured)
Technology (μm, CMOS)	0.18	0.18/0.18
Supply voltage (V)	1.8	1.8/1.8
Max. frequency (GHz)	4.5	4.5/4
Power consumption (mW)	3.1	2.5/3.2

7.4 Summary

At the beginning of this chapter, the basic concepts of the prescaler were discussed. This was followed by a brief discussion of various implementations of a DFF, which is the most important basic building of the prescaler. Then the detailed descriptions of a high-speed prescaler were presented. This low power divide-by-2/3 prescaler example was implemented using the CMOS technology. Compared to existing design, a 20% power consumption reduction was achieved while operating at the same operating frequency from the simulation. Measurement results showed that a divide-by-8/9 dual-modulus prescaler implemented with this divide-by-2/3 unit example using 0.18μm CMOS process was capable of operating up to 4GHz at a 1.8V supply voltage with a power consumption of 3.2mW.

References

[1] J. Craninckx and M. Steyaert, "Wireless CMOS Frequency Synthesizer Design", *1st Edition, Boston Kluwer Academic Publishers*, 1998.

[2] C. Lam and B. Razavi, "A 2.6GHz/5.2GHz CMOS Voltage-Controlled Oscillator", *ISSCC Digest of Technical Papers*, pp.402-403, Feb. 1999.

[3] K. Murata, et al, "A novel high speed latching operation flip-flop (HLO-FF) circuit and its application to a 19-Gb/s decision circuit using a 0.2-μm GaAs MESFET," *IEEE J. Solid-State Circuits*, vol. 30 , pp. 1101 – 1108, Oct. 1995.

[4] M. Mizuno, et al, "A GHz MOS adaptive pipeline technique using MOS current-mode logic," *IEEE J. Solid-State Circuits*, vol. 31, pp. 784-791, Jun. 1996.

[5] Tanabe, et al, "0.18-μm CMOS 10-Gb/s multiplexer/demultiplexer ICs using current mode logic with tolerance to threshold voltage fluctuation," *IEEE J. Solid-State Circuits*, vol. 36, pp. 988–996, Jun. 2001.

[6] Cicero S. Vaucher, *Architectures for RF Frequency Synthesizers*, Kluwer Academic Publishers, 2002.

[7] J. M. Rabaey, et al, *Digital Integrated Circuits, a Design Perspective, 2nd ed.*, Prentice-Hall Electronics and VLSI series, Upper Saddle River, New Jersey 07458, 2003.

[8] Dirk Pfaff, *Frequency Synthesizer for Wireless Transceivers, Swiss Federal Institute of Technology*, Zurich (ETH), Doctor of technical sciences dissertation, 2003.

[9] A.P. Chandrakasan, R.W. Brodersen, "Minimizing power consumption in digital CMOS circuits," *Proc. IEEE*, vol. 83, pp. 498–523, Apr. 1995.

[10] K. S. Yeo, S. S. Rofail, and W. L. Goh, *CMOS/BiCMOS ULSI: Low-Voltage Low-Power*, Prentice-Hall, Upper Saddle River, New Jersey 07458, Professional Technical Reference, International Edition, 2002.

[11] J. Navarro Soares, Jr., W.A.M. Van Noije, "A 1.6-GHz dual modulus prescaler using the extended true-single-phase-clock CMOS circuit technique (E-TSPC)," *IEEE J. Solid-State Circuits*, vol. 34, pp. 97 –102, Jan. 1999.

[12] P. R. Gray and R. G. Meyer, "Analysis and Design of Analog Integrated Circuits", *2nd Edition, New York Wiley*, 1984.

[13] S. Pellerano, S. Levantino, C. Samori, and A. L. Lacaita, "A 13.5-mW 5-GHz frequency synthesizer with dynamic-logic frequency divider," *IEEE J. Solid-State Circuits*, vol. 39, pp. 378 –383, Feb. 2004.

[14] J. Yuan, and C. Svensson, "High speed CMOS circuit technique," *IEEE J. Solid-State Circuits*, vol. 24, no. 1, pp. 62 – 70, Feb. 1989.

[15] R. Rogenmoser, *The Design of High Speed Dynamic CMOS Circuits for VLSI (Series in microelectronics)*, Hartung-Gorre; 1. Aufl edition, 1996.

[16] C. Lam, and B. Razavi, "A 2.6-GHz/5.2-GHz frequency synthesizer in 0.4-μm CMOS technology," *IEEE J. Solid-State Circuits*, vol. 35, pp. 788 –794, May. 2000.

Big things come in small ICs.

Kiat Seng YEO

INDEX